AQUACULTURE

PRINCIPLES AND PRACTICES

AQUACULTURE

PRINCIPLES AND PRACTICES

Skylar Bayer
Henderite Ohee

Kruger Brentt
Publishers
2025

Kruger Brentt Publishers UK. LTD.
Company Number 9728962

Regd. Office: 68 St Margarets Road, Edgware, Middlesex HA8 9UU

ISBN: 978-1-78715-281-6

For information on all our publications visit our website at http://krugerbrentt.com/

PREFACE

"Aquaculture" immerses readers into the diverse and dynamic world of aquatic farming, offering a comprehensive exploration of the principles, practices, and applications that define the field of aquaculture. This book serves as an essential resource for students, researchers, practitioners, policymakers, and aquaculture enthusiasts seeking to understand the science, technology, and sustainability of aquatic food production.

Aquaculture, or the farming of aquatic organisms such as fish, shellfish, and aquatic plants, plays a crucial role in meeting the growing global demand for seafood while alleviating pressure on wild fish stocks. From freshwater ponds and marine cages to land-based recirculating systems and aquaponics, aquaculture offers a diverse array of production methods tailored to different species, environments, and market demands.

The primary objective of this book is to provide readers with a comprehensive overview of aquaculture, covering key topics such as aquaculture systems and practices, species selection and breeding, nutrition and feed management, health and disease management, environmental sustainability, and socio-economic considerations. Each chapter offers insights, case studies, and practical guidance to help readers navigate the complexities of aquaculture and make informed decisions in their aquaculture endeavors.

Contributions from leading experts in the field explore various aspects of aquaculture, including finfish, shellfish, crustaceans, algae, and integrated multi-trophic aquaculture (IMTA) systems. Special emphasis is placed on emerging trends and innovations in aquaculture technology, such as recirculating aquaculture systems (RAS), genetic improvement programs, precision aquaculture, and digital technologies for monitoring and management.

As we confront the challenges of feeding a growing global population while preserving the health of aquatic ecosystems, it is essential to embrace sustainable aquaculture practices that minimize environmental impact, promote animal welfare, and support local communities. By fostering collaboration, innovation, and knowledge sharing in the aquaculture community, we can build a more resilient, equitable, and sustainable future for aquatic food production.

We extend our sincere gratitude to the contributors whose expertise and dedication have enriched this book. It is our hope that "Aquaculture" will serve as a valuable resource and catalyst for dialogue, innovation, and collaboration in advancing the science, technology, and sustainability of aquaculture worldwide.

Skylar Bayer
Henderite Ohee

CONTENTS

01 RESEARCHING BIOPOLYMERS AS BINDERS FOR FEED FOR AQUACULTURE

1. INTRODUCTION: AN OVERVIEW

There are many diets available for aquatic animals. They vary among species and can adapt to different nutritional needs at different stages of the life cycle, tailored for larvae, juveniles, adults, and breeders. This review will specifically examine binders in aquaculture feed. Binders can exist in liquid or solid form and have the ability to create connections, coverings, or layers that result in robust bonding between particles. Binders are utilized to enhance feed production and to solidify diets in water. Aquaculture feed needs to undergo a higher amount of processing compared to livestock feed to ensure it remains stable in water for an extended period, allowing animals to ingest it. The job of the binder is essential in determining varied amounts of stiffness suitable for specific feeding behavior. Feed stability is a more critical issue with crustaceans than with fish. However, certain fish are benthic and prefer small pellets that settle quickly to the bottom, making them easily detectable by the fish's chemoreceptors. Typically, commercial fish feed remains stable after extrusion and does not require binders to enhance water stability. Recent research have included binders in fish diets to produce harder feces, which helps reduce water pollution (Brinker, 2007). Crayfish, prawns, and shrimps are sluggish feeders that have a common inclination to manipulate food with their mouth appendages before eating. Therefore, it is vital to stabilize feed pellets and minimize nutrient leaching and disintegration in the manufacture of aquatic animal feed. Stability in feeding is essential in echinoculture. Sea urchins are grazers that require time to consume their food. It is important to keep the feed intact for several days to minimize nutrient loss and simplify the management of the rearing structure (Caltagirone et al., 1992; Mortensen et al., 2004; Pearce et al., 2004). Prepared diets often result in substandard gonad quality in terms of texture, firmness, color, and flavor, leading to reduced marketability of the product (Pearce et al., 2002a). Research should prioritize selecting suitable binders that maintain consistency in experimental feed while considering their impact on feed stability, gonad yield, and sensory quality. Many studies have been carried out to assess the type of binder, the ideal inclusion level, and its impact on growth and digestibility in various species of aquatic animals due to the variability of feeding behavior within species and the potential suboptimal nature of binders for some species. This review will analyze feed in terms of binder selection, preparation method, water stability assessment, nutrient leaching evaluation, and the impact of binders on animal development and consumption. This review intentionally does not address the dietary composition for aquatic species. The study of dietary needs of aquatic species is a competitive and rapidly evolving area, but it is not covered in this article. Readers interested in this topic are encouraged to consult the various reviews and books that are accessible.

2. FEED BINDERS

Food is the substance that can be digested, absorbed, and utilized by an animal after consumption. Some components of food are indigestible. The primary constituents of food are as follows:

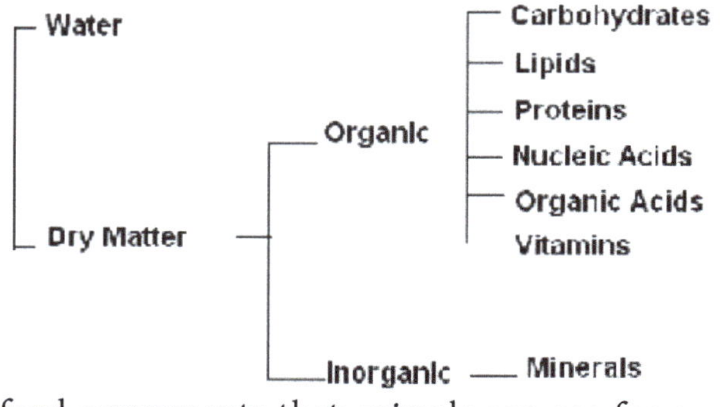

Nutrients are food components that animals can use for energy or metabolic functions. Proteins, lipids, and carbohydrates are the primary energy source components, whereas vitamins and minerals are required in modest quantities for metabolism and life maintenance. Several factors influence the selection of feed for farmed fish. The economic factor is typically prioritized among them. Feeding expenses can account for around 50% of the operational expenditures in agriculture. Key considerations in producing aquatic animal feeds are the feed's water stability and palatability. Adding a binder is essential to guarantee feed stability in water. Several chemicals have been chosen for their ability to connect with other molecules. Various natural binders have been used to produce solid pellets to enhance water stability and reduce nutrient loss. Some binders like wheat gluten or starches provide sustenance for animals, whereas others are inert raw materials with little nutritional value (Sinha et al., 2011). Over 50 organic and inorganic binders have been used in the feed business according to Kalian & Morey's research in 2009. Biopolymers are the most extensively researched and utilized natural binders among those explored by researchers. Common biopolymers are primarily sourced from plants and marine and agricultural animals. Examples include cellulose, starch, pectin, chitin, casein, whey, collagen, and soy proteins. This evaluation will concentrate on the most commonly used ingredients in aquatic feed.

According to De Silva and Anderson (1995) binders can be grouped in three different classes:

1. Binders of protein origin
2. Carbohydrate source binders
3. Binders with no nutritional value

For a carnivorous species, a protein binder is more suitable nutritionally than a carbohydrate binder, whereas the opposite is true for a herbivorous species. Carbohydrates are commonly added to crustacean artificial diets to help conserve protein, as demonstrated by Shiau & Peng in 1992 and Rosas et al. in 2000. As to the Minimal Total Discomfort theory proposed by Forbes in 2001, consuming plant-based carbohydrates in wild animals leads to a protein sparing effect, promoting growth without protein wastage. Food consumption has physiological effects through physical (such as distension) and chemical (like glucose) stimulation of receptors in the internal organs, and over time, through alterations in signals from adipose tissue (such as leptin), which are processed by the central nervous system (Paolucci, 2010). The consequences are linked to the sensory properties of the food, where repeated exposure to a food leads to a conditioned acceptance or rejection response, with the physiological effects of eating as the unconditioned stimulus (US) and the sensory traits of the food as the conditioned stimulus (CS). Learned preferences and aversions are present in various species across the animal kingdom, including nematodes, insects, laboratory animals, and farm animals. Food preferences and dislikes are evident in selective decisions between different foods and can impact the amount consumed when only one food option is present. Theory of Minimal Total Discomfort (Forbes, 2001) suggests that animals adjust their daily food intake and food choices to minimize signals of excesses or deficiencies in food components. Adjusting food intake and selection can accommodate variations in dietary composition and nutrient requirements.

2.1 Carbohydrates

Polysaccharides are biopolymers made up of large carbohydrates with high molecular weight (Aspinall, 1982). The polymers are biodegradable, biocompatible, and non-toxic, exhibiting unique physical-chemical properties and environmentally sustainable characteristics. Polysaccharides have macromolecular structures with multiple polar functional groups that enable them to retain large amounts of water or biological fluids. This property allows them to form hydrogels, which are three-dimensional networks that can resist water under various chemical or physical conditions like gel formation, retrogradation process, pH changes, and cross-linking processes. The wide range of carbohydrates' molecular diversity enables numerous significant roles. Polysaccharides in the food business serve as phytocolloids and emulsifying agents (Farris et al., 2009). Non-digestible carbohydrates can be either insoluble, such as cellulose, or soluble, such pectin, gum, β-glucan, mucilage, and algal polysaccharides (Atkins, 1985; Clark & Ross-Murphy, 1987). Technologically, hydrocolloids are essential for modifying the texture and consistency of meals (Drochner et al., 2004) and possess characteristics that make them appropriate for extending the shelf-life of food products (Volpe et al., 2010). Carbohydrates are the most prevalent biological molecules and play various essential roles in living organisms. They are involved in storing and transporting energy (starch and glycogen), forming structural components (cellulose in plants, chitin in animals), and connecting intercellular walls (hemicellulose) (Kennedy & White, 1983). Polysaccharides are large carbohydrates with high molecular weight. They can be seen as condensation polymers formed by glycosidically joining carbohydrates, resulting in the removal of water molecules, as described by the empirical equation.

$$n\, C_6H_{12}O_6 \rightarrow (C_6H_{10}O_5)n + (n-1)H_2O$$

The diverse macromolecular architectures and chemical compositions of polysaccharides contribute to their wide range of physical and biological uses. Various polysaccharides like agar, alginate, chitin, and pectin can absorb water in both cold and high temperatures, resulting in the formation of viscous solutions, dispersions, and gels. The high interest in these polysaccharides for aquatic animal feed is mostly due to their gelling capabilities.

2.1.1 Agar

Agar is a polymer derived from agarophyte, a type of seaweed, usually a red alga (Usov, 1998). The substance is composed of a minimum of two distinct polymers that can be separated into agarose and agaropectin. Agarose has a high molecular weight and a low sulfate content. Agaropectin has a low molecular weight and a high sulfate content. Agarose is the component in agar responsible for gel formation. Agarose is a linear polymer composed of alternating D-galactose and 3,6 anhydro-L-galactose units, as illustrated in Figure 1.

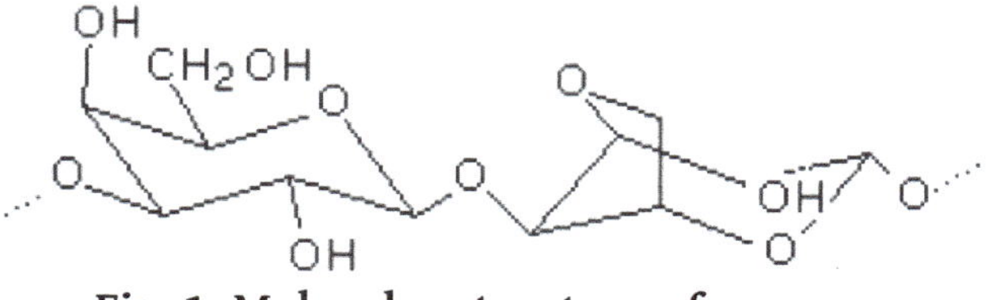

Fig. 1. Molecular structure of agarose

The many types of agarose influence the physical and chemical properties of agar, including its gelling and melting temperature and reactivity. Agar is a blend of agarose and agaropectin in varying ratios determined by the source material and manufacturing method used. Agar gelation is solely induced by its agarose component and is exclusively facilitated by hydrogen bonding. Agar has the ability to gel without requiring any other substances, making it highly versatile for various applications such as food component, biotechnology, cell and tissue cultures, and as a support for electrophoresis or chromatography. Agarose forms "physical gels," where aqueous gels are created by polymer molecules connected only by hydrogen bonds. Because of this distinctive gelling feature, these gels retain a large amount of water within their internal network, allowing water to flow freely across the macroreticulum. Each molecule retains its structure autonomously, resulting in electrostatic attraction rather than polymerization. Agar is a very intricate polysaccharide that exhibits significant variability based on its source. In 1991, eleven distinct agarose structures were discovered in several agar-producing weeds, influenced by factors such as gender, species, climatic circumstances, and time of year. Agar can be altered by replacing sulphate, pyruvate, uronate, or methoxyl groups. Contemporary alkali treatment techniques typically enhance the anhydrous bridging within the molecule, leading to a subsequent enhancement in gel strength.

2.1.2 Alginate

Sodium Alginate is the sodium salt of alginic acid, which is a complex mixture of oligo-polymers, polymannuronic acid (MM), polyguluronic acid (GG), and a mixed polymer (MG) with sequences like GGM and MMG (Smidsrod et al., 1966). Mannuronic acid generates β (1-4) connections, resulting in linear and flexible conformation of M-block segments. Guluronic acid forms α (1-4) connections, creating steric hindrance around the carboxyl groups. This results in G-block segments adopting a folded and hard structure, contributing to the polymer's stiffness. Fig. 2 shows the M and G blocks of the alginic acid salts.

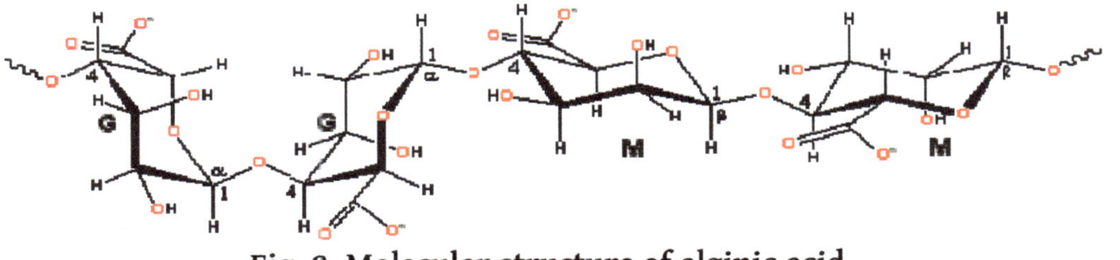

Fig. 2. Molecular structure of alginic acid

In a water solution and in the presence of divalent cations, such as calcium ions, the peculiar buckled backbone of G segments gives rise to water insoluble gel due to the strong interactions between the divalent cations and the COO- groups of the base residual of guluronic acid. Cations can be trapped in a stable, continuous and thermo-irreversible three- dimensional network, whose conformation is typical of an "egg box" (Grant et al., 1973; Grasdalen, et al., 1981) as shown in Fig. 3.

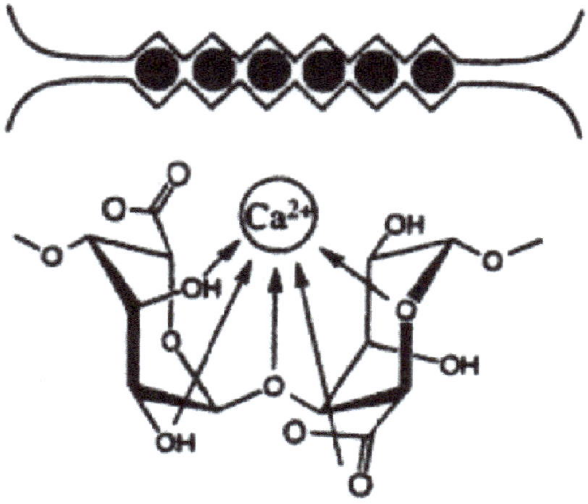

Fig. 3. Alginic acid "egg box" structure.

Therefore, calcium alginate gel is insoluble in water. However, a rehydration process takes place, causing the gel to expand and control the release of trapped chemicals.

2.1.3 Pectin

Pectin, a polysaccharide derived from plant cell walls, serves as a thickening or gelling agent in several industrial applications (Stephen, 1983). Pectin is a polymer of α- galacturonic acid with a variable number of methyl ester groups (Fig. 4).

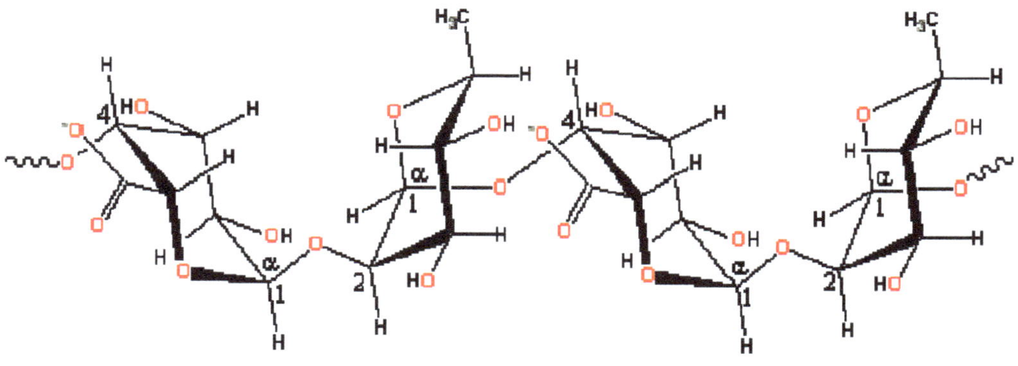

Fig. 4. Molecular structure of pectin

Pectin lacks a defined structure, resulting in a flexible and bent shape in solution rather than a straight one. Methylating the carboxylic acid groups produces methyl esters that are significantly more hydrophobic, leading to a distinct impact on the structure of the surrounding water. Pectin's characteristics are determined by its degree of esterification, typically about 70%. Gel strength rises with higher $Ca2+$ levels but decreases with elevated temperature and increased acidity ($pH < 3$). Low methoxy-pectin gels (~35% esterified) create clear gels at low temperatures (~10°C) by cooperative "zipped" connections in the absence of additional cations.

This hydrogen-bonded interaction is expected to resemble that of alginate. High methoxyl-pectin with an esterification level above 43%, often about 67%, forms gels through hydrogen bonding and hydrophobic interactions. This process occurs in acidic conditions with a pH of approximately 3.0 to minimize electrostatic repulsions, and with the addition of sugars, such as around 62% sucrose by weight, to decrease polymer-water interactions. The gel produced by low methoxyl-pectin is thermoreversible and unstable in water, but the gel generated by high methoxyl-pectin is thermoirreversible, especially in the presence of high sugar content (sucrose) and at low pH. Under these circumstances, the re-hydration process of the pectin gel is partly prevented by the strong hydrophobic structural framework.

2.1.4 Chitosan

Chitosan is a positively charged carbohydrate polymer derived from chitin, which is the second most common polysaccharide in nature after cellulose. Chitin primarily comes from the discarded shells of shrimps, lobsters, and crabs (Johnson & Peniston, 1982). Chitosan is obtained by treating chitin with high alkali to remove most of the acetyl groups. Chitosan can be described as a group of linear binary copolymers consisting of (1-4)-linked 2-acetamido-2-deoxy-β-D-glucopyranose (GlcNAc) and 2-amino-2-deoxy-β-D-glucopyranose (GlcN) molecules (Fig.5).

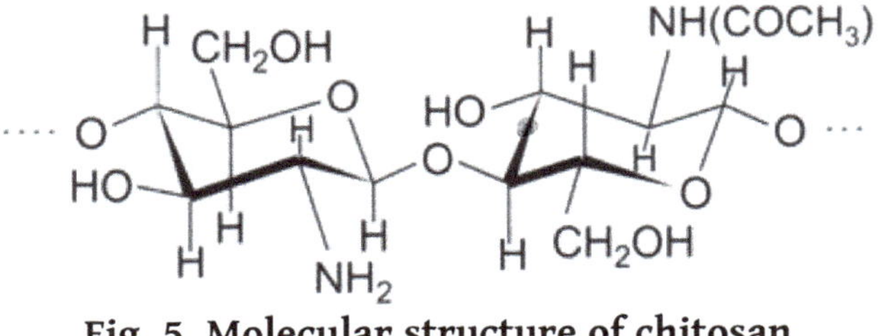

Fig. 5. Molecular structure of chitosan

Chitosan's polycationic characteristics make it stand out among polysaccharides and natural polymers, leading to a wide range of established and potential applications. Chitosan is recognized for its ability to form complexes with specific metal ions, which is utilized to eliminate small amounts of heavy metals or radioactive isotopes from wastewater. Chitosan is well-documented for lowering cholesterol levels and is also controversially used as a weight-reducing agent. It is also utilized as a possible method for delivering controlled-release medications orally. This material is suggested for creating microspheres and microcapsules (El-hefian et al., 2011). These are just a handful of the uses of chitosan. Chitin is not soluble in most solvents, however the characteristics of chitosan are connected to its nature as a polyelectrolyte polymeric carbohydrate. It is not soluble in water but easily dissolves in weak solutions of various organic acids as acetic, citric, and tartaric acids. Chitosan gel is stable in water as it does not swell much due to the polymer's insolubility in water (Cerver et al., 2004).

2.1.5 Carrageenan

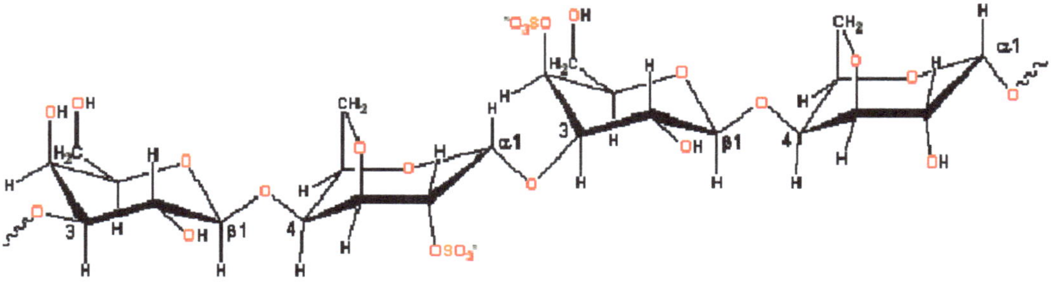

Fig. 6. Molecular structure of carrageenan

Table 1. Comparative properties of types of carrageenan

Kappa carrageenan	Iota carrageenan	Lambda carrageenan
Soluble in hot water. The addition of potassium ions induces the formation of a durable, brittle gel; it also increases the gelling and melting temperatures. Strong, rigid gel, some syneresis, forms helix with K+ ions. Ca++ causes helices to aggregate and the gel to contract and become brittle. Slightly opaque gel. Becomes clear with sugar. Approximately 25% ester sulfate and 34% 3,6-AG. Compatible with water miscible solvents. Insoluble in most organic solvents. Typical use levels — 0.02 to 2.0%.	Dilute solutions exhibit thixotropic characteristics. Soluble in hot water; sodium iota carrageenan is soluble in cold and hot water. The addition of calcium ions will induce the formation of a durable, elastic gel, and increase gelling and melting temperatures. Elastic gels, forms helix with Ca++. Limited aggregation contributes to elasticity, no syneresis. Clear gel. Freeze/thaw stable. Insoluble in most organic solvents. Approximately 32% ester sulfate and 30% 3,6-AG. Typical use levels — 0.2 to 2.0%.	Free flowing, non-gelling pseudo-plastic solutions in water. Partially soluble in cold water, fully soluble in hot water. No gel, random distribution of polymer chains. Range from low to high viscosity. Addition of cations has little effect on viscosity. Compatible with water miscible solvents. Insoluble in most organic solvents. Stable over a wide range of temperatures, including freeze/thaw cycles. Soluble in 5% salt solution, hot or cold. Approximately 35% ester sulfate and little or no 3,6-AG. Typical use level — 0.1 to 1.0%.

Carrageenan is a hydrocolloid extracted from certain red seaweeds using water or aqueous alkali. Carrageenan is mostly composed of potassium, sodium, calcium, magnesium, and ammonium sulfate esters of galactose and 3,6-anhydrogalactose copolymers. The hexoses are connected in a polymer structure by alternating α-1,3 and β-1,4 linkages (Fig. 6). The three primary copolymers in the hydrocolloid are kappa-carrageenan (k-carrageenan), iota-carrageenan (ι-carrageenan), and lambda-carrageenan (λ-carrageenan). K-carrageenan is mostly composed of a polymer that alternates between D-galactose-4-sulfate and 3,6-anhydro-D-galactose. I-carrageenan differs in that the 3,6-anhydrogalactose is sulfated at carbon 2. There is a range of compositions between k-carrageenan and ι-carrageenan that vary in the level of sulfation at carbon 2. The alternating monomeric units of λ-carrageenan consist mostly of D-galactose-2-sulfate (1,3-linked) and D-galactose-2,6-disulfate (1,4-linked). Carrageenan typically contains between 18% to 40% ester sulfate concentration. Carrageenan contains inorganic salts derived from seaweed and the extraction process. Carrageenan is extracted by methods such as alcohol precipitation, drum drying, or freezing. The main factors that affect the characteristics of k-, ι- and λ-carrageenan are the quantity and location of the ester sulfate groups on the repeating galactose units (Kennedy & White, 1983). Elevated concentrations of ester sulfate decrease the solubility temperature of carrageenan and result in weaker gel formation, or hinder gel formation. Table 1 presents the comparative features of the three kinds of carrageenan.

2.1.6 Carboxymethylcellulose

Carboxymethylcellulose (CMC) is a cellulose derivative created through a reaction between cellulose and alkali along with chloroacetic acid. The CMC structure is derived from the cellulose polymer composed of β-(14)-D-glucopyranose units (Fig. 7).

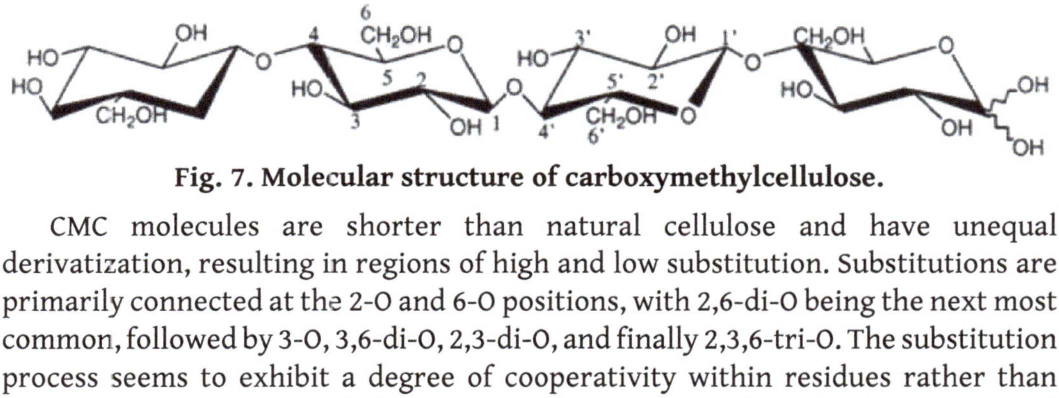

Fig. 7. Molecular structure of carboxymethylcellulose.

CMC molecules are shorter than natural cellulose and have unequal derivatization, resulting in regions of high and low substitution. Substitutions are primarily connected at the 2-O and 6-O positions, with 2,6-di-O being the next most common, followed by 3-O, 3,6-di-O, 2,3-di-O, and finally 2,3,6-tri-O. The substitution process seems to exhibit a degree of cooperativity within residues rather than randomness, resulting in slightly greater than anticipated levels of unsubstituted and trisubstituted areas. CMC molecules exhibit a rod-like conformation at low concentrations, but at greater concentrations, they overlap and coil up, eventually entangling to form a thermoreversible gel. Higher ionic strength and lower pH levels decrease viscosity by inducing the polymer to coil more tightly. Carboxymethyl cellulose dissolves quickly in cold water and is mostly utilized to manage viscosity without forming a gel. As it becomes less viscous when heated, it can enhance the volume yield in baking by promoting the development of gas bubbles. Its ability to modulate viscosity makes it suitable for use as a thickening, phase or emulsion stabilizer (such as with milk casein), and suspending agent. CMC can also be utilized for its great water-holding ability, especially when combined with Ca2+ salt, even at low viscosity (Arisz et al., 1995).

2.1.7 Guar gum

Guar has been farmed throughout India for ages. India produces 80% of the world's guar. Historically, guar has been a significant source of sustenance for both people and animals. Guar gum, also known as galactomannan, is a large carbohydrate molecule obtained from the seeds of the guar plant (Cyampopistetragonolobus). Guar gum is a polysaccharide made up of a mannose backbone and a galactose side chain. The galactose is distributed randomly on the mannose backbone with an average ratio of 1:2 galactose to mannose. The polymerized composition of guar gum, which includes several hydroxyl groups, has been utilized by industry for a considerable period of time (Goldstein et al., 1973). Figure 8 displays the chemical structure of guar gum. Guar gum's key characteristic is its quick hydration in cold or hot water, resulting in a consistent and high viscosity at low concentrations. Guar gum has the benefit of maintaining high viscosity even when dissolved in cold water. Guar gum is a highly cost-effective stabilizer and emulsifier widely used in the food industry for its ability to improve texture, mouthfeel, and regulate crystal formation through its exceptional water-binding capabilities (Castillo-Garcia et al., 2005). It is chemically inactive and shows resistance to oil, greases, and solvents. It exhibits strong compatibility with several hydrocolloids, especially xanthan gum (Ahmed et al., 2005). Even in little amounts, it exhibits a remarkably high viscosity.

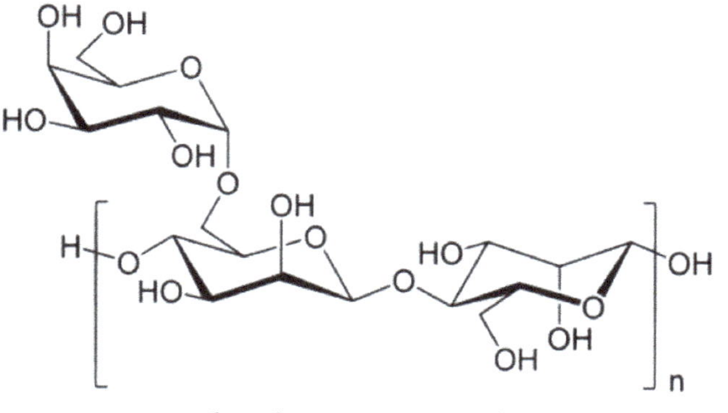

Fig. 8. Molecular structure of guar gum

2.2 Proteins

The most often utilized proteins as feed binders are urea formaldehyde, wheat gluten, and gelatin. Gelatin-based binders are a suitable substitute for gluten and urea formaldehyde. Gelatin and gluten are both easily digestible and rich in proteins (De Muylder et al., 2008). However, the use of gluten may have disadvantages because of its antigenic properties (Shewry, 2009).

2.2.1 Gelatin

Gelatin is a blend of water-soluble proteins obtained from collagen by hydrolysis. The protein fraction is predominantly composed of amino acids. The amino acids are connected by amide bonds to create a linear polymer with a molecular weight ranging from 15,000 to 250,000. Gelatin is a favorable choice for creating hydrogel packaging due to its cost-effectiveness, biodegradability, and ability to facilitate various molecular interactions (Farris et al., 2009). There are two varieties of gelatins distinguished by their method of production. Gelatin Type A, with a pH range of 3.8–6.0 and an isoelectric point of 6–8, is produced by acidic hydrolysis of pork skin. It enhances the flexibility and elasticity of the blend. Type B gelatin, with a pH range of 5.0–7.4 and an isoelectric point of 4.7–5.3, is produced through alkaline hydrolysis of bones and animal skin, enhancing the gel strength of the mixture. Commercially, gelatin is available in different purity degrees in the form of translucent sheets, granules, or powders. Gelatin is virtually devoid of odor and flavor. It is not soluble in acetone, chloroform, ethanol (95%), ether, and methanol. It dissolves in glycerin, acids, and alkalis, but precipitates when exposed to strong acids or alkalis. It expands and becomes more pliable when in contact with water, eventually soaking up 5 to 10 times its original weight in water. It dissolves in hot water. When cooled to 35–40 °C, it solidifies into a jelly or gel. The system occurs as a sol at temperatures equal to or greater than 40 °C. In alkaline conditions, a gel with greater viscosity is produced compared to acidic conditions (Robinson et al., 1975). Gelatin, being a protein, has chemical features typical of proteins, such as being broken down by proteolytic processes to release its amino components. Gelatin interacts with acids, bases, aldehydes, aldehydic sugars, anionic and cationic polymers, electrolytes, metal ions, plasticizers, preservatives, and surfactants.

3. FEED MANUFACTURE

Feed utilized in small-scale research is typically created by uncomplicated procedures. Ingredients are manually ground and blended or processed through extrusion at room temperature (cold-extrusion). Grinding feed materials is essential since pellets produced from finely crushed ingredients will be more resilient and tough (Lim & Cuzon, 1994). Cold extrusion is advantageous for maintaining the nutritional qualities of the meal when compared to high-temperature extrusion. Extrusion technique is commonly used in aquaculture feed manufacturing, although the high temperature and pressure involved can potentially reduce the nutritional quality of the components. Heat-extrusion of macroalgae-based diet enhances the digestibility of carbohydrates compared to fresh algae, resulting in improved gonad production in farmed sea urchins (Akiyama et al., 2001). Binders are typically dissolved in warm or hot water at a concentration between 1 to 10% to get the desired final percentage. As the temperature drops and the solution solidifies into a gel, diet components are incorporated (Ruscoe et al., 2005; Volpe et al., 2008; Simon, 2009). Binders in powdered form are added to the diet ingredients, mixed, and hot water is poured over the combination. The mixture is then churned vigorously to create a dough (Orire et al., 2010). Zein binders are initially dissolved in ethanol since they are not soluble in water (Genodepa et al., 2004; Johnston & Johntson, 2007). The result is a diet consisting of little portions. Pellets are created by extruding the dough through a mincer to generate noodles that are then cut into little pieces, or by spreading the dough on a tray and cutting it into bits. The mixture can also be placed in a marumerizer to form spheroid fed particles (Liu et al., 2008). Steam pellets can be produced by utilizing a laboratory pellet mill as demonstrated by Palma et al. in 2008. Pellets are stored at low temperatures (4°C to -20°C) or dried to reduce moisture content before storage (Ruscoe et al., 2005). The former can be described as "moist pellets" and the latter as "dry pellets." Gelatinous pellets are a type of pellet formed by combining nutritional components with oil, as described by Johnston & Johnstson (2007) and Liu et al. (2008). Pellets can be coated with binders to create a microcoated diet, as described by Bautista et al. in 1989.

4. ASSESSMENT OF FEED STABILITY IN WATER

Ensuring the water stability of feed is crucial when producing aquaculture diets. The stability of water is significantly affected by the features of binders, which are in turn impacted by the components (Dominy & Lim, 1991). Durazo-Beltran and Viana (2001) found it challenging to link the components of fish silage to produce pellets for abalone farming, perhaps because to the elevated levels of hydrolyzed protein. The aquaculture sector is concerned about the water stability of aquatic feed, yet there is currently no established technique to assess this stability. The dry matter weight loss is often evaluated by placing a specific amount of feed, usually in pellet form, in a beaker of water for a variable period, sometimes shaking it. The pellets are poured off and dried. Literature varies in the factors related to pellet incubation, such as water amount, temperature, time intervals, and drying processes (Ruscoe et al., 2005; Johnston & Johnston, 2007; Palma et al., 2008; Orire et al., 2010). The American Feed Industry Association (AFIA 1999 in Liu et al., 2008) proposes a minor adjustment to the dry matter weight loss. AFIA defines water stability as the proportion of pellets that remain on a wire mesh sieve after being immersed in a shaking water bath for a specific period, extracted, and dried. Obaldo et al. (2002) created three techniques to assess the water durability of shrimp pelleted feeds: a static method, a horizontal method, and a vertical shaking method. Various parameters, including water and pellet agitation, direction of agitation, water temperature, salinity, leaching container, and filtering media, were considered. Among these factors, temperature and water salinity had the most significant impact on the rates of dry matter retention. Increased water temperatures and decreased salinity levels resulted in greater leaching and less dry matter retention. Overall, the three methods seemed to offer a reliable and exact way to measure the water stability of shrimp diets. Water shaking can replicate both indoor and outdoor cultivation settings accurately and can be easily adapted to different situations. Water stability can be measured in different ways: as a percentage of the immersed diet weight compared to the initial sample weight, as the percent loss of dry matter calculated by the difference in sample weight after reweighing, or as the percentage of dry matter remaining calculated using a specific formula. Caltagirone et al. (1992) propose considering a subjective criterion, along with the %LDM, when assessing binder efficiency, which is the uniformity of food, specified on a certain scale. 1) Inconsistency (feed disintegrates completely upon removal from water; 2) Weak consistency (feed partially disintegrates upon removal from water; 3) Good consistency (feed remains intact upon removal from water but breaks apart with slight pressure; 4) Very good consistency (feed remains intact upon removal from water and withstands slight pressure). Caltagirone et al. (1992) assessed six binding agents - agar-agar, cellulose, CMC, guar gum, gelatine, and sodium alginate - after 48 hours of water immersion based on specific evaluation criteria. Cellulose and CMC were ineffective as binders even at high concentrations; sodium alginate maintained decent consistency but exhibited significant weight loss, whereas guar gum, gelatine, and agar-agar proved to be effective binders for pellet stability. Pearce et al. (2002b) examined the efficacy of various binders (gelatin, guar gum, sodium alginate, and maize starch) in prepared feed by assessing their performance after 216 hours of immersion in sea water. Guar gum and starch-based pellets lost consistency after 24 hours in water; sodium alginate maintained consistency for 48 hours, and gelatine stayed firm even after 216 hours in sea water. Mortensen et al. (2004) conducted a study to assess the durability of a pellet made from fish-skin gelatine reinforced with transglutaminase enzyme, using the same evaluation method. The feed maintained a stable consistency for 7 days in seawater. The dry matter loss increased over time, reaching 17.6%, compared to 60.1% for the salmon feed used as a control. We used a unique methodology in our laboratory to test pellet water stability. This involved evaluating the diameter of released particles in water at different time intervals using a Low Angle Laser Light Scattering Technique. (Volpe et al., 2008; Coccia et al., 2010; Fabbrocini et al., 2011). The size of particles released by pellets is constantly measured over time to show how water stability changes based on particle size. Pellets that break down into smaller particles are less stable in water compared to pellets that break down into larger particles. The evaluation approach indicated that pectin exhibited superior water stability compared to alginate and agar, two algal polysaccharides, based on binder capacity (Volpe et al., 2008). Adding a coating on the microbound pellets enhanced their water stability.

Microbound pellets broke down into particles with a diameter about double the size of pellets made without coating (Coccia et al., 2010). Palma et al. (2008) found that pellets lacking a binder but having a microcoat of lignosol performed worse in water compared to pellets manufactured with lignosol as a binder. This suggests that an internal binder is essential for giving firmness to the feed. Research utilizing the Low Angle Laser Light Scattering Technique found that 3% and 6% agar-based round pellets made for sea urchin feeding demonstrated strong water stability. The diameter of the released particles increased notably after 5 days in sea water for 3% agar pellets and after 6 days for 6% agar pellets (Fabbrocini et al., 2011). Because of the wide range of feed ingredients, binder percentages, and manufacturing techniques, it is not possible to definitively determine if one binder is superior to another in terms of water stability. Experimental results in literature are sometimes presented as relative, indicating that one binder performs better than another under specific conditions, which hinders an objective assessment of binder performance. Literature data on feed water stability are included in table 2.

Table 2. Effect of binders on feed water stability.

WATER STABILITY TEST	MAX TIME INTERVAL CONSIDERED	% OF BINDER EMPLOYED	OUTCOME1	REFERENCE
Dry matter weight loss	10 min	3% CMC, corn starch, guar gum, wheat gluten	CMC, guar gum, wheat gluten > corn starch	Fagbenro and Jauncey, 1995
Dry matter weight loss	2 h	2% agar, lignin	Lignin>agar	Palma et al., 2008
Dry matter weight loss	12 h	5% carrageenan 3% zein, agar 2% alginate, gelatin	Alginate,gelatin> carrageenan, zein, agar	Johnston and Johnston, 2007
Dry matter weight loss	12 h	10% Wheat flour 2% Whole cassava meal, dried molasses, agar, Langobin°	Wheat flour>whole cassava meal> dried molasses>agar> langobin	Seixas Filho et al., 1997a
Dry matter weight loss	24 h	0.5-2.5% Agar, Alginate, Carrageenan	All effective at 0.5%	Durazo- Beltran and Viana, 2001
Dry matter weight loss	48h	Agar-agar, carboxyl-methyl-cellulose (CMC), guar gum, gelatin, sodium alginate, cellulose	Agar-agar, gelatin, guar gum > sodium alginate > CMC, cellulose	Caltagirone et al., 1992
Dry matter weight loss	168h	Fish-skin gelatin+0.13% transglutaminase (TG)	Gelatin + TG > commercial salmon feed	Mortensen et al., 2004

WATER STABILITY TEST	MAX TIME INTERVAL CONSIDERED	% OF BINDER EMPLOYED	OUTCOME1	REFERENCE
Dry matter weight loss	216h	Guar gum, gelatin, sodium alginate, corn starch	Gelatin > sodium alginate > guar gum, corn starch	Pearce et al., 2002b
Dry matter retention	30 min	7% starch 2% carrageenan, CMS, alginate, gelatin	Carrageenan> alginate> gelatin> CMS	Liu et al., 2008
Dry matter remaining after immersion in water	180 min	3 and 5% carrageenan, CMC, agar, gelatin	Carrageenan , CMC> agar, gelatin	Ruscoe et al., 2005
Variation in weight after soaking in water	24 hours	10% Gelatin 30% alginate	Alginate>gelatin	Valverde et al., 2008
Variation in weight after soaking in water	24 hours	2% alginate plus 1% sodium hexametaphosphate or D-gluconic acis as sequestrants 9 and 13% agar 6 and 15% gelatin 20% agar/gelatin mixture	Agar/gelatin > agar> alginate>gelatin	Knauer et al., 1993
Long Angle Laser Light Scattering Technique	24 hours	2.5% Agar, alginate, pectin	Pectin>agar, alginate	Volpe et al., 2008
Long Angle Laser Light Scattering Technique	24 hours	1% chitosan, pectin, agar, alginate	Chitosan, pectin> agar, alginate	Coccia et al., 2010
Long Angle Laser Light Scattering Technique	96h 144h	3% agar, 6% agar	No differences agar 6% > agar 3%	Fabbrocini et al., 2011

[1]The reported outcome refers to the maximum time interval tested . > means that the binder(s) on the left perform better that the binder(s) on the right. * a commercial binder.

5. ASSESSMENT OF NUTRIENT LEACHING

The type of binder can impact the stability of feed in water and influence the attractiveness of microbound diets through the leaching of attractant molecules.

According to Partridge and Southgate (1999), leaching rose as binder content decreased. O'Mahoney et al. (2011) conducted a study to examine the properties of potential binders for abalone farming. They observed that feed containing glucomannan-xanthan gum did not exhibit any dry matter leaching even after being immersed in water for up to 4 days. There is currently no standardized method for measuring nutrient leaching, despite its significance. Some authors equate nutrient leaching with dry matter loss following water immersion. Other authors utilize various methods to assess nutrient leaching, such as including radiolabelled chemicals into the feed, using internal markers, and detecting specific nutrients, proteins, or lipids. Genodepa et al. (2007) conducted an experiment where they added 14C-labelled rotifers to the diet and examined how zein, agar, alginate, and carrageenan affected nutrient leaching. Leaching was mostly observed in the initial 30 minutes of immersion. Longer immersion times (up to 240 min) led to only a slight rise in leaching rates. The leaching rate of diets bound with zein was notably decreased compared to diets bound with agar, alginate, and carrageenan. Fagbenro and Jauncey (1995) found that the leaching of total protein content was minimal when pellets produced with 3% CMC, maize starch, guar gum, or wheat gluten were immersed in water for 10 minutes. Simon (2009) discovered that there was no notable variation in chromic oxide (used as an internal standard) and nutritional levels following a 1-hour submersion in water of pellets containing gelatine (8%), alginate (7%), or agar (8%). Kovalenko et al. (2002) studied the levels of water soluble vitamins riboflavin and thiamine in feed before and after immersion in water with 5.38% alginate. They found a significant decrease in both vitamins after 90 minutes of immersion, with the highest percentage loss occurring within the first 1 to 15 minutes. Falayi et al. (2006) assessed the overall protein and fat levels in pelleted feed samples with 10% wheat grain starch and cassava tuber starch serving as binding agents and carbohydrate sources. After 1 hour of water immersion, protein retention decreased to approximately 80%, but lipid retention remained at around 97-95%.

6. BINDER EFFECT ON GROWTH

The correlation between food availability and composition and the growth rate of a specific aquatic species is vital for optimizing rearing circumstances. To comprehend this link, one must consider and assess the primary physiological processes of the organism, such as ingestion, assimilation, respiration, development, and reproduction, using integrated models (Van der Meer, 2006). In recent decades, theories of dynamic energy budgets (DEB) gained momentum, surpassing the previously dominant static approach. The Kooijman κ-rule DEB theory (Kooijman, 2001) is one of the most comprehensive models among those proposed. According to Figure 9, this model posits that a portion of consumed food is absorbed, stored in reserves, and then divided for maintenance and development (primary flow), with the remainder allocated for maturation and reproduction. Since all DEB models are rather simplistic, they do not incorporate the species-specific physiological features of the single species being studied, which must be taken into consideration each time. Considering the significant role binders play in prepared pellets, it is crucial to carefully assess their impact on the ingestion, digestion, and assimilation processes of each species in order to optimize feed efficiency in experiments.

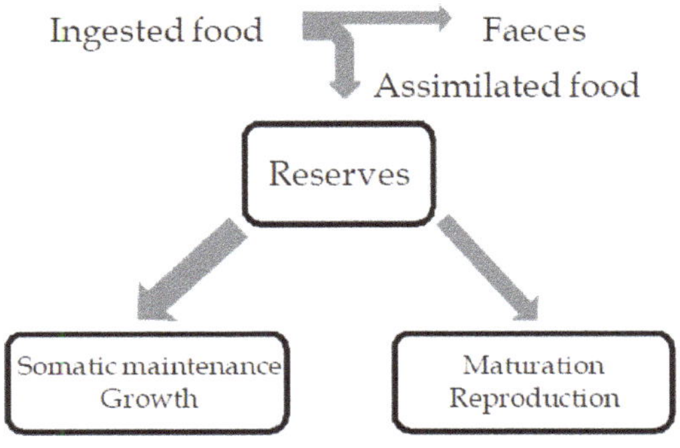

Fig. 9. Schematic representation (modified from van der Meer, 2006) of the Koijman Dynamic Energy Budget model.

6.1 Fish

The availability and nutritional value of live food can be a limiting issue in larviculture of many fish and crustaceans. Many species of larvae have been successfully produced utilizing Artemia salina nauplii and rotifers. The nutritional value of the cysts may vary depending on their source and the timing of harvest. Formulated feed is more appealing since it can be customized to match the nutritional requirements of the species by adjusting its content and size. Many commercial larval diets utilize microbound diets, incorporating binders to stabilize feed pellets and minimize leaching and disintegration (Hashim and Saat, 1992). Carrageenan pellets are appropriate for ayu fish but not for postlarval tongue sole. Alginate and zein are found to be appropriate for cultivating sea bass larvae, but not barramundi larvae. However, a previous study indicated that barramundi larvae could digest both alginate and gelatine bound diets. The addition of sodium carboxymethylcellulose, sodium alginate, and gelatin to the meal for postlarval tongue sole resulted in a growth rate comparable to the control group (Liu et al., 2008). A patented microbound diet consisting of soy lecithin, wheat gluten, and alginate has been developed for feeding larval fish and crustaceans (D'Abramo, 2003).

6.2 Crustaceans

Dietary binders have been evaluated for growth performance in many species, yielding inconsistent outcomes. Wheat flour, whole cassava meal, and dried molasses were shown to be excellent binders that resulted in the maximum weight gain in crayfish Macrobrachium rosenbergii post-larvae compared to agar. However, pellets containing agar had the best texture. The study did not include a comparison with a control group that was fed a natural diet. Kovalenko et al. (2002) found that larvae fed alginate-containing feed had a growth rate that was only 90% of those fed a natural diet (newly hatched nauplii of Artemia). Lignosol and agar both caused a notable weight gain in juvenile shrimp Palaemonetes varians and Palaemon elegans (Palma et al., 2008). Studies by Volpe et al. (2008) and Coccia et al. (2010) found that alginate, agar, and pectin led to significant growth improvements in both young and adult Cherax albidus compared to those fed a natural diet. Pectin was the most effective among the three substances for both age groups.

6.3 Molluscs

Alginate and gelatin have been examined for their effects on cephalopod development. Alginate, a food binder in Octopus maya, has a detrimental impact on growth rate and survival even at a low concentration. The survival rate of Octopus maya fed alginate-bound crab was just 15% compared to the control group fed with crabs (Rosas et al., 2008). Octopus maya and Octopus vulgaris that were fed diets with alginate and gelatin as binders showed significantly decreased growth compared to those provided natural diets, according to studies by Rosas et al. (2008), Valverde et al. (2008), and Garcia et al. (2011). Konjac glucomannan and xanthan gum were used in varying ratios (1:1; 1:3) in abalone feed. While there were no observed changes in growth between the treatments, animals that consumed a natural diet exhibited much higher growth characteristics (O'Mahoney et al., 2011).

6.4 Sea urchins

The swift growth of sea urchin culture has raised the need for specialized formulated feed tailored to meet the dietary requirements of echinoids. Studies have shown that prepared diets enhance test growth and gonad yield more than natural food. Additionally, feed designed for species other than echinoids has been found to effectively promote the growth and maturation of urchins' gonads. However, these diets often result in inferior gonad quality in terms of texture, firmness, color, and flavor, leading to reduced marketability of the product. Research on formulating appropriate feed for sea urchins should consider the impact on gonad yield and sensory quality of the experimental diet components and binders used to impart consistency to the pellets. Various binders have been effectively used in creating feed for various types of sea urchins. Gelatin-based pellets in many echinoid species have been found to promote greater urchin growth and higher gonad yield compared to urchins fed on fresh macroalgae. Additionally, pellets strengthened with the enzyme transglutaminase have shown better results in terms of gonad index compared to a diet of Laminaria hyperboreus in Strongylocentrotus droebachiensis. A diet with a combination of binder gelatin and sodium alginate was more effective in promoting the growth of Strongylocentrotus droebachiensis juveniles compared to fresh Laminaria, as demonstrated by Kennedy et al. in 2007. Agar, sodium alginate, guar gum, and corn starch-based pellets outperformed a natural diet in terms of gonad indices (Barker et al., 1998; Akiyama et al., 2001; Pearce et al., 2002b). Although most research on sea urchin farming in controlled environments has concentrated on developing specialized feed, few studies have investigated how the type and amount of binder used affect the growth of urchins' gonads and their sensory characteristics. In 1997, Akiyama and colleagues examined the binding capabilities of 30% sodium alginate and 15% Curdlan™, a polysaccharide derived from the bacteria Alcaligenes faecalis var. myxogenes and frequently utilized in human food products. Gonad growth in Pseudocentrotus depressus was greater when fed with alginate-based feed compared to Curdlan™-based pellets. Pearce et al. (2002b) conducted an experiment testing various types of binders (gelatin, guar gum, sodium alginate, and maize starch) at concentrations of 3% and 5% to assess their impact on gonad growth in Strongylocentrotus droebachiensis. Although gonad growth was quicker with gelatin and alginate-based pellets, after the 12-week study, gonad indices (GI) were comparable for the different binders at the higher concentration (5%), but somewhat lower with 3% guar gum and corn starch. Fabbrocini et al. (2011) examined the efficacy of agar-based pellets in promoting gonad growth and gamete generation in Paracentrotus lividus. Agar did not inhibit gonad growth in urchins fed on pellets made of Classic© commercial feed and 3% agar. Urchins fed on Ulva-agar pellets in a 4-week trial showed similar progress in the reproductive cycle and comparable gonad index levels to those fed on commercial feed alone. A meal rich in gelatin yielded positive outcomes in terms of gonad sensory quality in Evechinus chloroticus (Phillips et al., 2009). Conversely, Mortensen et al. (2004) found that a meal containing gelatin-transglutaminase resulted in unsatisfactory gonad taste in Strongylocentrotus droebachiensis. On the other hand, Pearce et al. (2002b) achieved the best gonad color outcomes using a starch-based pellet. When selecting a binder, it is important to examine the specific species of urchin being raised and the desired outcome, such as urchin growth, gonad yield, gamete production, or gonad sensory quality.

7. BINDER EFFECT ON FEED DIGESTIBILITY

Multiple findings suggest a correlation between binders and the digestion of nutrients. Various methods are employed to assess feed digestibility, with the complete collection technique being the most dependable while being laborious and taxing for the animal (Khan et al., 2003). The animal is given a measured amount of feed, and precise records of both feed consumption and fecal excretion are maintained. Digestibility trials in aquatic animals may be prone to mistakes due to challenges in accurately quantifying feed intake and the procedures used to collect feces, such as settlement and filtering, which can lead to leaching losses (Irvin & Tabrett, 2005). The settlement collection approach leads to an overestimation of protein digestibility by up to 6% in shrimp when feces are collected every 6 hours instead of hourly post-feeding (Smith & Tabrett, 2004). In vitro methods can be utilized as a rapid and cost-effective way to predict nutrient digestibility. Researchers estimated the digestibility of different carbohydrate sources for young spiny lobsters by measuring how quickly they were broken down in a lab setting using enzyme mixtures. They also looked at the levels of glucose in the lobsters' blood after they consumed specific diets. Various factors impact the effectiveness of digestion. The influence of binder type on feed digestibility is one of the aspects to consider when constructing a formulated feed for optimal success. Excessive amounts of binder can actually decrease the digestibility of the meal. Assessing the apparent digestibility of binders is essential to comprehend their bioavailability in formulated diets and their impact on the digestibility of other main macro-ingredients like protein. The impact of binders on food digestibility has been widely ignored. Simon (2009) discovered that gelatin, at a 7-8% inclusion level, improved digestibility in juvenile lobsters of the species Jasus edwardsii, while agar and alginate did not have the same effect. Octopus maya fed alginate-containing pellets had weight loss and mortality, indicating that this binder hinders nutrient absorption. Conversely, gelatin was found to enhance nutritional absorption (Rosas et al., 2008). Binders in fish feed can reduce nutritional digestibility by speeding up the passage of food through the gastrointestinal tract (Storebakken, 1985). Adding guar gum to tilapia feed results in poorer nutritional digestibility coefficients compared to diets with CMC, corn starch, and wheat gluten (Fagbenro & Jauncey, 1995). Consistent with prior findings, fish feed with potent binders was found to have adverse effects on macronutrient digestibility, likely attributed to physical factors including alterations in viscoelastic characteristics (Storebakken & Austreng, 1987; Storebakken et al., 1998). The impact of two natural binders, fava and kidney beans, on protein and fat digestibility in Atlantic salmon has been studied recently. Salmo salar did not exhibit any notable variations among the dietary treatments (Pratoomyot et al., 2011), and the inclusion of guar gum in the rainbow trout diet enhanced digestibility when combined with plant meal (Brinker & Reiter, 2011). The distribution of food within the gut remained consistent regardless of diet or size of the sea urchin Psammechinus miliaris (Otero-Villanueva et al., 2004). Lawrence et al. (1989) discovered that in Paracentrotus lividus, the time it takes for food to pass through the stomach and the digestibility of dry matter were comparable when fed agar-based diets with either soybean or fish meals. Klinger et al. (1994) found comparable outcomes for Litechinus variegatus. Evechinus chloroticus showed no variation in absorption efficiency when fed extruded or agar-based pellets, as reported by Barker et al. (1998). Research by Akiyama et al. (2001) found that in Pseudocentrotus depressus, alginate-based pellets had higher feed efficiency compared to kelp. Additionally, Akiyama et al. (1997) observed that alginate-based pellets had lower feed efficiency than Curdlan™-based pellets.

Table 3. Effect of feed binders on growth

BINDER	% Binder employed	SPECIES	GROWTH	Reference
CARBOHYDRATES				
Agar	2	*Palemonetes varians* (juveniles)	Positive	Palma et al., 2008
	2	*Palemon elegans* (juveniles)	Positive	
	2.5	*Cherax albidus* (adults)	Positive	Volpe et al., 2008
	2	*Cherax albidus* (juveniles)	Positive	Coccia et al., 2010
	3	*Paracentrotus lividus* (adults)	Same as control	Fabbrocini et al., 2011
Alginate	2.5	*Cherax albidus* (adults)	Positive	Volpe et al., 2008
	5.38	*Macrobrachium rosembergi* (larvae)	Negative	Kovalenko et al., 2002
	10	*Octopus vulgaris*	Negative	Garcia et al., 2011
	1	*Octopus Maya*	Negative	Rosas et al., 2008
	30	*Octopus vulgaris*	Negative	Valverde et al., 2008
	2	*Cynoglossus semilaevis* (postlarvae)	Same as control	Liu et al., 2008
	2	*Cherax albidus* (juveniles)	Positive	Coccia et al., 2010
	30	*Pseudocentrotus depressus* (juveniles)	Negative vs control Positive vs curdlan®	Akiyama et al., 1997
	30	*Pseudocentrotus depressus* (juveniles)	Positive	Akiyama et al., 2001
	3; 5	*Strongylocentrotus droebachiensis* (adults)	Positive	Pearce et al., 2002b
Carboxymethyl cellulose	2	*Cynoglossus semilaevis*	Same as control	Liu et al., 2008
Carrageenan	2	*Cynoglossus semilaevis*	Negative	
Chitosan	1	Cherax albidus (juveniles)	Positive	Coccia et al., 2010
Konjac glucomannan-xanthan gum	0.1; 7,5	*Haliotis discus hannai*	Negative	O'Mahoney et al., 2011
Guar gum	3; 5	*Strongylocentrotus droebachiensis* (adults)	Positive	Pearce et al., 2002b

BINDER	% Binder employed	SPECIES	GROWTH	Reference
Lignosol	2	*Palemonetes varians* (juveniles)	Positive	Palma et al., 2008
	2	*Palemon elegans* (juveniles)	Positive	
Corn starch	3; 5	*Strongylocentrotus droebachiensis* (adults)	Positive	Pearce et al., 2002b
Pectin	2.5	Cherax albidus (adults)	Positive	Volpe et al., 2008
	1	Cherax albidus (juveniles)	Positive	Coccia et al., 2010
PROTEINS				
gelatin	10	*Octopus vulgaris*	Negative	Garcia et al., 2011
	2	*Octopus Maya*	Same as control	Rosas et al., 2008
	10	*Octopus vulgaris*	Negative	Valverde et al., 2008
	2	*Cynoglossus semilaevis* (postlarvae)	Same as control	Liu et al., 2008
	3; 5	*Strongylocentrotus droebachiensis* (adults)	Positive	Pearce et al., 2002b
	5	*Strongylocentrotus droebachiensis* (juveniles)	Positive	Pearce et al., 2002b
	5	*Strongylocentrotus droebachiensis* (juveniles)	Positive	Daggett et al., 2005
MIXED BINDERS				
Gelatin+ transglutaminase	? 0.13%	*Strongylocentrotus droebachiensis* (juveniles)	Positive	Mortensen et al., 2004
Gelatin+ alginate	5 2	*Strongylocentrotus droebachiensis* (juveniles)	Positive	Kennedy et al., 2007

8. BINDER EFFECT ON DIGESTIVE ENZYMES

It is well accepted that species possess a specific set of digestive enzymes that correspond to their distinct life histories (Figueiredo & Anderson, 2009). Therefore, by analyzing the digestive enzyme profile, one can forecast the species' capacity to utilize certain nutrients. Higher digestive enzyme activity is observed for substrates that are frequently consumed in the diet, leading to the assumption that the activity of digestive enzymes can be enhanced by introducing specific nutrients into the diet for a sufficient duration. Crustaceans are known to adjust their digestive enzymes to match the composition of their diet, allowing them to break down a diverse range of food sources they encounter in their natural environment. Changes in digestive enzyme activity in several penaeid shrimp larvae have been linked to the amount or type of certain dietary components (Le-Vay et al., 1993; Rodriguez et al., 1994; Le Moullac et al., 1997; Lemos & Rodriguez, 1998; Brito et al., 2000). The diet's protein composition enhanced protease activity in Marsopenaeus japonicas juveniles and Litopenaeus vannamei larvae, as reported by Van Wormhoudt et al. (1986), Le Moullac & Van Wormhoudt (1994), and Le Moullac et al. (1997). Some authors propose that the rise in amylase activity observed in Penaeus monodon (Fang & Lee, 1992) and Litopenaeus vannamei (Brito et al., 2000) could be linked to the limited carbohydrate content in the diet (Le Vay et al., 1993; Rodriguez et al., 1994; Kumlu & Jones, 1995; Lemos & Rodriguez, 1998). Protease, amylase, and cellulase activities are positively associated with dietary protein levels and adversely associated with high cellulose levels (Pavasovic et al., 2007b, 2006). However, the dietary protein intake in white shrimp Penaeus vannamei did not affect overall protease activity, while changes in trypsin- and chymotrypsin-like activities have been observed (Muhlia-Almazan et al., 2003). No substantial difference was found in protease activity between Litopenaeus vannamei and Scylla serrata under various dietary regimens. The diet composition of Cherax quadricarinatus affects the secretion of digestive enzymes, which can be influenced by the presence of animal- or plant-derived substances (Lopez-Lopez et al., 2005; Pavasovic et al., 2007a). Zhou et al. (2007) observed a favorable association between dietary lipid level and lipase activity in the mollusk Babylonia areolata. Proteolytic activity in teleosts does not seem to be affected by the protein content of the diet in Pseudoplatystoma corruscans and early weaned sea bass larvae, as shown in studies by Lundstedt et al. (2004) and Cahu & Zambonino-Infante (1994) respectively. However, in Colossoma macropomum, digestive protease levels increased with higher dietary protein content, as observed in the study by De Almeida et al. (2006). Conversely, Dentex dentex exhibited increased alkaline protease activity when fed a meal with lower protein content and greater carbohydrate content (Perez-Jimenez et al., 2009). Amylase activity was elevated in Pseudoplatystoma corruscans when fed diets with 13-25% starch. Similarly, this activity could be regulated by adjusting the starch content of the diet in Colossoma macropomum and Dentex dentex. De Almeida et al. (2006) found a positive correlation between lipase and dietary lipid levels in Colossoma macropomum, while Luo et al. (2010) identified a similar correlation in Lateolabrax japonicus. The developing picture in teleosts is characterized by considerable complexity and variety, influenced by the diverse environments and life histories of these species. There is a lack of research specifically focused on studying how feed binders impact digestive enzyme performance. We have investigated the impact of several binders on the profile of digestive enzymes in Cherax albidus (Volpe et al., submitted). We conducted a study to investigate how the presence of binders affects the enzymatic profile in the digestive system of juvenile Cherax albidus. We analyzed carbohydratases, lipases, and proteases in the digestive tract under natural diet conditions and experimental diets with pellets containing various binders but similar proximate composition. Our study revealed that digestive enzyme activity were not significantly different in the digestive tract, except for amylase, which was notably greater in the intestine compared to the gastric juice and hepatopancreas of rats that had pectin-containing pellets for 12 weeks.

9. AT A GLANCE

High-quality water-resistant pellets are a significant issue for the aquaculture sector. Binder agents are commonly used to enhance the stability of aquatic animal feed. Binders are thought to help avoid or minimize significant nutrient loss and maintain the texture of feed when pellets are submerged in water for extended durations. Preventing or minimizing this inconvenience would have positive effects on the aquaculture business. The cost of the diet can be greatly impacted by binder prices, thus using natural components that are biodegradable and renewable may be beneficial from both an environmental and economic perspective. Ongoing research in the field of binding agents acknowledges the variability in eating behaviors and nutritional requirements among various species and even within the same species at different life stages. Although there has been improvement in recent years, understanding of microbound diets and their effects on digestive physiology and growth of farmed aquatic species is still insufficient to meet the industry's requirements, necessitating further research.

REFERENCES

1. Ahmed, J.; Ramaswamy, H.S. ; Ngadi, M.O.(2005). Rheological characteristics of Arabic gum in combination with guar and xanthan gum using response surface methodology: Effect of temperature and concentration. *International Journal of Food Properties*, Vol. 8, No. 2, (April 2005), pp. 179-192, ISSN 1094-2912

2. Akiyama, T.; Unuma, T.; Yamamoto, T.; Furuita, H. & Konishi, K. (1997). An evaluation of amino acid sources and binders in semipurified diet for red sea urchin *Pseudocentrotus depressus*. *Fisheries Science* , Vol. 63, No. 6, pp. 881-886, (December 1997), ISSN 1444-2906

3. Akiyama, T.; Unuma T. & Yamamoto, T. (2001). Optimum protein level in a purified diet for young red sea urchin *Pseudocentrotus depressus*. *Fisheries Science*, Vol. 67, No. 2, (April 2001), pp. 361-363, ISSN 1444-2906

4. Aspinall, G.O. (1982) Chemical characterization and structure determination of polysaccharides. In: *The Polysaccharides*, Aspinall G.O. (ed.), Vol. 1, pp. 35-131, Academic Press, Inc., New York

5. Arisz, P.W.; Kauw, H.J.J. & Boon, J.J. (1995) Substituent distribution along cellulose backbone in O-methylcelluloses using GC and FAB-MS for monomer and oligomer analysis. *Carbohydrate Research*, Vol. 271, No. 1, (July 1995), pp. 1-14, ISSN 0008-6215

6. Atkins, E. D. T. (Ed.). (1985). *Polysaccharides: Topics in structure and morphology.* Wiley-VCH Verlag GmbH, ISBN 9783527264100, Germany

7. Azad, A.K.; Pearce, C.M. & McKinley R.S. (2011). Effects of diet and temperature on ingestion, absorption, assimilation, gonad yield, and gonad quality of the purple sea urchin (*Strongylocentrotus purpuratus*). *Aquaculture*, Vol. 317, No. 1-4, (July 2011), ISSN 0044-8486

8. Barker, M.F.; Keogh, J.A.; Lawrence, J.M. & Lawrence, A.L. (1998). Feeding rate, absorption efficiencies, growth, and enhancement of gonad production in the New Zealand sea urchin *Evechinus chloroticus* Valenciennes (Echinoidea: Echinometridae) fed prepared and natural diets. *Journal of Shellfish Research*, Vol. 17, No. 5, (December 1998), pp. 1583-1590, ISSN 0730-8000

9. Barker, M.F.; Bremer, P.; Silcock, P.; Delahunty, C. & Sewell, M. (2006). Enhancement of yield and quality of *Evechinus chloroticus* roe through controlled diet. *Journal of Shellfish Research*, Vol. 25, No. 2, (August 2006), p. 711, ISSN 0730-8000

10. Bautista M.N.; Millamena, O.M. & Kanazawa, A. (1989). Use of kappa-carrageenan microbound diet (C-MBD) as feed for *Penaeus monodon* larvae. *Marine Biology*, Vol. 103, pp. 169-173, ISSN 0025-3162

11. Brinker, A. (2007). Guar gum in rainbow trout (*Oncorhynchus mykiss*) feed: the influence of quality and dose on stabilization of faecal solids. *Aquaculture*, Vol. 267, No.1-4, (July 2007), pp. 315-327, ISSN 0044-8486

12. Brinker, A. & Reiter, R. (2011). Fish meal replacement by plant protein substitution and guar gum addition in trout feed, Part I: Effects on feed utilization and fish quality. *Aquaculture*, Vol. 310, No. 3-4, (January 2011), pp.350-360, ISSN 0044-8486

13. Brito R.; Chimal, M.E.; Gaxiola, G. & Rosas, C. (2000). Growth, metabolic rate, and digestive enzyme activity in the white shrimp *Litopenaeus setiferus* early postlarvae fed different diets. *Journal of Experimental Marine Biology and Ecology*, Vol. 255, No. 1, (December 2000), pp. 21-36. ISSN 0022-0981

14. Cahu, C.L. & Zambonino-Infante, J.L. (1994). Early weaning of sea bass (*Dicentrarchus labrax*) larvae with a compound diet: effect on digestive enzymes. *Comparative Biochemistry and Physiology part A*, Vol. 109, No. 2 (October 1994), pp. 213–222, ISSN 1095-6433

15. Caltagirone, A.; Francour, P. & Fernandez C. (1992). Formulation of an artificial diet for the rearing of the urchin *Paracentrotus lividus*: I. Comparison of different binding agents. In: *Echinoderm Research* L. Scalera-Liaci, C. Canicattì (Eds), pp. 115-119, ISBN 90-5410-049-4, Balkema, Rotterdam

16. Castillo-García, E.; López-Carballo, G. & López-Castellano, A. (2005). Guar gum characteristics and applications. *Ciencia y Tecnologia Pharmaceutica*, Vol. 15, No. 1, (March 2005), pp. 3-10, ISSN 1575-3409

17. Cerver, M.F.; Heinämäki, J.; Räsänen, M.; Maunu, S.L.; Karjalainen, M.; Acosta, O.M., Colart,

18. A. I. & Yliruusi, J. (2004). Solid-state characterization of chitosan derived from lobster chitin. *Carbohydrate Polymers*, Vol. 58, No. 4-7, (December 2004), pp. 401-408, ISSN 0144-8617

19. Chevanan, N.; Muthukumarappan, K. & Rosentrater, K. A. (2009). Extrusion Studies of Aquaculture Feed using Distillers Dried Grains with Solubles and Whey. *Food and Bioprocess Technology*, Vol. 2, No. 2, (March 2009), pp. 177–185, ISSN 1935-5130

20. Clark, A. H. & Ross-Murphy, S. B. (1987). Structural and mechanical properties of biopolymers gels. In: *Advances in Polymer Science,* Vol. 83 (October 1987), pp. 57– 192, ISSN 0065-3195

21. Coccia, E.; Santagata, G.; Malinconico, M.; Volpe, M.G.; Di Stasio, M. & Paolucci, M. (2010) *Cherax albidus* juveniles fed polysaccharide-based pellets: rheological behavior and effect on growth. *Freshwater crayfish*, Vol. 17, pp.13-18, ISBN 978-951-27-1322-6

22. Coccia, E.; Varricchio, E. & Paolucci, M. (2011). Digestive enzymes in the crayfish *Cherax albidus*: polymorphism and partial characterization. *International Journal of Zoology*, Vol. 2011, (March 2011), pp.1-9, ISSN 1687-8477

23. Coviello, T.; Alhaique, F.; Dorigo, A.; Matricardi, P. & Grassi, M. (2007). Two galactomannans and scleroglucan as matrices for drug delivery: preparation and release studies. *European Journal of Pharmaceutics and Biopharmaceutics,* Vol. 66, No. 2, (May 2007), pp. 200-209, ISSN 0939-6411

24. Cuzon, G.; Guillaume, J. & Cahu, C. (1994). Composition, preparation and utilization of feeds for Crustacea. *Aquaculture*, Vol. 124, No.1-4, (June 1994), pp. 253–267, ISSN 0044-8486

25. D'Abramo, L.R. (2003). Microparticulate microbound diet for the culture of larval fish and crustaceans. US 6645536 (2003)

26. Daggett, T.L.; Pearce, C.M.; Tingley, M.; Robinson, S.M.C. & Chopin, T. (2005). Effect of prepared and macroalgal diets and seed stock source on somatic

growth of juvenile green sea urchins (*Strongylocentrotus droebachiensis*). *Aquaculture,* Vol. 244, No. 1-4, (February 2005), pp. 263-281, ISSN 0044-8486

27. De Almeida, L.C.; Lundstedt, L.M. & Moraes, G. (2006). Digestive enzyme responses of tambaqui (*Colossoma macropomum*) fed on different levels of protein and lipid. *Aquaculture Nutrition*, Vol. 12, No. 6, (December 2006), pp. 443-450, ISSN 1353-5773
28. De Muylder, E.; Hage, H. & van der Velden, G. (2008). Binders: Gelatin as alternative for urea formaldehyde and wheat gluten in the production of water stable shrimp feeds. *Aquafeed International magazine*, Volume 11, No. 2, (April 2008), ISSN 1464-0058
29. De Silva, S.S. & T.A. Anderson. (1995). *Fish Nutrition in Aquaculture*. Chapman & Hall, ISBN 0412 55030X, London
30. Dominy, W.G. & Lim, C. (1991). Performance of binder in pelleted shrimp diets. *Proceedings of the aquaculture feed processing and nutrition workshop*, pp. 149-157, Thailand and Indonesia, September 1991
31. Drochner, H.; Kerler, A. & Zazharias, B. (2004). Pectin in pig nutrition, a comparative review. *Journal of Animal Physiology and Animal Nutrition*, Vol. 88, No. 11-12, (April 2004), pp. 367-380, ISSN 0931-2439
32. Durazo-Beltran, E. & Viana, M.T. (2001). Effect of the concentration of agar, alginate and carrageenan on the stability, toughness and nutrient leaching in artificial diets for abalone. *Ciencias marinas*, vol. 27, No. 1, pp. 1-19, ISSN 0185-3880
33. El-hefian, E.A.; Nasef, M.M. & Yahaya, A.H. (2011). Chitosan physical forms: A short review. *Australian Journal of Basic and Applied Sciences* Vol.5, No. 5, (May 2011), pp. 670-677, ISSN 19918178677
34. Fabbrocini, A. & D'Adamo, R. (2010). Gamete maturation and gonad growth in fed and starved sea urchin *Paracentrotus lividus* (Lamarck, 1816). *Journal of Shellfish Research,* Vol. 29, No. 4, (December 2010), pp. 1051-1059, ISSN 0730-8000
35. Fabbrocini, A.; Volpe, M.G.; Di Stasio, M.; D'Adamo, R.; Maurizio, D.; Coccia, E. & Paolucci,
36. M. (2011). Agar-based pellet as feed for sea urchins (*Paracentrotus lividus*): rheological behaviour, digestive enzymes and gonad growth. *Aquaculture Research,* (in press, DOI: 10.1111/j.1365-2109.2011.02831.x), ISSN 1365-2109
37. Fagbenro, O. & Jauncey, K. (1995). Water stability, nutrient leaching and nutritional properties of moist fermented fish silage diets. *Aquaculture Engineering*, Vol. 14, No. 4, pp. 143–153, ISSN 0144-8609

38. Falayi, B.A.; Sadiku, S.O.; Eyo, A.A. & Okaeme, A.N. (2006). Water stability and floatation of fish diets bound with different agents and *Saccharomyces cerivisiae*. *Nigerian Journal of Fisheries*, Vol 2, No. 3, pp. 181-196, ISSN 1597-443.

39. Fang, L.S. & Lee, B.N. (1992). Ontogenic change of digestive enzymes in *Penaeus monodon*. *Comparative Biochemistry and Physiology Part B: Comparative Biochemistry*, Vol. 103, No. 4, (December 1992), pp. 1033-1037, ISSN 1096-4959

40. Farris, S.; Schaich, K.M.; Liu, L.; Piergiovanni, L. & Yam, K.L. (2009). Development of polyion-complex hydrogels as an alternative approach for the production of bio- based polymers for food packaging applications: a review. *Trends in Food Science and Technology*, Vol. 20, No. 2, (February 2009), pp. 316-332, ISSN 0924-2244

41. Fernandez, C. & Pergent, G. (1998). Effect of different formulated diets and rearing conditions on growth parameters in the sea urchin *Paracentrotus lividus*. *Journal of Shellfish Research*, Vol. 17, No. 5, (December 1998), pp. 1571-1581, ISSN 0730-8000

42. Figueiredo, M.S.R.B., & Anderson, A.J. (2009). Digestive enzyme spectra in crustacean decapods (Paleomonidae, Portunidae and Penaeidae) feeding in the natural habitat. *Aquaculture Research*, Vol. 40, No. 3, (February 2009), pp. 282–291, ISSN 1365-2109

43. Forbes, J.M. (2001). Consequences of feeding for future feeding. *Comparative Biochemistry and Physiology Part A: Molecular & Integrative Physiology*, Vol. 128, No. 3, (March 2001), pp. 463-70, ISSN 1095-6433

44. Garcia, S.; Domingues, P.; Navarro, J.C.; Hachero, I.; Garrido, D. & Rosas, C. (2011). Growth, partial energy balance, mantle and digestive gland lipid composition of *Octopus vulgaris* (Cuvier, 1797) fed with two artificial diets. *Aquaculture Nutrition*, Vol. 17, No. 2 (April 2011), pp. e174-e 187, ISSN 1353-5773

45. Genodepa, J.; Zeng, C. & Southgate, P. C. (2004). Preliminary assessment of a microbound diet as an *Artemia* replacement for mud crab, *Scylla serrata*, megalopa. *Aquaculture*, Vol. 236, No. 1-4 (June 2004), pp. 497-509, ISSN 0044-8486

46. Genodepa, J.; Zeng, C. & uence of binder type on leaching rate and ingestion of microbound diets by mud crab,⇓Southgate, P. C. (2007). In *Scylla serrata* (Forsskal), larvae. *Aquaculture Research*, Vol. 38, No. 14, (October 2007), pp. 1486-1494, ISSN 1365-2109

47. Goldstein, A.M.; Alter, E.N. & Seaman, J.K. (1973). Guar gum, *In: Industrial Gums, Polysaccharides and their Derivatives* Whistler R.L. Ed., pp. 303–321, Academic Press, New York

48. Grant, N. H.; Clark, D. E. & Rosanoff, E. I. (1973). *Biochemical and Biophysical Research Communications*, Vol. 51, pp. 100-106, ISSN 0006-291X

49. Grasdalen, H.; Larsen, B. & Snidsrod, O. (1981). 13C-N.M.R. studies of monomeric composition and sequence in alginate *Carbohydrate Research*, Vol. 89, No. 2, (January 1981), pp. 179-191, ISSN 0008-6215

50. Hashim, R. & Saat, M.A.M. (1992). The utilization of seaweed meals as binding agents in pelleted feeds for snakehead (*Channa striatus*) fry and their effects on growth. *Aquaculture*, Vol. 108, No. 3-4, (December 1992), pp. 299-308, ISSN 0044-8486

51. Holdich, D.M. (2002). *Biology of freshwater crayfish.* Blackwell Science, ISBN 978-0-632-05431- 2, United Kingdom.

52. Irvin, S.J. & Tabrett, S.J. (2005). A novel method of collecting fecal samples from spiny lobsters. *Aquaculture*, Vol. 243, No. 1-4, (January 2005), pp. 269-272, ISSN 0044-8486

53. Johnson, E.L. & Peniston, Q.P. (1982). Utilization of Shellfish Waste for Chitin, Chitosan Production, In: *Chemistry and Bio-chemistry of Marine Food Products*, Martin, R.E.; Flick, G.J.; Hebard C.E. & Ward D.R. eds, pp. 415-428, AVI Publishing Co, ISBN 0870554085, 9780870554087, Westport, CT, USA

54. Johnston, D.J. & Freeman, J., (2005). Dietary preference and digestive enzyme activities as indicators of trophic resource utilization by six species of crab. *The Biological Bulletin*, Vol. 208, No. 1, (February 2005), pp. 36–46, ISSN 0006-3185

55. Johnston, M. D. & Johnston, D.L. (2007). Stability of Formulated Diets and Feeding Response of Stage I Western Spiny Lobster, *Panulirus cygnus*, Phyllosomata. *Journal of the World Aquaculture Society*, Vol. 38, No.2, (June 2007), pp. 262–271, ISSN 1749-7345

56. Kalia, A.N. (2005). Textbook of Industrial Pharmacognosy. New Delhi CBS Publishers and Distributors, ISBN 8123912099, India

57. Kalian, N. & Morey, R.V. (2009) Factors affecting strength and durability of densified biomass products. *Biomass and Bioenergy*, Vol. 33, No. 3, (March 2009), pp. 337-359,

58. ISSN 0961-9534

59. Kennedy, J.F. & White, C.A. (1983). Bioactive Carbohydrates, In: *Chemistry, Biochemistry and Biology*, Kennedy, J.F. & White, C.A. Eds., pp. 163, Ellis Horwood Publishers, Chichester, United Kingdom

60. Kennedy, E.J.; Robinson, S.M.C.; Parsons, G.J. &. Castell, J.D. (2007). Effects of dietary minerals and pigment on somatic growth of juvenile green sea urchins,

Strongylocentrotus droebachiensis. *Journal of the World Aquaculture Society*, Vol. 38, No. 1, (March 2007), pp. 36-48, ISSN 1749-7345

61. Khan, M.A.; Nisa, M.U. & Sarwar, M., (2003). Techniques Measuring Digestibility for the Nutritional Evaluation of Feeds. A Review in *International Journal of Agriculture and Biology*, Vol. 5, No. 1, pp. 91-94, ISSN 1560–8530.
62. Klinger, T.S.; Lawrence, J.M. & Lawrence A.L. (1994). Digestive characteristics of the sea- urchin *Lytechinus variegatus* (Lamarck) (Echinodermata: Echinoidea) fed prepared feeds. *Journal of the World Aquaculture Society*, Vol. 25, No. 4, (December 1994), pp. 489-496, ISSN 1749-7345
63. Knauer, J.; Britz, P.J. & Hecht, T. (1993). The effect of seven binding agents on 24-hour water stability of an artificial weaning diet for the South African abalone, *Haliotis midae* (Haliotidae, Gastropoda). *Aquaculture*, Vol. 115, No. 3-4, (September 1993), pp. 327-334, ISSN 0044-8486
64. Koojiman, S.A.L. (2001) Quantitative aspects of metabolic organization: A discussion of concepts. *Philosophical Transactions of the Royal Society B: Biological Sciences*, Vol. 356, No. 1407, (March 2001), pp. 331-349, ISSN 0962-8436
65. Kovalenko, E. E.; D'Abramo, L. R.; Ohs, C. L. & Buddington, R. K. (2002). A successful microbound diet for the larval culture of freshwater prawn *Macrobrachium rosenbergii*. *Aquaculture*, Vol. 210, No. 1-4 (July 2002), pp. 385-395. ISSN 0044-8486
66. Kumlu, M. & cial diets on growth, survival, and trypsin activity in larvae of Penaeus indicus.⇒Jones, D.A. (1995). The effect of live and arti *Journal of World Aquaculture Society*, Vol.26, No. 4, (December 1995), pp 406–415, ISSN 0893-8849
67. Lawrence, J.M.; Regis, M.B.; Delmas, P.; Gras, G. & Klinger T.S. (1989). The effect of quantity and quality of food on feeding and digestion in *Paracentrotus lividus* (Lamarck) (Echinodermata: Echinoidea). *Marine Behaviour and Physiology*, Vol. 15, No. 2, (August 1989), pp. 137-144, ISSN 0091-181X
68. Lee, P.S.; Southgate, P.C. & Fielder, D.S. (1996). Assessment of two microbound artificial diets for weaning Asian sea bass (*Lates calcarifer*, Bloch). *Asian Fisheries Science*, Vol. 9, No.1, pp. 115-120
69. Lemos, D. & Rodriguez, A. (1998). Nutritional effects on body composition, energy content and trypsin activity of *Penaeus japonicus* during early postlarval development. *Aquaculture*, Vol. 160, No. 1-5 (January 1998), pp. 103–116, ISSN 0044-8486
70. Le Moullac, G.; Klein, G.B.; Sellos, D. & Van Wormhoudt, A. (1997). Adaptation of trypsin, chymotrypsin anda-amylase to casein level and protein source in *Penaeus vannamei* (Crustacea, Decapoda). *Journal of Experimental Marine Biology and Ecology*, Vol. 208, No. 1-2 ,(January 1997), pp. 107–125, ISSN 0022-0981

71. Le Moullac, G. & uence of protein and carbohydrate quality in⇓bre levels and in⇒Van Wormhoudt, A. (1994). Adaptation of digestive enzymes to dietary protein,carbohydrate and *Penaeus vannamei* larvae(Crustacea, Decapoda), *Aquatic Living Resources*, Vol. 7, No. 3 (July 1994), pp. 203–210, ISSN 0990-7440

72. Le-Vay, L.; Rodriguez, A.; Kamarudin, M.S. & cial diets on tissue composition and trypsin activity in⇒uence of live and arti⇓Jones, D.A. (1993). In *Penaeus japonicus* larvae. *Aquaculture*, Vol. 118, No. 3-4 (December 1993), pp. 287–297, ISSN 0044-8486

73. Lim C. & Cuzon, G. (1994). Water stability of shrimp pellet: a review. *Asian Fisheries Science*, Vol 7, No. 2-3, (October 1994), pp. 115-126, ISSN 0116-6514

74. Linton, S.M.; Allardyce, B.J.; Hagen, W.; Wencke, P. & Saborowski, R. (2009). Food utilization and digestive ability of aquatic and semi-terrestrial crayfishes, *Cherax destructor* and *Engaeus sericatus* (Astacidae, Parastacidae). *Jounal of Comparative Physiology, part B*, Vol. 179, No. 4, (April 2009), pp. 493-507, ISBN 0174-1578

75. Liu, F.; Ai, Q.; Mai, K.; Tan, B.; Ma, H.; Xu, W.; Zhang, W. & LiuFu, Z. (2008). Effects of Dietary Binders on Survival and Growth Performance of Postlarval Tongue Sole, *Cynoglossus semilaevis* (Günther). *Journal of the World Aquaculture Society*, Vol. 39, No. 4, (August 2008), pp. 441–571, ISSN 1749-7345

76. Lopez-Lopez, S.; Nolasco, H.; Villarreal-Colmenares, H. & Civera-Cerecedo, R. (2005). Digestive enzyme response to supplemental ingredients in practical diets for juvenile freshwater crayfish *Cherax quadricarinatus. Aquaculture nutrition,* Vol. 11, No. 2, (April 2005), pp. 79–85, ISSN 1353-5773

77. Lundstedt, L.M.; Melo, J.F. & Moraes, G. (2004). Digestive enzymes and metabolic profile of *Pseudoplatystoma corruscans* (Teleostei: Siluriformes) in response to diet composition. *Comparative Biochemistry and Physiology, Part B Biochemistry and Molecular Biology,* Vol. 137, No. 3 (March 2004), pp. 331-339, ISSN 1096-4959

78. Luo, G.; Xu, J.; Teng, Y.; Ding, C. & Yan, B. (2010). Effects of dietary lipid levels on the growth, digestive enzyme, feed utilization and fatty acid composition of Japanese sea bass (*Lateolabrax japonicus* L.) reared in freshwater. *Aquaculture research*, Vol. 41, No. 2 (January 2010), pp. 210-219, ISSN 1365-2109

79. Mortensen, A.; Siikavuopio, S.I. & Raa, J. (2004). Use of transglutaminase to produce a stabile sea urchin feed. In: *Sea Urchins-Fisheries and Ecology* Lawrence, J.D. Ed., pp. 203-213. DEStech Publications Inc., ISBN 13 978-1-932078-32-9, ,Landcaster, PA, USA.

80. Moss, S.M.; Divakaran, S. & Kim, B.K. (2001). Stimulating effects of the pond water on digestive enzyme activity in the Pacific white shrimp, *Litopenaeus*

vannamei (Bonne). *Aquaculture Research,* Vol. 32, No. 2, (February 2001), pp. 125–131, ISSN 1365-2109

81. Muhlia-Almazan, A.; Garcìa-Carreño, F.L.; Sanchez-Paz, J.A.; Yepiz-Plascencia, G. & Peregrino-Uriarte, A.B. (2003). Effects of dietary protein on the activity and mRNA level of trypsin in the midgut gland of the white shrimp *Penaeus vannamei. Comparative Biochemistry and Physiology, part B*, Vol. 135, No. 2, (June 2003), pp. 373– 383, ISSN 1096-4959

82. Murano, E.; Toffanin, R.; Zanetti, F.; . Knutsen, S.H.; Paoletti, S. & Rizzo,R. (1992). Chemical and macromolecular characterisation of agar polymers from Gracilariadura (C. Agardh) J. Agardh (Gracilariaceae, Rhodophyta). *Carbohydrate Polymers,*Vol. 18, No. 3, (October 1992), pp. 171-178, ISSN 0144-8617

83. No, H.K. & Meyers, S.P. (1995). Preparation and Characterization of Chitin and Chitosan-A Review. *Journal of Aquatic Food Product Technology*, Vol. 4, No. 2, (June 1995), pp. 27- 52, ISSN 1049-8850

84. Obaldo, L. G.; Divakaran, S. & Tacon, A. G. (2002). Method for determining the physical stability of shrimp feeds in water. *Aquaculture Research*, Vol. 33, No. 5, (April 2002), pp. 369-377, ISSN 1365-2109

85. O' Mahoney, M.; Mouzakitis, G.; Doyle, J. & Burnell, G. (2011). A novel konjac glucomannan–xanthan gum binder for aquaculture feeds: the effect of binder configuration on formulated feed stability, feed palatability and growth performance of the Japanese abalone, *Haliotis discus hannai. Aquaculture Nutrition*, Vol. 17, No. 4, (August 2011), pp. 395–407, ISSN 1353-5773

86. Orire, A.M.; Sadiku, S.O.E. & Tiamiyu, L.O. (2010). Evaluation of Yam Starch (Discorea rotundata) as Aquatic Feed Binder. *Pakistan Journal of Nutrition,* Vol. 9, No. 7, (July 2010), pp. 668-671, ISSN 1680-5194

87. Otero-Villanueva, M.; Kelly, M.S. & Burnell, G. (2004). How diet influences energy partitioning in the regular echinoid *Psammechinus miliaris*; constructing an energy budget. *Journal of Experimental Marine Biology and Ecology*, Vol. 304, No. 2, (June 2004), pp. 159-181, ISSN 0022-0981

88. Palma, J.; Bureau, P. & Andrade, J. P. (2008). Effects of binder type and binder addition on the growth of juvenile *Palaemonetes varians* and *Palaemon elegans* (Crustacea: Palaemonidae). *Aquaculture International*, Vol. 16, No.5, (October 2008), pp. 427-436, ISSN 0967-6120

89. Paolucci, M. Ed. (2010). *Leptin in non mammalian vertebrates.* Transworld Research Network. ISBN 978-81-7895-436-3, Trivandrum-695 023, Kerala, India.

90. Partridge, G.J. & Southgate, P.C. (1999). The effect of binder composition on ingestion and assimilation of microbound diets (MBD) by barramundi *Lates calcarifer* Bloch larvae. *Aquaculture Research*, Vol. 30, No. 11-12, (November 1999), pp. 879–886, ISSN 1365- 2095

91. Pavasovic, M.; Richardson, N.A.; Anderson, A.J.; Mann, D. & Mather, P.B. (2004). Effect of pH, temperature and diet on digestive enzyme profiles in the mud crab, *Scylla serrata*. *Aquaculture*, Vol. 242, No. 1-4 , (December 2004), pp. 641–654, ISSN 0044- 8486

92. Pavasovic, A.; Richardson, N.A.; Mather, P.B. & Anderson, A.J. (2006). Influence of insoluble dietary cellulose on digestive enzyme activity, feed digestibility and survival in the red claw crayfish, *Cherax quadricarinatus* (von Martens). *Aquaculture Research*, Vol. 37, No. 1,, (January 2006), pp. 25–32, ISSN 1365-2109

93. Pavasovic, A.; Anderson, A.J.; Mather, P.B. & Richardson, N.A. (2007a). Effect of a variety of animal, plant, and single cell-based feed ingredients on diet digestibility and digestive enzyme activity in redclaw crayfish, *Cherax quadricarinatus* (Von Marten 1868). *Aquaculture*, Vol. 272, No. 1-4, (November 2007), pp. 564-572, ISSN 0044-8486

94. Pavasovic, A.; Anderson, A.J.; Mather, P.B. & Richardson, N.A. (2007b). Influence of dietary protein on digestive enzyme activity, growth and tail muscle composition in redclaw crayfish, *Cherax quadricarinatus* (von Martens). *Aquaculture Research*, Vol. 38, No. 6, (April 2007), pp. 644-652, ISSN 1365-2109

95. Pearce, C.M.; Daggett, T.L. & Robinson, S.M.C. (2002a). Effect of protein source ratio and protein concentration in prepared diets on gonad yield and quality of the green sea urchin, *Strongylocentrotus droebachiensis*. *Aquaculture*, Vol. 214, No. 1-4, (November 2002), pp. 307-332, ISSN 0044-8486

96. Pearce, C.M.; Daggett, T.L. & Robinson, S.M.C. (2002b). Effect of binder type and concentration of prepared feed stability and gonad yield and quality of the green sea urchin, *Strongylocentrotus droebachiensis*. *Aquaculture*, Vol. 205, No. 3-4, (March 2002), pp. 301-323, ISSN 0044-8486

97. Pearce, C.M.; Daggett, T.L. & Robinson S.M.C. (2004). Effect of urchin size and diet on gonad yield and quality in the green sea urchin (*Strongylocentrotus droebachiensis*). *Aquaculture*, Vol. 233, No. 1-4, (April 2004), pp. 337-367, ISSN 0044-8486

98. Perez-Jimenez, A.; Cardenete, G.; Morales, A. E.; Garcia-Alcazar, A.; Abellan, E. & Hidalgo, C. (2009). Digestive enzymatic profile of Dentex dentex and response to different dietary formulations. *Comparative Biochemistry and Physiology - Part A: Molecular & Integrative Physiology*, Vol. 154, No. 1 (September 2009), pp 157-164, ISSN 1095-6433

99. Person Le Ruyet, J.P.; Alexander, J.C.; Thebaud, L. & Mugnier, C. (1993). Marine fish larvae feeding: formulated diets or live prey? *Journal of World Aquaculture Society*, Vol. 24, No.2, (June 1993), pp. 211– 224, ISSN 1749-7345

100. Phillips, K.; Bremer, P.; Silkock, P.; Hamid, N.; Delahunty, C.; Barker, M. & Kissick, J. (2009). Effect of gender, diet and storage time on the physical properties and the sensory quality of sea urchin (*Evechinus chloroticus*) gonads. *Aquaculture,* Vol. 288, No. 3-4, (March 2009), pp. 208-215, ISSN 0044-8486

101. Pratoomyot, J.; Bendiksen, E.Å; Campbell, P. J.; Jauncey, K. J.; Bell, J.G. & Tocher, D.R. (2011). Effects of different blends of protein sources as alternatives to dietary fishmeal on growth performance and body lipid composition of Atlantic salmon (*Salmo salar L.*). *Aquaculture*, Vol. 316, No. 1-4, (June 2011), pp. 44-52, ISSN 0044- 8486

102. Rivas-Vega, M.E.; Goytortúa-Bores, E.; Ezquerra-Brauer, J.M.; Salazar-García, M.G.; Cruz- Suárez, L.E.; Nolasco, H. & Civera-Cerecedo, R. (2006). Nutritional value of cowpea (*Vigna unguiculata* L. Walp) meals as ingredients in diets for Pacific white shrimp (*Litopenaeus vannamei* Boone). *Food Chemistry*, Vol. 97, No. 1, (July 2006), pp. 41–49, ISSN 0308-8146

103. Robinson, J.A.J.; Kellaway, I.W. & Marriott, C. (1975). Effects of Blending on the Rheological Properties of Gelatin Gels. *Journal of Pharmacy and Pharmacology*, Vol. 27, No 11, (November 1975), pp. 188–824, ISSN 0022-3573

104. Rodriguez, A.; Le Vay, L.; Mourente, G. & Jones, D.A. (1994). Biochemical composition and digestive enzyme ′activity in larvae and postlarvae of Penaeus japonicus during herbivorous and carnivorous feeding, Marine Biology, Vol. 118, No. 1, (March 1994), pp. 45–51, ISSN 0025-3162

105. Rosas, C.; Cuzon, G.; Gaxiola, G.; Arena, L.; Lemaire, P.; Soyez, C. & uence of dietary carbohydrate on the metabolism of juvenile Litopenaeusstylirostris.⇓Van Wormhoudt , A. (2000). In *Journal of Experimental Marine Biology and Ecology* , Vol. 249, No. 2, (June 2000), pp. 181–198, ISSN 0022-0981

106. Rosas, C.; Tut, J.; Baeza, J.; Sanchez, A.; Sosa, V.; Pascual, C.; Arena, A.; Domingues, P. & Cuzon, G. (2008). Effect of type of binder on growth, digestibility, and energetic balance of *Octopus maya*. Aquaculture, Vol. 275, No. 1-4, (March 2008) pp. 291-297, ISSN 0044-8486

107. Ruscoe, I.M; Jones, C.M.; Jones, P.L. & Caley, P. (2005) The effects of various binders and moisture content on pellet stability of research diets for freshwater crayfish . *Aquaculture Nutrition,* Vol 11, No.2, (April 2005), pp. 87-93, ISSN 1353-5773

108. Schlosser, S.C.; Lupatsch, I.; Lawrence, J.M.; Lawrence, A.L. & Shpigel, M. (2005). Protein and energy digestibility and gonad development of the European sea urchin *Paracentrotus lividus* (Lamarck) fed algal and prepared diets during spring and fall. *Aquaculture Research*, Vol. 36, No. 10, (July 2005), pp. 972-982, ISSN 1365-2109

109. Seixas Filho J.T.; Rostagno, H.S.; Queiroz, A.C.; Euclydes, R.F. & Barabarino, P. Jr. (1997a). Evaluation of the performance of post-larvae of the freshwater prawn *Macrobrachium rosenbergii* fed with balanced diets containing different binders. *Revista Brasiliana de Zootecnia*. Vol. 26, No. 4, pp. 638-644, ISSN 1806-9290

110. Seixas Filho J.T.; Rostagno, H.S.; Euclydes, R.F.; Queiroz, A.C. & Barabarino, P. Jr. (1997b). Effect of binders on the hydrosolubility of balanced diets for freshwater prawns (*Macrobrachium rosenbergii*, de Man) in the post-larval stage. *Revista Brasiliana de Zootecnia*. Vol. 26, No. 4, pp. 638-644, ISSN 1806-9290

111. Shewry, P.R. (2009). Wheat. *Journal of Experimental Botany*, Vol. 60, No. 6, (April 2009) pp. 1537-53, ISSN 0022-0957

112. Shiau, S.Y. & Peng, C.Y. (1992). Utilization of different carbohydrates at different dietary protein levels in grass prawn, *Penaeus monodon*, reared in seawater. *Aquaculture*, Vol. 101, No. 3-4, (February 1992), pp. 241–250, ISSN 0044-8486

113. Simon, C.J. (2009). The effect of carbohydrate source, inclusion level of gelatinised starch, feed binder and fishmeal particle size on the apparent digestibility of formulated diets for spiny lobster juveniles, *Jasus edwardii*. *Aquaculture*, Vol. 296, No. 3-4 , (November 2009), pp. 329-336, ISSN 0044-8486

114. Sinha, A.K.; Kumar, V.; Makkar, H.P.S.; De Boeck, G. & Becker, K. (2011). Non-starch polysaccharides and their role in fish nutrition - A review *Food Chemistry*, Vol. 127, No. 4, (August 2011) pp. 1409-1426, ISSN 03088146

115. Smidsrod, O.; Haug, A. & Larsen, B. (1966). The influence of pH on the rate of hydrolysis of acidic polysaccharides. *Acta chemical Scandinavica*, Vol. 20, No. 4, pp. 1026-1034, ISSN 0302-4377

116. Smith, D.M. & Tabrett, S.J. (2004). Accurate measurement of in vivo digestibility of shrimp feeds. *Aquaculture*, Vol. 232, No.1-4, (April 2004), pp. 563-580, ISSN 0044-8486

117. Stephen, A.M. (1983). Other plant polysaccharides, In: The polysaccharides. Aspinall, G.O., pp. 97–193, Academic Press, New York

118. Storebakken, T. (1985). Binders in fish feeds I. Effect of alginate and guar gum on growth, digestibility, feed intake and passage through the gastrointestinal tract of rainbow trout.⇒Storebakken, T. (1985). Binders in *Aquaculture*, Vol. 47, No. 1, (July 1985), pp. 11–26, ISSN 0044-8486

119. Storebakken, T. & sh feeds II. Effect of different alginates on the digestibility of macronutrients in rainbow trout.⇒Austreng, E. (1987). Binders in *Aquaculture*, Vol. 60, No. 2, (February 1987), pp. 121–131, ISSN 0044-8486

120. Storebakken, T.; Shearer, K.D. & sh meal, soy-protein concentrate and phytase-treated soy protein-concentrate-based diets to Atlantic salmon,⇒Roem, A.J. (1998). Availability of protein, phosphorus another elements in *Salmo salar*. *Aquaculture*, Vol. 161 No. 1-4, (February 1998), pp. 365–379, ISSN 0044-8486

121. Teshima. S.; Kanazawa, A. & Sakamoto, M. (1982). Microparticulate diets for the larvae of aquatic animals. Mini review and data file of Fisheries Research, Kagoshima University 2, pp. 67-86

122. Usov, A I. (1998). Structure analysis of red seaweed galactan of agar and carrageenan groups. *Food Hydrocolloids*, Vol. 12, No 3, (July 1998), pp. 301-308, ISSN0268-005X

123. Valverde, J. C.; Hernández M. D.; Aguado-Gimenez, F. & Garciá Garciá, B. (2008). Growth, efficiency and condition of common octopus (*Octopus vulgaris*) fed on two formulated moist diets, Aquaculture, Vol. 275, No. 1-4, (March 2008), pp. 266-273, ISSN 0044-8486

124. Van der Meer, J. (2006). Metabolic theories in ecology. *Trends in Ecology and Evolution*. Vol. 21, No. 3, (November 2006), pp. 136-140, ISSN 0169-5347

125. Van Wormhoudt, A.; Cruz, E.; Guillaume, J. & Favrel, P. (1986). Action de l'inhibiteur trypsique de soja sur lacroissance et l'activite des enzymes digestives chez ′ *Penaeus japonicus* (Crustacea, Decapoda): role ^eventuel des hormones gastro-intestinales. *Oceanis*, Vol. 12, pp. 305–319, ISSN 0182-0745

126. Volpe, M.G.; Monetta, M.; Di Stasio, M. & Paolucci, M. (2008). Rheological behavior of polysaccharide based pellets for crayfish feeding tested on growth in the crayfish *Cherax albidus*. *Aquaculture*, Vol. 274, No. 2-4, (February 2008), pp. 339–346, ISSN 0044-8486

127. Volpe, M.G.; Malinconico, M.; Varricchio, E. & Paolucci, M. (2010). Polysaccharides as biopolymers for food shelf-life extension: Recent patents. *Recent Patents on Food, Nutrition & Agriculture*, Vol. 2, No. 2, (June 2010), pp. 129-139, ISSN 1876-1429

128. Watanabe, T. (2002). Strategies for further development of aquatic feeds. *Fisheries Science*, Vol. 68, No. 2, (May 2002), pp. 242-252, ISSN 1444-2906

129. Zhou, Q-C.; Zhou, J-B.; Chi, S-Y.; Yang, Q-H; Liu, C-W. (2007). Effect of dietary lipid level on growth performance, feed utilization and digestive enzyme of juvenile ivory shell, *Babylonia areolate*. *Aquaculture*, Vol. 272, No. 1-4 (November 2007), pp. 535-540, ISSN 0044-8486.

02 ZEBRAFISH EXPERIENCE DIMINISHED GROWTH AND JAW DYSMORPHOGENESIS WHEN FED AN ARTEMIA DIET WITHOUT SUPPLEMENTATION

1. INTRODUCTION: AN OVERVIEW

The utilization of zebrafish as an experimental animal model has significantly increased in laboratories during the last twenty years (Craig et al., 2006). The number of zebrafish facilities worldwide has significantly risen to accommodate the rising need for quality animal care and maintenance. To meet this need, aquaculture facilities need to implement husbandry methods that aim to consistently provide a supply of healthy and viable eggs. Many husbandry practices, especially feeding protocols, are often shared informally between lab members without thorough experimental confirmation. An optimal diet should include a limited selection of food items that are nutritionally balanced, easy to cook, non-messy, and affordable. Previous research focused on optimizing feeding strategies for adult zebrafish in large aquaculture facilities has highlighted the importance of cost-effective, single-food diets. However, these diets have been found to result in reduced survival rates and reproductive abilities, indicating a deficiency in essential nutrients. Limiting adult fish meals to one type of food may save time and money but could lead to an inadequate balance of trace minerals necessary for hormone and enzyme production, skeletal construction, and other physiological functions. However, due to the demanding breeding schedules in many institutions to fulfill research requirements, a high-protein diet is sometimes used to increase egg output. Flake food diets that have been processed are typically abundant in proteins and vital trace elements necessary for proper bodily functions. Recent studies focusing on finding new diets to enhance the growth and reproductive abilities of zebrafish have not given much consideration to the minimum amount of iodine needed in their diet, despite its importance in regulating growth and metabolic rate in warm-blooded animals. Zebrafish can tolerate temporary iodine shortage because they store a significant amount of T4 in the follicular colloid, as suggested by Brown et al. (2004). Iodized salts in water systems and iodine-fortified processed foods are frequently utilized to prevent iodine-deficiency disorders including hyperthyroidism and goiter. However, adverse environmental factors like low water quality or malnutrition can cause significant growth of thyroid tissue in certain fish species, suggesting the importance of ensuring a sufficient iodine level in controlled environments like zebrafish aquaculture facilities. Previous research has demonstrated the effectiveness of dietary supplements in preventing thyroid hyperplasia in saltwater cyprinodont Fundulus and goiters in Chinook salmon. However, there is currently no study that has determined the minimum iodine level required to maintain normal development and thyroid health in zebrafish. During our dietary experiments, we found that decreased iodine levels led to slower physical growth and a higher occurrence of jaw deformities. We decided to assess the developmental effects of raising zebrafish on either a diet of just live food or a mixed food diet containing different amounts of iodine. We show that feeding zebrafish just Artemia can lead to changes in their jaw structure due to tissue swelling and increased blood vessel formation, despite the diet being rich in protein needed for egg production.

2. EXPERIMENTAL METHODS

2.1 Animal housing and maintenance

Adult wild-type zebrafish were bred and cared for in the Zebrafish Model Organism Laboratory at the University of Cincinnati Genome Research Institute using established procedures (Craig et al., 2006; Kimmel et al., 1995; Westerfield, 2000). Zebrafish fry were raised in a multi-rack system from Aquatic Habitats in Apopka, FL, with automated regulation of water chemistry using a YSI Model 5200 control system from Yellow Springs, OH, under a 14-hour light and 10-hour dark cycle. The water chemistry was controlled to maintain optimal circumstances for fish health. The temperature is 28.5 ± 1 oC, pH is 7.5 ± 0.2, ammonia is 0 ± 0.25 mg/l, nitrite is 0 ± 0.20 mg/l, nitrate is 5 ± 5 mg/l, total hardness is 102±17 mg/l, and alkalinity is 50 ± 17 mg/l. The conductivity was kept at 750 ± 50 µS. Adult breeding fish were fed a regular diet consisting of 3-day-old HUFA-enriched Artemia (Super SELCO, Aquatic Ecosystems) in the morning and a balanced flake food (Aquatox, Aquatic Ecosystems) in the afternoon.

2.2 Dietary paradigm

The fry were given a varied diet consisting of Paramecium, rotifers, and Zeigler's larval diet until they were able to consume Artemia at around 1 month old to promote their early development. From the age of 1 month, fish were fed either a live study diet consisting of three Artemia feedings or a mixed control diet consisting of two Aquatox flake feedings and one Artemia feeding each day. The amount of Artemia consumed by fish was determined by giving them 17.2 mg of Artemia per tank after an overnight fast, with a feeding time of 15 minutes to reach satiety. The fish were euthanized using a concentrated solution of 250 mg/l MS-222, then gutted and had their digestive tract removed. The gut was turned inside out, and the complete quantity of consumed shrimp was tallied using a stereomicroscope. Aquatox flake food consumption was assessed by feeding fish until they were full (4.7 mg per tank, with a 15-minute feeding period), then removing the leftover food with a siphon, drying it, and weighing the uneaten portion.

2.3 Iodine analyses

The quantification of iodine in the food items was conducted using ICP-MS, following the methodology outlined in a recent study on heavy metals by Figueroa et al. (2008). Artemia samples were cultivated for 72 hours, then harvested and freeze-dried overnight using a lyophilizer (FreeZone 4.5 Liter Benchtop System, Labconco Corporation, Kansas City, MI). Three dried Artemia samples and system water samples (~30 mg each) were placed in septum-sealed glass tubes. They were treated with HNO3 and digested using a microwave with the following settings: power of 130 W, ramp time of 1 minute, hold time of 1 minute, and temperature of 75 °C. Then, 0.1 mL of 30% hydrogen peroxide (H2O2) was added, and the samples underwent the second digestion stage with the following parameters: power 130 W, ramp 1 minute, hold 1 minute, temperature 110 °C. The samples were diluted with deionized water to a volume of 10 mL and then analyzed for total iodine concentration using an Agilent 7500ce ICP-MS instrument with a quadrupole mass analyzer in continuous flow mode.

2.4 Histology

Hematoxylin and eosin staining was conducted at Oregon State University on both phenotypic and control juvenile zebrafish to determine the characteristics of the jaw dysmorphogenesis. 4-micron thick sagittal jaw sections were prepared by cutting and fixed in paraformaldehyde. The sections were then stained with Alizarin Red following established techniques (Cailliet et al., 1986). Adult fish were treated with Alizarin Red dye to enhance the visibility of bone and pigment structure. The fish were immersed in 10% phosphate buffered formalin for 3 days. The fish were initially washed in deionized water, treated with a solution of 3% hydrogen peroxide and 2% potassium hydroxide to eliminate pigment, washed again in deionized water, subjected to trypsin digestion for eight days (3375 U/ml in 30% Sodium Borate) until most of the spine was exposed, immersed in 0.5% KOH, soaked in Alizarin Red for a few hours, and then destained in 2% KOH overnight. The stained fish were gradually moved into higher glycerol concentrations (reaching 90% in the end) for imaging and long-term storage.

2.5 Morphometrics

The impact of nutrition on the typical growth of fish was assessed using standard growth measurements. Body mass measurements were taken on fish that were sedated with 125 mg/l MS-222. Water accumulated on the body surface can result in significant overestimation of body mass and so it was removed by gently blotting the animal with absorbent paper (Kimwipe). The fish were individually weighed three times using a 5-place analytical balance from Mettler-Toledo, Inc., Columbus, Ohio, and the average weight was determined. The total lengths were measured from still photos using a calibrated measuring tool in Rincon 7.1 software or a hand dial caliper. Craniofacial development was evaluated by measuring the lengths and widths of the Meckel's and ceratohyal complexes in adult fish using still photos of Alizarin Red bone stained specimens exposed to various food regimes.

2.6 Statistics

The statistical analyses were conducted using SigmaStat XI software developed by Systat Software Inc. in San Jose, California. The data was analyzed using Student's two-tailed unpaired t-tests or nonparametric rank sum tests as required. Significance was attributed to probability (P) values less than 0.05. Incidence levels were compared using a Yates-corrected Chi-square test.

3. RESULTS AND DISCUSSION

3.1 Dietary paradigm results in a 6-fold difference between *Artemia*-only and mixed food diets

Fish were fed Artemia three times a day and consumed an average of 1.2 mg of Artemia per feeding, totaling 3.6 mg of dry mass. The fish fed a mixed diet consumed 4.8mg of dry mass daily, consisting of 1.2mg of Artemia twice and 2.37mg of Aquatox flakes once. ICP-MS analysis of iodine levels in diets showed that the Artemia-only diet had 11.48ng of iodine, which is approximately one-sixth of the 72.6ng found in the mixed diet (Table 1).

Table 1. Iodide concentrations of study-related salts, feed and supplements. Mean values are listed ± 1 s.d. a(Bidwell and Spotte, 1985), bICP-MS quantified.

Source	Concentration (ppm)
Crystal Sea Marine Mixa	0.12
Instant Ocean Sea Salta	0.26
Seawaterb	0.49-0.58
System Water (Instant Ocean + salts)	0.73 ± 0.26
SuperSelco Enriched *Artemia*b	3.19 ± 1.04
Aquatox Flakesb	27.38 ± 1.56

The decrease in physical growth could be due to reduced hormone production, potentially triggered by a lack of iodine or other essential nutrients (Ikeda et al., 1973). The reduced body size was unlikely a result of insufficient eating as the fish were given three filling meals of Artemia each day. Additionally, there were no signs of reduced feeding capacity or rapid deterioration in the overall appearance of the fish, even though we anticipated that the movement of the lower jaw bones would impact their ability to feed. Throughout the trial, the fish fed with Artemia were noticeably smaller.

3.2 *Artemia*-only diet results in smaller, less sexually dimorphic individuals compared with mixed diets

Fish fed on the Artemia diet exhibited significantly lower body mass and length compared to fish raised on a mixed diet, as indicated in Table 2 (P<0.0001). These measurements were taken well after the fish were expected to have attained their terminal length (~150 dpf, unpublished data). Sexual dimorphism was noticeable in the population that had a mixed diet by 4 months of age, but the fish fed with Artemia showed negligible dimorphism at 10 months of age. Notably, fish given exclusively processed flake food had shorter terminal lengths compared to those fed Artemia or mixed food diets (data not shown).

Table 2. Body morphometrics of 160-180 day old fish on study diets. Mean values listed ± 1 s.d.

	Artemia Only	Mixed Diet	P-value
Mass (mg)	139 ± 64 (78)	278.80 (15)	0.079
Length (mm)	25.4 ± 1.9 (78)	60.4 ± 2.1 (15)	<0.001
Meckle's Length (mm)	1.23 ± 0.29 (4)	1.52 ± 0.75 (2)	0.499
Ceratohyal Length (mm)	1.43 ± 0.53 (4)	1.79 ± 0.84 (2)	0.541
Iodine Content (mg/day)	0.11	0.28	n/a

3.3 Enlarged thyroid causes displacement of the jaw bone in fish that are given low-iodine diets

The protruded mandibular jaw phenotype was noticed one week after starting the Artemia diet, with the occurrence rate growing logarithmically to 6.3% at 180dpf and stabilizing at 10.5% by the time the fish were 10 months old (n=828 total fish). The occurrence rate in the group consuming a varied diet was consistently lower and decreased to 3.4% after 10 months (n = 507 total fish, Yates-corrected Chi-square stat: 21.4, $\alpha = 0.05$, 95% CI: 4.3 – 10.1%). Over the 10-month test period, there was a slow expansion in the mandibular protrusion and localized vascularization in 78% of the fish studied. These growths protruded an average of 2.1% of the fish's total body length (Fig. 1A). Upon histologic examination, it was observed that there was evident glandular participation (Fig. 1B) along with a noticeable repositioning of the basihyal jawbone from a sagittal to a more transverse alignment (Fig. 1C). The jaw deformity seemed to be caused by the thyroid gland pushing the bone out of place near the basihyal bone, rather than a lack of bone formation. However, the lengths of Meckel's cartilage and ceratohyal bone were slightly shorter in fish fed the mixed diet compared to those in Table 2 (reduced by 19% and 20% for Meckel's and ceratohyal lengths).

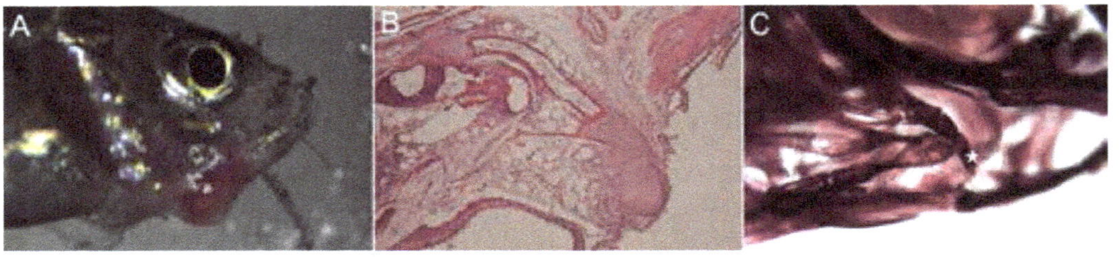

Fig. 1. Phenotypic characterization of low iodine diet-reared fish. Gross phenotype (A), histological section stained with hematoxilin and eosin (B), and Alizarin Red bone stain (C). Ventral rotation of the basihyal bone is indicated by an *.

We suspected that the enlarged, hypervascularized tissue mass around the thyroid gland was caused by an iodine deficit, leading to the observed phenotype. The classification of piscine thyroid morphologies is typically challenging due to the teleost thyroid being an unencapsulated colloid-filled follicle, first described by Gundernatsch in 1911. However, previous research in different fish species has demonstrated lower jaw abnormalities resembling those observed in the zebrafish given Artemia in this study (Gaylord & Marsh, 1912; Marine & Lenhart, 1911; Marine, 1914). The swollen, highly vascular tumors growing under the jaw of New York hatchery salmon were identified as thyroid carcinomas (Gaylord & Marsh, 1912). Brook trout in a Pennsylvania hatchery were observed to have goiters that looked identical to the ones seen in our study. No signs of abnormal cell growth, nuclear abnormalities, or extra layers characteristic of carcinomas were found in fish raised on our low-iodine diet. Instead, the analysis revealed features of simple goiter such as pushing borders, organ involvement, and hypervascularization (Gaylord & Marsh, 1912). The histological study showed a bone displacement towards the ventral side, which was linked to the protrusion of the lower jaw. This indicated that the glandular tissue was pressing against the jawbones and changing their usual position. Our research ended after six months, but if chronic iodine deficiency had continued, all fish would have developed thyroid abnormalities. However, the steady increase in blood vessel creation seemed to stabilize at 10.5% of the fish studied. Additional factors including environmental pressures and the health of individual fish may have influenced the reported characteristics.

4. AT A GLANCE

Freshwater fish typically need a minimum of 1 – 4 mg total iodine per kilogram (Watanabe et al., 1997) and a dietary minimum of 2.8 mg iodine per kilogram (Lovell, 1979). However, plasma iodide levels in freshwater fish show significant variation, suggesting a broad range in the effectiveness of iodine absorption and use. Freshwater fish have low iodine concentration ranging from 0.1 to 10 µg/l, which is insufficient to meet their metabolic needs. Therefore, they primarily depend on food sources of iodine rather than the iodine present in their environment. Freshwater fish acquire iodine from food by absorption in the stomach, uptake from the environment through the gills, and a minor amount through hormone recycling. Our data indicates that the minimum iodine level needed to prevent reduced growth and other iodine-related effects in zebrafish is slightly lower compared to other freshwater fish. This level falls between the iodine content found in our Artemia and mixed food diets, ranging from 0.11 to 0.28 mg I kg-1 per day above the minimal 0.7 ppm environmental exposure. It is uncertain if providing more iodine than what is naturally found in the diet can reverse the gland enlargement and bone displacement observed. Previous research indicates that trout goiters can be reduced with a potassium iodide supplement in just 10 days (Marine & Lenhart, 1911; Marine, 1914). We believe that the minor costs of incorporating iodine-rich flake food into the fish's diet are justified by the improved fish health and accelerated sexual maturation.

REFERENCES

1. Barnard, E. & Bagatto, B. (2002). Effects of maternal diet on reproductive strategies and embryo development in the zebrafish, Danio rerio. *Integrative and Comparative Biology* 42:1191
2. Bidwell, J. P. & Spotte, S. (1985). *Artificial Seawaters: Formulas and Methods*. Jones and Bartlett, Boston, Massachusetts.
3. Brown, S.B.; Adams, B.A.; Cyr, D.G.; & Eales, J.G. (2004). Contamination effects on the teleost fish thyroid. *Environmental Toxicology and Chemistry* 23:1680-1701.
4. Cailliet, G. M.; Love, M.S.; & Ebeling, A.W. (1986). *Fishes. a field and laboratory manual on their structure, identification, and natural history*. Wadsworth Publishing Co, Belmont, California.
5. Craig, M. P.; Gilday, S.D.; & Hove, J.R. (2006). Dose-dependent effects of chemical immobilization on the heart rate of embryonic zebrafish. *Lab Animal (NY)* 35:41-7.
6. Eales, J. G. (1979). *Thyroid function in cyclostomes and fishes*. Academic Press, London.
7. Eales, J.G.; & Brown, S.B. (1993). Measurement and regulation of thyroidal status in teleost fish. *Rev. Fish Biol Fish* 3:299-347.

8. Figueroa, J. A.; Wrobel, K.; Afton, S.; Caruso, J.A.; & Corona Felix Gutierrez, J. (2008). Effect of some heavy metals and soil humic substances on the phytochelatin production in wild plants from silver mine areas of Guanajuato, Mexico. *Chemosphere* 70:2084-91.
9. Gaylord, H. R.; & Marsh, M.C. (1912). Carcinoma of the thyroid in salmonid fishes. *Bureau of Fisheries* 32:363-524.
10. Goolish, E. M.; Okutake, K; & Lesure, S. (1999). Growth and survivorship of larval zebrafish Danio rerio on processed diets. *North American Journal of Aquaculture* 61:189-198.
11. Gundernatsch, J. F. (1911). The thyroid gland of teleosts. Journal of Morphology 21:709-782. Hoover, K.L. (1984). Hyperplastic thyroid lesions in fish. *National Cancer Institute Monograph* 65:275-289.
12. Ikeda, Y., H. Ozaki, & H. Yasuda. 1973. Effects of potassium iodide on growth of body and scale in goldfish. *Journal of Tokyo University of Fisheries* 59:333-342.
13. Kimmel, C. B.; Ballard, W.W.; Kimmel, S.R.; Ullmann, B.; & Schilling, T.F. (1995). Stages of embryonic development of the zebrafish. *Developmental Dynamics* 203:253-310.
14. Lovell, R. T. (1979). Formulating diets for aquaculture species. *Feedstuffs* 51:29-32.
15. Marine, D. (1914). Further observations and experiments on goitre (so-called thyroid carcinoma) in brook trout (Salvelinus fontinalis): III. Its prevention and cure. *Journal of Experimental Medicine* 19:70-88.
16. Marine, D.; & Lenhart, C.H. (1911). Further observations and experiments on the so-called thyroid carcinoma of the brook trout (Salvelinus fontinalis) and its relations to endemic goitre. *Journal of Experimental Medicine* 13:455-475.
17. Meinelt, T.; Schulz, C.; Wirth, M.; Kurzinger, H.; & Steinberg, C. (1999). Dietary fatty acid composition influences the fertilization rate of zebrafish (Danio rerio Hamilton- Buchanan). *Journal of Applied Ichthyology* 15:19-23.
18. Pickford, G.E. (1954). The response of hypophysectomized male killifish to prolonged treatment with small doses of thyrotropin. *Endocrinology* 55:274-287
19. Schlumberger, H. G. (1954). Spontaneous hyperplasia and neoplasia in the thyroid of animals. *Brookhaven Symposia in Biology* 7:169-191.
20. Watanabe, T.; Kiron, V.; & Satoh, S. (1997). Trace minerals in fish nutrition. *Aquaculture* 151:185-207.
21. Westerfield, M. (2000). *The zebrafish book: a guide for the laboratory use of zebrafish.* University of Oregon Press, Eugene, OR. Woodall, A.N.; & La Roche, G. (1964). The nutrition of salmonoid fishes. XI. Iodide requirements of Chinook salmon. *Journal of Nutrition* 82:475-482.

03 SUBSTITUTION OF FISHMEAL WITH PLANT PROTEIN INGREDIENTS IN DIETS FOR ATLANTIC SALMON (SALMO SALAR) - IMPACT ON WEIGHT GAIN AND ACCRETION

1. INTRODUCTION: AN OVERVIEW

Cultivating carnivorous fish, particularly, relies on high-protein diets. Atlantic salmon (Salmo salar) typically need 55-60% dietary protein throughout their juvenile stages, with a decrease in protein need as they mature (NRC, 1993). In the past, fishmeal from wild caught fish has been used as the protein source, while fish oil from the same source has been used as the lipid source. Over the past decade, farmed Atlantic salmon production in Norway has risen consistently from 440,061 tons in 2000 to 927,876 metric tons in 2010. Meanwhile, wild fish catch has remained stable during this period, according to statistics from the Directorate of Fisheries, Norway. This also applies to the global wild-caught fish and aquaculture production as documented by the Food and Agriculture Organization (FAO) and many researchers (Tacon, 1995; FAO, 2006; Tacon & Metian, 2008). If farmed fish production is to rise, other protein ingredients and fat sources must be evaluated as feed components. This involves promoting the growth of farmed fish while ensuring their health is not compromised. In future aquaculture, these innovative elements must serve as the primary protein and lipid sources in the fish diets. The primary sources are expected to be of plant origin, while some animal by-products and microbiologically manufactured proteins may be permitted by national and/or international laws. Animal by-products can come from chicken and swine processing, as well as from microbiologically produced protein sources such single-cell proteins. Plant proteins are the most probable choices due to their prevalence and comparatively affordable price. Substituting fishmeal with plant protein ingredients can lead to reduced growth performance in Atlantic salmon, as well as in other commercially important fish species such as Atlantic cod, sea bream, rainbow trout, and turbot. Substituting fishmeal with plant components raises concerns about higher feed conversion ratio (FCR) and lower protein utilization, as highlighted in studies by Robaina et al. (1995), Refstie et al. (1998, 2000), and Opstvedt as al. (2003). The decreased performance could be attributed to an unbalanced amino acid profile or reduced voluntary feed intake in fish fed diets containing plant components. Protein synthesis necessitates the presence of all required amino acids in the cell. If an amino acid is missing or not present in sufficient amounts for synthesis in a specific tissue, the protein cannot be produced. The amino acids will then be metabolized and used as fuel because free amino acids are not stored in tissues (Geiger, 1947). The primary storage site for amino acids is the muscle, however they are not accessible for protein metabolism until the proteins are broken down into free amino acids. Free amino acids often make up less than 0.1% of the protein-bound amino acids in mammals, as reported by Njaa (1990), Millward (1988, 1999), and Espe (2008). Fish requires a continual supply of essential amino acids (IAAs) because only around 50% of the protein-bound amino acids are reused for protein synthesis, despite the ongoing turnover of proteins (Houlihan et al., 1995). Recent research discovered that fish need not just indispensable amino acids (IAAs) but also dispensable amino acids (DAAs) (Abbouti et al., 2009). Ensuring sufficient quantities of amino acids with a well-balanced profile is crucial for achieving maximum protein accumulation and growth in fish (El-Mowafi et al., 2010). Animal-based protein components offer a well-rounded amino acid composition, regardless of whether they come from beef or fish muscle, and supply the essential amino acids needed by consumers or animals. The diets are formulated based on a balanced amino acid profile utilizing the ideal protein concept, where the quantity of each essential amino acid is compared to the amount of lysine (Wang & Fuller, 1987; 1989; Rollin et al., 2003). Moreover, a decrease in digestibility or the existence of potential anti-nutrients could contribute to the diminished nutrient absorption in fish fed plant-based diets, potentially leading to decreased performance. The decreased feed intake and absorption in fish using these diets could alter the availability of the balanced amino acid profile in tissue compartments. Imbalances of amino acids in tissues can adversely influence metabolism and have a substantial impact on the health and protein/lipid deposition in farmed fish and animals (Wu, 2009). Feed intake directly influences weight gain in fish, similar to domestic animals. In the past, feed consumption was not a major worry when fishmeal made up a large part of the diet because fish naturally ate enough to promote maximum growth. Substituting fishmeal with plant protein components may decrease the attractiveness of the feed, leading to lower growth performance (Kaushik et al., 1995; Fournier et al., 2004; Glencross et al., 2004; Dias et al., 2005). One of the main issues in creating high plant protein diets for farmed Atlantic salmon is ensuring that the fish readily absorb the feed as they do with fishmeal-based diets. When creating diets, it is important to look for elements that can help maintain feed consumption. Both fish protein hydrolysate (FPC) and stick water include several non-protein nitrogen molecules that are present in plant components in varied amounts or are absent. Examples of NPN include taurine, betaine, trimethylamine oxide, spermine, and spermidine, as documented by Espe & Lied (1999), Liaset & Espe (2008), and historical data from NIFES. Both FPC and stick water have been shown to enhance voluntary feed intake in poultry, rats, and Atlantic salmon. These components have the potential to support the consumption of fishmeal replacement diets for Atlantic salmon. Hydrolyzed squid protein has a high level of non-protein nitrogen (NPN) molecules, around 6%, with roughly 1/6th being taurine based on previous data from NIFES. This makes it a potential option for enhancing the taurine content in diets. Squid also has a significant amount of betaine (trimethylglycine, Treberg & Driedze, 2007). Betaine is essential for the remethylation of homocysteine to produce methionine in the choline-betaine pathway, playing a critical role in the methylation process in animals (Finkelstein, 1990; Mato et al., 2002) Initially, diets were created without fishmeal or with low fishmeal content to study the impact on feed intake and growth. Stick water (5%) or hydrolyzed squid (3%) were included in the diets for this purpose. Furthermore, 5% FPC was included. The fish fed the whole replacement diet ate somewhat less feed than the fishmeal-fed Atlantic salmon. Fish fed the diet with hydrolyzed squid ate more feed than those fed the diets containing stick water. Fish that were not given meals containing fishmeal grew less than fish that were provided reference diets with a high fishmeal content. The decreased growth was attributed to a general reduction in lipid accumulation (Espe et al., 2006). The inclusion of 5% fishmeal in the meals resulted in lipid accretion levels comparable to those of the control group fed with fish meal (Espe et al., 2007). It is advisable not to create diets for Atlantic salmon without any fishmeal. However, incorporating a small amount of fishmeal (5%) along with marine foods such as squid or stick water and FPC appears to be effective. Fournier et al. (2003) included FPC in turbot diets without fishmeal, resulting in identical feed consumption compared to fish fed fishmeal, but growth was decreased. Table 1 summarizes experiments involving various species fed diets with complete or significant substitution of fishmeal, showing that replacing marine protein with plant proteins typically decreased performance. Atlantic salmon fed FPC and squid meal showed better performance compared to those fed stick water. In our trials with high fishmeal replacement diets for Atlantic salmon, the test diets always included FPC and hydrolyzed squid meal or krill meal, with fishmeal never accounting for less than 5%. Replacing fish oil depends on using a combination of plant oils to ensure the necessary intake of polyunsaturated fatty acids, as indicated by Torstensen et al. (2004; 2005; 2008). Our research with Atlantic salmon exclusively used fish oil as the lipid source, so any mention of alternate lipid sources in fish feed is not part of this chapter. Leaver et al. (2009) recently studied the impact of replacing fish oil with plant oil.

2. IMPACT OF SUBSTITUTING FISH MEAL ON GROWTH AND ACCUMULATION

Comparison was made among six experiments including high fishmeal replacement diets fed to Atlantic salmon to assess wet weight gain, protein gain, and lipid gain. The diets examined had balanced amino acid profiles (Figure 1) but were formulated using various plant protein sources (Table 2) and were enhanced with a low concentration of crystalline amino acids to maintain the balance without affecting growth (Espe & Njaa 1991; Espe et al., 1992a; Espe & Lied 1994; Berge et al., 1994; Cowey 1995; Dabrowski & Guderley 2002; Rollin et al., 2003). Daily voluntary feed intake was closely monitored to ensure that any variations in feed intake would affect growth or accumulation. Previous studies have reported growth data on the amino acid requirements of Atlantic salmon, but this analysis only includes fish that were fed diets with comparable amino acid profiles in each experiment. Furthermore, certain unpublished experiments are incorporated. The results were organized based on the replacement of high protein low fishmeal (HPLF) and medium protein medium fishmeal (MPMF). Each answer was statistically modeled using a mixed-effects model, with the level of fishmeal (HPLF or MPMF) as a fixed effect and trial as a random effect.

Table 1. The effect on voluntary feed intake, growth performance, protein accretion and feed utilisation in fish fed high plant protein diets compared to fishmeal control fed fish in some studies testing total or high fishmeal replacement diets. -No information given, Total fishmeal replacement trials are marked with bold letters

Species	Fishmeal protein (% of total)	Marine protein (% of total)	Voluntary feed intake	Growth	PER	PPV	FCR	References
Atlantic salmon (Salmon salar)	0	7	**Less or equal**	**Reduced**	**Equal**	**Equal**	**Equal**	**Espe et al., 2006**
	7	17	Equal	Equal	Equal	Equal	Equal	Espe et al., 2007
	11-14	17-18	Equal	Equal	Equal	-	Equal	Kousoulaki et al., 2009
	20	28	-	Equal	Equal	Equal	Equal	Torstensen at al., 2008

Species	Fishmeal protein (% of total)	Marine protein (% of total)	Voluntary feed intake	Growth	PER	PPV	FCR	References
Sea bass (Dicentrachus labrax) **Seabream** (Sparus aurata)	**0**	**17**	**Reduced**	**Reduced**	**Reduced**	**Reduced**	**Increased**	**Dias et al., 2005**
	7	7	Equal	Equal	Reduced	-	Equal	Kaushik et al., 2004
	27	27	Reduced	Reduced	Reduced	-	Increased	Gomez-Requeni et al., 2004
	0	**0**	**Reduced**	**Reduced**	**Reduced**	-	**Increased**	**Gomez-Requeni et al., 2004**
Atlantic cod (Gadus morlwa) **Turbot** (Psetta maxima)	24	24	-	Reduced	Reduced	Reduced	Increased	Hansen et al., 2007
	12	17	Reduced	Reduced	Equal	-	-	Fournier et al., 2004
	0	6	**Reduced**	**Reduced**	**Equal**	-	-	**Fournier et al., 2004**
	0	6	**Equal**	**Reduced**	**Reduced**	**Reduced**	**Increased**	**Fournier et al., 2003**
Rainbow trout (Oncorhynchus mykiss)	26	26	Reduced	Reduced	-	-	Increased	Glencross et al., 2004
	0	8	**Equal**	**Reduced**	**Equal**	**Reduced**	**Equal**	**Fournier et al., 2003**
	0	0	**Reduced**	**Reduced**	**Equal**	-	**Equal**	**Gomes et al., 1995**
	34	34	Equal	Equal	Equal	-	Equal	Gomes et al., 1995
	10	34	Reduced	Reduced	Equal	reduced	Increased	Aksnes et al., 2006

IAA (g/16g N):

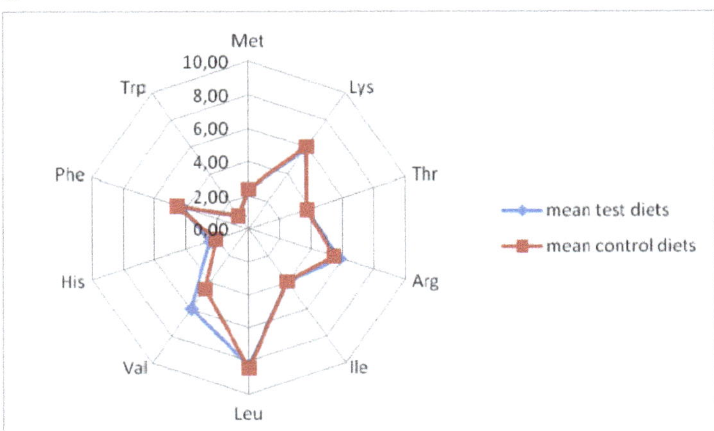

DAA (g/16gN):

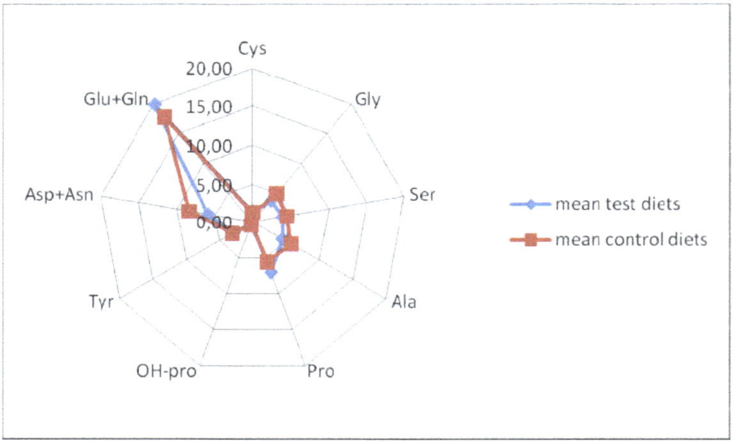

Fig. 1. The diets were balanced in both indispensable (IAAs) and dispensable (DAAs) amino acids towards the controls used in each trial and equalled the requirement of amino acids. Here is shown the mean values over all trials (g amino acid/16gN).

Table 2. The compositions (g/kg) in the diets used in the different trials (I-VI). For each trial the control (C) contained more fishmeal, while the test (T) diets are based on different blends of plant proteins containing lower fishmeal inclusion.

	Trial I		Trial II		Trial III		Trial IV		Trial V		Trial VI	
Diets	C	T	C	T	C	T	C	T	C	T	C	T
Fishmeal (Norse LT-94)	490	50	400	50	260	50	250	50	354	87.5	280	50
FPC (C.P.S.P.	0	50	0	50	0	50	0	50	0	112.5	0	59
Wheat gluten	0	151	0	135	80	238	70	75.3	40	120	50	150
Corn gluten	100	222	170	190	140	0	150	250	0	0	0	0
Soy protein	55	0	35	0	123	11	96	0	0	0	0	0
Wheat grain	106.4	112.4	156.4	190.4	106.4	225.8	115.4	146.1	0	0	144.4	0
Peal/soy protein mix	0	0	0	0	0	0	0	0	323	382	0	0
Pea protein conc, 50	0	0	0	0	0	0	0	0	0	0	120	130
Tapioca	0	0	0	0	0	0	0	0	0	0	0	155.9
Squid hydrolysates	0	30	0	30	0	30	0	30	0	0	0	30
Krill meal	0	0	0	0	0	0	0	0	0	0	0	50
Fish oil	240	258	230	253	272	280	310	325	267	273	312	293.9

	Trial I		Trial II		Trial III		Trial IV		Trial V		Trial VI	
AA mix	0	126.6	0	71	0	71.6	0	50	0	0	0	4.4
Micronutrients*	8.6	30.5	8.6	30.6	19.5	43.6	8.6	23.6	16	25	13.6	42.1
Crude protein	477	504	424	446	433	430	389	400	440	418	373	360
Crude fat	248	224	241	239	274	289	332	331	287	290	373	367
Gross energy (MJ kg)	23.6	24.3	23.7	23.8	24.0	24.3	25.6	26.2	24.0	23.9	24.7	25.8
% fishmeal protein of diet protein	69.8	6.7	64.1	7.6	40.9	7.9	43.7	8.0	54.8	14.1	50.9	9.4
% marine protein of total diet protein	69.8	17.0	64.0	18.5	40.9	20.0	43.7	21.5	54.8	21.7	50.9	18.6

*crystalline amino acid to balance the dietary amino acids and vitamin + mineral mixtures to fulfil the requirement (for more details see Espe et al., 2007; 2008). Percentage of marine protein in the diets was calculated assuming fishmeal, krill meal and FPC contained 68% protein and squid 60% protei

In order to compare the growth and accretion data from the six studies, growth data was computed as g/kg average BW/day, accounting for variations in both the duration of the trials and the initial body weight of the fish. The protein and fat increases were computed as gain per kilogram of average body weight per day. Table 3 displays the average wet weight gain, protein gain, and lipid gain from the six studies. Table 2 summarizes the replacement ratios of fishmeal protein and marine protein components with plant proteins. The experimental diets had a protein content ranging from 6.7% to 14.1% derived from fishmeal, whereas the control diets (MPMF) had a protein content ranging from 40.9% to 69.8% sourced from fishmeal. The test diets contained marine protein ranging from 17.0 to 21.5% of total protein when FPC and squid or krill meal were added. The medium fishmeal replacement diets had marine protein levels ranging from 40.9 to 69.8% of the total dietary protein. The weight increase ranged from 6-12g per kilogram of average body weight per day in the trials included in Table 3, with the highest rates observed in trials III and V. The protein intake ranged from 0.6 to 1.8g/kg of Average Body Weight per day, with the greatest values observed in trials III and V as shown in Table 3. The lipid accumulation ranged from 0.8 to 3.2g/kg average body weight per day, with the highest levels observed in trials I, III, and V as shown in Table 3. Feeding Atlantic salmon high plant protein diets significantly reduced lipid gain with a high likelihood (>99%). The 95% credible interval (CI) for this difference is between -3% and -16%, indicating a clear negative effect. Protein gain was highly likely to decrease by more than 99% when fishmeal was substituted with plant components, with the effects ranging from roughly -3% to -16% (95% CI, Table 4). In the MPMF, the estimated average wet weight growth was 8.1g per kilogram of body weight per day. There was a strong likelihood (>99%) that wet weight gain reduced in the HPLF fed groups, with a 95% confidence interval for the effect ranging from approximately -2% to -13% (Table 4). The impacts on protein and fat accumulation, as well as wet weight growth, are evidently adverse but all remain below 10% (i.e., the median size effect; Table 4). A decrease in performance of approximately 10% is not catastrophic and indicates that the adjustment of dietary amino acids and inclusion of marine components likely enhanced the performance of the fish fed the HPLF diets. Weight increase typically decreases when fish are fed high plant protein diets, as shown in various studies (Gomes et al., 1995; Kaushik et al., 1995, 2004; Mambrini et al., 1999; Dias et al., 2005; Torstensen et al., 2008). The decrease in performance may be attributed to the unbalanced amino acid profile of the meals. However, even when the amino acid profile is balanced, performance still drops as the proportion of plant ingredients increases (Fournier et al., 2003, 2004). When juvenile Senegalese sole were fed diets with low fishmeal content but balanced in all dietary amino acids and supplemented with taurine, their wet weight increase matched that of fish on a diet with high fishmeal content (Silva et al., 2009). Taurine supplementation in high fishmeal replacement diets administered to juvenile Atlantic salmon was discovered to decrease the whole body lipid to protein ratio (Espe et al., 2011). Greater protein accumulation leads to increased cost efficiency in feed production. Reduced nitrogen leakage into the ocean occurs when sediment accumulation is substantial (Kaushik et al., 2004). Hence, the new fishmeal replacement diets must lead to protein accretion similar to that of traditional fishmeal-based diets, considering the environmental perspective. Feeding rainbow trout meals high in plant proteins results in decreased protein accretion (Fournier et al., 2003; Aksnes et al., 2006). This is typically true when any marine species is provided with higher quantities of plant proteins in their diet. Atlantic cod decreases protein growth significantly when plant protein makes up more than 50% of the diet's protein, according to Hansen et al. (2007). On the other hand, the European seabass does not decrease protein growth until plant protein approaches 80% of the diet's protein, as found by Kaushik et al. (2004). Seabream experienced decreased protein gain when fishmeal was substituted with plant proteins, attributed to an unbalanced amino acid profile (Robaina et al., 1995; Gomez-Requeni et al., 2004). Despite having a similar amino acid composition to the fishmeal-based control, turbot fed diets with less than 20% fishmeal showed lower protein accretion compared to the control group (Fournier et al., 2004).

Table 3. The mean initial and end body wt (BW, g), mean wet wt gain (g/kg ABW/day), mean apparent digestibility of nitrogen and mean protein and lipid gain (g/kg ABW/day) in the six trials included in the statistics to test the effects of high or medium fishmeal replacement diets fed to Atlantic salmon.

	Trial I		Trial II		Trial III		Trial IV		Trial V		Trial VI	
Diets	C	T	C	T	C	T	C	T	C	T	C	T
Initial BW	333±4	327±8	633±31	619±50	481±25	495±5	1123±9	1128±55	371±11	383±6	1441±25	1402±30
End BW	763±37	735±15	1108±23	986±48	1235±94	1125±64	1891±25	1928±77	1065±93	965±91	2435±88	2497±72
Weight gain	8.7±0.4	8.3±0.2	6.7±0.7	5.3±0.4	10.3±0.3	9.5±0.7	6.0±0.3	6.0±0.8	12.3±0.9	10.8±1.0	5.8±0.3	6.3±0.4
ADC-N	89±1	92±2	88±1	93±1	93±2	93±1	86±5	92±1	88±1	89±1	92±1	94±1
Protein gain	1.13±0.1	1.18±0.1	1.14±0.1	0.84±0.1	1.72±0.1	1.43±0.1	0.77±0.1	0.75±0.1	1.85±0.1	1.73±0.1	0.62±0.1	0.74±0.1
Lipid gain	3.20±0.2	2.81±0.2	1.23±0.1	1.02±0.1	2.40±0.1	1.90±0.2	1.30±0.1	1.24±0.3	1.71±0.2	1.53±0.1	0.75±0.1	0.84±0.1

Arbitrary BW (ABW) is the mean wt during the trials (end BW+initial BW/2). In trials I to IV is mean of 3 tanks per treatment while trials V and VI is the mean of 4 tanks per treatment. Values are the tank means±standard deviation

Table 4. Statistical estimates of the mean wet weight gain, protein and lipid gain of the MPMF feeds together with the estimated differences of the HPLF feeds (g/Kg ABW/day). Median effect size and its 95% credible intervals are given as percentage of the MPMF estimate. The estimated probability that HPLF is smaller than MPMF is also given in the last column.

	MPMF	Diff with HPLF	SE diff	Effect size median	Effect size 95% CI low	Effect size 95% CI high % of MPMF	Probability HPLF<MPMF
	g/Kg ABW/day	g/Kg ABW/day	g/Kg ABW/day	% of MPMF	% of MPMF	% of MPMF	
Wet wt gain	8.07	-0.59	0.23	-7.24	-13.14	-1.87	0.995
Protein gain	1.19	-0.11	0.04	-9.47	-16.10	-2.77	0.998
Lipid gain	1.64	-0.15	0.06	-9.47	-16.48	-3.01	0.997

MPMF medium protein, medium fishmeal

HPLF high protein, low fishmeal

All living beings, including humans, possess a remarkable capacity to adjust to a reduced protein consumption (Millward, 1999). In Atlantic salmon, protein gain remained unaffected despite a decrease in feed intake and a reduction in lipid gain when the fish was fed fishmeal-free diets (Espe et al., 2006). Dias et al. (2005) found that European seabass fed plant protein-based diets had lower lipid accumulation compared to those fed a fishmeal diet. The meals contained a mixture of plant proteins as the primary protein source, with consistent amino acid profiles throughout all trials, exceeding the expected amino acid needs. The decreased performance cannot be attributed to a lack of dietary amino acids since the digestibility was consistently good in all trials (Table 3). The decreased performance of Atlantic salmon when fed plant protein-based diets is not due to a reduced supply of amino acids. The metabolic cost of high replacement diets may increase because NPN metabolites found in fishmeal may not be provided or need to be synthesized by the fish. Previous research showed lower fat content in whole fish, muscle, and liver of Atlantic salmon that were fed hydrolyzed proteins compared to fish fed diets containing intact control protein, as reported by Espe et al. (1992c). The lower fat deposition in the test diets compared to the control diet, which did not contain FPC, may be attributed to the presence of 5-11.25% FPC in the test diets. The decreased lipid accumulation in diets with complete fishmeal substitution is likely a result of metabolic alterations or increased metabolic expenses. To develop diets for Atlantic salmon that optimize the burning of lipids and deposition of dietary protein, it is crucial to investigate how fish detect energy and select between protein and fat as fuel sources. Recent findings indicate that FPC influences the internal organ mass of Atlantic salmon without impacting the overall weight when fed diets with medium fishmeal content partially substituted with FPC. This suggests a change in how Atlantic salmon deposits nutrients when FPC is included in their diets. Moreover, a study by Torstensen et al. (2011) found that Atlantic salmon had increased adipose tissue when both the protein and lipid sources were of plant origin. This suggests that the deposition pattern should be considered when evaluating the impact of high fishmeal replacement diets.

3. FUTURE ASPECTS

Despite Atlantic salmon successfully adapting to diets with complete substitution of fish meal with plant ingredients or high replacement levels, changes in compartmentalization and fat deposition may occur even when feed consumption remains consistent. This could be caused by changes in the concentration of non-protein nitrogen (NPN) in the new diets. In trials III and VI, the replacement diets showed greater NPN levels, as depicted in Figure 2. However, wet weight increase was most favorable in trials III and V. Therefore, NPN alone cannot account for the decreased performance. Atlantic salmon that were fed diets deficient in methionine showed decreased levels of taurine in both muscle and liver, as reported by Espe et al. (2008). Decreased methionine intake and lowered liver taurine levels led to increased lipid accumulation in the liver, as shown by Espe et al. (2010). Additionally, lysine restrictions caused greater lipid concentration in the entire fish, as demonstrated by Espe et al. (2007). The study by Gaylord et al. (2007) revealed that methionine, not taurine, had a greater impact on lipid deposition in rainbow trout. Taurine supplementation in juvenile Atlantic salmon fed plant protein-based diets decreased whole-body lipid accumulation without impacting protein deposition (Espe et al., 2011). The changed lipid distribution resulting from amino acid restrictions is likely influenced by metabolite interactions occurring when they are either abundant or scarce in the diets. The studies on Atlantic salmon used diets with balanced amino acids. However, future research should investigate how amino acid requirements may change when fishmeal is substituted. Research should focus on the metabolic effects of feeding Atlantic salmon diets with total or nearly total fish meal replacement, and determine if NPN compounds are conditionally essential. The need for essential and non-essential amino acids should be considered when synthesizing metabolites that are not included in the diet, as this may raise the demand for precursor amino acids.

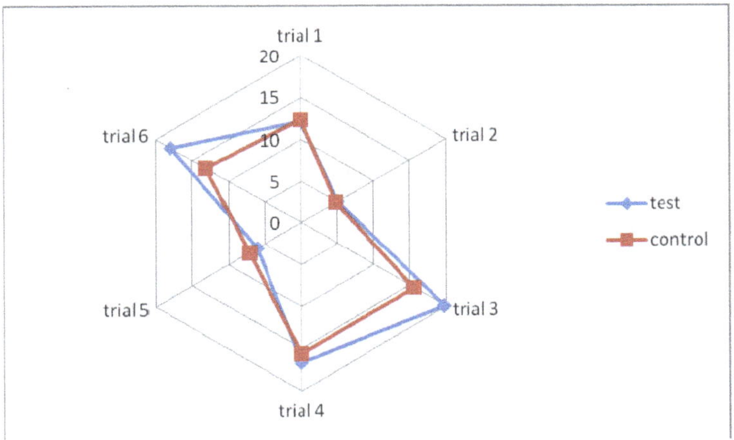

Fig. 2. The non amino acid nitrogen (NPN) differed between the diets in trial III and VI, otherwise the NPN was pretty similar. Here is shown the NPN in the relative high fishmeal diets (control) vs the NPN in the replacement diets.

Previously, it was found that replacing all protein with hydrolyzed proteins decreased lipid accumulation in various parts of the fish. Therefore, optimizing the inclusion of FPC could be a viable strategy to produce leaner Atlantic salmon in fish farming. The visceral lipid in rats decreased when they were fed a diet with FPC as the dietary protein source compared to diets with casein and soy protein (Liaset et al., 2009). The recent findings on substituting fishmeal with plant protein in Atlantic salmon diets demonstrate the feasibility of achieving sustainable aquaculture in the future. Utilizing marine ingredients, such as fishmeal, to enhance feed intake, growth, metabolism, and overall health of farmed fish can boost production without relying heavily on limited marine resources.

REFERENCES

1. Abbouti, T., Mambrini, M., Larondelle, Y. & Rollin, X. (2009). The effect of dispensable amino acids on nitrogen and amino acid losses in Atlantic salmon (*Salmo salar*) fry fed a protein-free diet. *Aquaculture* 289, 327-333
2. Aksnes, A., Hope, B. & Albrektsen, S. (2006). Size-fractionated fish hydrolysate as feed ingredient for rainbow trout (*Oncorhynchus mykiss*) fed high plant protein diets. II: Flesh quality, absorption, retention and fillet levels of taurine and anserine. *Aquaculture*, 261, 318-326
3. Berge, G.E., Lied, E. &, Espe, M. (1994). Absorption and incorporation of dietary free and protein bound (U14C)-lysine in Atlantic cod (*Gadus morhua*). *Comparative Biochemistry and Physiology* 109A, 681-688
4. Berge, G.M., Baeverfjord, G., Skrede, A. & Storebakken, T. (2005). Bacterial protein grown on natural gas as protein source for Atlantic salmon, *Salmo salar*, in saltwater. *Aquaculture* 244, 233-240
5. Carter, C.G. & Hauler, R.C. (2000). Fish meal replacement by plant meals in extruded feeds for Atlantic salmon, *Salmo salar* L. *Aquaculture*, 185, 299-311.
6. Cowey. C.B. (1995). Protein and amino acid requirements: A critique of methods. *Journal of Applied Ichthyology* 11, 199-204
7. Dabrowski, K, & Guderley, H.Y. (2002). Intermediary metabolism. In: *Fish nutrition* (Ed: Halver J & Hardy R), Academic press, pp 309-365
8. Dias, J., Alvarez, M.J., Arzel, J., Corraze, G., Diez, A., Bautista, J.M. & Kaushik, S.J. (2005). Dietary protein source affects lipid metabolism in the European seabass (*Dicentrarchus labrax*). *Comparative Biochemistry and Physiology* 142, 19-31.
9. El-Mowafi, A., Ruohonen, K., Hevrøy, E.M. & Espe, M. (2010). Impact of three amino acid profiles on growth performance in Atlantic salmon. *Aquaculture Research* 41, 373-384
10. Espe, M. (2008). Understanding factors affecting flesh quality in farmed fish. In: *Improving farmed fish for the consumer* (Ed: Ø Lie), Woodhead Publishing, Cambridge UK pp 241-269
11. Espe M & Njaa LR (1991) Growth and chemical composition of Atlantic salmon (*Salmo salar*) given a fish meal diet or a corresponding free amino acid diet. *Fiskeridirektoratets Skrifter Serie Ernaering* 4, 103-110

12. Espe, M. & Lied, E. (1994). Do Atlantic salmon (*Salmo salar*) utilize mixtures of free amino acids to the same extent as intact protein sources for synthetic purposes? *Comparative Biochemistry and Physiology* 107A, 249-254.
13. Espe, M. & Lied, E. (1999). Fish silage prepared from different cooked and uncooked raw materials. Chemical changes during storage at different temperatures. *Journal of Food Science and Agriculture* 79, 327-332.
14. Espe, M., Raa, J. & Njaa, L.R. (1989). Nutritional value of stored fish silage as a protein source for young rats. *Journal of Food Science and Agriculture* 49, 259-270
15. Espe, M., Haaland, H. & Njaa, L.R. (1992a). Substitution of fish silage protein and a free amino acid mixture for fish meal protein in a chicken diet. *Journal of Food science and Agriculture* 58, 315-320
16. Espe, M., Haaland, H. & Njaa, L.R. (1992b). Growth of young rats on diets based on fish silage with different degrees of hydrolysis. *Food Chemistry* 44, 195-200.
17. Espe, M., Haaland, H. & Njaa, L.R. (1992c). Autolysed fish silage as a feed ingredient for Atlantic salmon (*Salmo salar*). *Comparative Biochemistry and Physiology* 103A, 369-372
18. Espe, M., Sveier, H., Høgøy, I. & Lied, E. (1999). Nutrient absorption and growth of Atlantic salmon (*Salmo salar* L.) fed fish protein concentrate. *Aquaculture* 174, 119-137
19. Espe, M., Lemme, A., Petri, A. & El-Mowafi, A. (2006). Can Atlantic salmon (*Salmo salar*) grow on diets devoid of fish meal? *Aquaculture* 255, 255-262
20. Espe, M., Lemme, A., Petri, A. & El-Mowafi, A. (2007). Assessment of lysine requirement for maximal protein accretion in Atlantic salmon using plant protein diets. *Aquaculture* 263, 168-178
21. Espe, M., Hevrøy, E.M., Liaset, B., Lemme, A. & El-Mowafi, A. (2008). Methionine intake affect hepatic sulphur metabolism in Atlantic salmon, *Salmo salar*. *Aquaculture* 274, 132-141.
22. Espe, M., Rathore, R.M., Due, Z-Y., Liaset, B. & El-Mowafi, A. (2010). Methionine limitation results in increased hepatic FAS activity, higher liver 18:1 to 18:0 fatty acid ratio and hepatic TAG accumulation in Atlantic salmon, *Salmo salar*. *Amino Acids* 39, 449- 460
23. Espe, M., Ruohonen, K. & El.Mowafi, A. (2011). Effect of taurine supplementation on the metabolism and body lipid-to-protein ratio in juvenile Atlantic salmon (*Salmo salar*). *Aquaculture Research* In press Doi: 10.1111/j.1365-2109.2011.02837.x
24. FAO. (2006). The State of World Fisheries and Aquaculture. FAO, Rome, 3 pp.
25. Finkelstein, J.D. (1990). Methionine metabolism in mammals. *Journal of Nutritional Biochemistry* 1, 228-237.

26. Fournier, V., Gouillou-Coustans, M.F., Metailler, R., Vachot, C., Moriceau, J., LeDelliou, H., Huelvan, C., Desbrugeres, E. & Kaushik, S.J. (2003). Excess arginine affects urea excretion but does not improve N utilisation in rainbow trout *Oncorhynchus mykiss* and turbot *Psetta maxima*. *Aquaculture* 217, 559-576

27. Fournier, V., Huelvan, C. & Desbruyeres, E. (2004). Incorporation of a mixture of plant feedstuffs as substitute for fish meal in diets of juvenile turbot (*Psetta maxima*). *Aquaculture*, 236, 451–465.

28. Gaylord, T.G., Barrows, F.T., Teague, A.M., Johansen, K.A., Overturf, K.E. & Shepherd, B. (2007). Supplementation of taurine and methionine to all-plant protein diets for rainbow trout (*Oncorhynchus mykiss*). *Aquaculture* 269, 514-524

29. Glencross, B., Ewans, D., Hawkins, W. & Jones, B. (2004). Evaluation of dietary inclusion of yellow lupin (*Lupinus luteus*) kernel meal on the growth, feed utilisation and tissue histology in rainbow trout (*Oncorhynchus mykiss*). *Aquaculture* 235, 411-422

30. Geiger, E. (1947). Experiments with delayed supplementation of incomplete amino acid mixtures. *Journal of Nutrition* 34, 97-111

31. Gomes, E.F., Remam P., Gouveia, A. & Teles, A.O. (1995). Replacement of Fish-Meal by Plant-Proteins in Diets for Rainbow-Trout (*Oncorhynchus-mykiss*) - Effect of the Quality of the Fish-Meal Based Control Diets on Digestibility and Nutrient Balances. *Water Science and Technology* 31, 205-211.

32. Gomez-Requeni, P., Mingarro, M., Calduch-Giner, J.A., Medale, F., Martin, S., Houlihan, D.F., Kaushik, S. & Perez-Sanchez, J. (2004). Protein growth performance, amino acid utilisation and somatotropic axis responsiveness to fish meal replacement by plant protein sources in gilthead sea bream (*Sparus aurata*). *Aquaculture* 232, 493- 510.

33. Hansen, A.C., Karlsen, O., Rosenlund, G., Rimbach, M. & Hemre, G-I. (2007). Dietary plant protein utilization in Atlantic cod, *Gadus morhua* L. *Aquaculture Nutrition* 13, 200- 215.

34. Hevrøy, E.M., Espe, M., Waagbø, R., Sandnes, K., Ruud, M. & Hemre, G-I. (2005). Nutrient utilization n Atlantic salmon (*Salmo salar* L.) fed increased levels of fish protein hydrolysate during a period of fast growth. *Aquaculture Nutrition* 11, 301-313.

35. Houlihan, D.F., Carter, C.G. & McCarthy, I.D. (1995). Protein synthesis in fish. In: *Biochemistry and Molecular Biology in Fishes* (Eds: Hochachka, PW & Mommsen ID), Elsevier Science, Amsterdam, pp 191-219.

36. Kousoulaki, K., Albrektsen, S., Langmyhr, E., Olsen, H.J., Campbell, P. & Aksnes, A. (2009). The water soluble fraction in fish meal (stickwater) stimulates growth

in Atlantic salmon (*Salmo salar*, L) given high plant protein diets. *Aquaculture* 289, 74-83.

37. Kaushik, S.J., Cravedi, J.P., Lalles, J.P., Sumpter, J., Fauconneau, B. & Laroche, M. (1995). Partial or total replacement of fish meal by soybean protein on growth, protein utilization, potential estrogenic or antigenic effects, cholesterolemia and flesh quality in rainbow trout (*Oncorhynchus mykiss*). *Aquaculture* 133, 257 - 274.
38. Kaushik, S.J., Covès, D., Dutto, G. & Blanc, D. (2004). Almost total replacement of fish meal by plant protein sources in the diet of a marine teleost, the European sea bass (*Dicentrarchus labrax*). *Aquaculture* 230, 391 – 404.
39. Krogdahl, A., Bakke-McKellup, A.M. & Baeverfjord, G. (2003). Effects of graded levels of standard soybean meal on intestinal structure, mucosal enzyme activities, and pancreatic response in Atlantic salmon (*Salmo salar* L.). *Aquaculture Nutrition* 9, 361- 371.
40. Laksesvela, B. (1958). Protein value and amino-acid balance of condensed herring soluble and spontaneously heated herring meal. Chick experiments. *Journal of Agricultural Science* 51, 164-176
41. Laksesvela, B. (1960). The potency of balancing interactions between dietary proteins. Chick experiments on the significance for prior judgment and use of herring proteins. *Journal of Agricultural Science* 55, 215-223
42. Leaver, M.J., Bautista, J.M., Bjornsson, B.T., Jonsson, E., Krey, G., Tocher, D.R. & Torstensen, B.E. (2009). Towards fish lipid nutrigenomics: Current state and prospects for finfish aquaculture. *Reviews in Fisheries Science* 16, 73-94
43. Liaset, B. & Espe, M. (2008). Nutritional composition of two main peptide fractions obtained in the enzymatic hydrolysis of fish raw material. *Process Biochemistry* 43, 42-48
44. Liaset, B., Lied, E. & Espe, M. (2000). Enzymatic hydrolysis of by-products from the fish- filleting industry; chemical characterisation and nutritional evaluation. *Journal of Food Science and Agriculture* 80, 581-589.
45. Liaset, B., Madsen, L., Qin, H., Criales, G., Mellgren, G., Marschall, H.U., Hallenborg, P., Espe, M., Frøyland, L. & Kristiansen, K. (2009). Fish protein hydrolysate elevates plasma bile acids and reduces visceral adipose tissue mass in rats. *Biochimica et Biophysica Acta - Molecular and Cell Biology of Lipids* 1791, 254-262
46. Mambrini, M., Roem, A.J., Cravèdi, J.P., Lallès, J.P. & Kaushik, S.J. (1999). Effects of replacing fish meal with soy protein concentrate and of DL-methionine supplementation in high-energy extruded diets on the growth and nutrient

utilization of rainbow trout, *Oncorhynchus mykiss*. *Journal of Animal Science* 77, 2990- 2999.

47. Mato, J.M., Corrales, F.J., Lu, S.C. & Avila, M.A. (2002). S-adenosylmethionine: a control switch that regulates liver function. *FASEB Journal* 16, 15-26.
48. Millward, D.J. (1988). Metabolic demand for amino acids and the human dietary requirement: Millward and Rivers (1988) revisited. *Journal of Nutrition* 128, 2563S- 2576S
49. Millward, D.J. (1999). The nutritional value of plant based diets in relation to human amino acid and protein requirements. *Proceedings of the Nutritional Society* 58, 249-260.
50. Njaa, L.R. (1990). Amino acid contents of fillet protein from 13 species of fish.
51. *Fiskeridirektoratets Skrifter Serie Ernaering*, III, 43-45. National Research Council (NRC) (1993). Nutritional Requirements of Fish, National Academy Press, Washington DC, USA 114 pp.
52. Olli, J.J., Krogdahl, Å. & Våbenø, A. (1995). Dehulled solvent-extracted soybean meal as a protein source in diets for Atlantic salmon, *Salmo salar*, L. *Aquaculture Research* 26, 167-174.
53. Opstvedt, J., Aksnes, A., Hope, B. & Pike, I.H. (2003). Efficiency of feed utilization in Atlantic salmon (*Salmo salar* L.) fed diets with increasing substitution of fish meal with vegetable proteins. *Aquaculture* 221, 365-379.
54. Ozyurt, M. & Deveci, U.D. (2004). Conversion of agricultural and industrial wastes for single cell production and pollution potential reduction: A review. *Fresenius Environmental Bulletin* 13, 693-699
55. Pongmaneerat, J. & Watanabe, T. (1992). Utilization of soybean-meal as protein-source in diets for rainbow trout. *Nippon Suisan Gakkaishi* 58, 1983-1990
56. Rahnema, S., Borton, R. & Shaw, E. (2005). Determination of the effects of fish vs plant vs meat protein based diets on the growth and health of rainbow trout. *Journal of Applied Animal Research* 27, 77-80
57. Refstie, S., Storebakken, T. & Roem, A.J. (1998). Feed consumption and conversion in Atlantic salmon (*Salmo salar*) fed diets with fish meal, extracted soybean meal or soybean meal with reduced content of oligosaccharides, trypsin inhibitors, lectins and soya antigens. *Aquaculture* 162, 301-312.
58. Refstie, S., Olli, J.J.& Standal, H. (2004). Feed intake, growth and protein utilisation by post- smolt Atlantic salmon (*Salmo salar*) in response to graded levels of fish protein hydrolysate in the diet. *Aquaculture* 239, 331 – 349.

59. Refstie, S., Korsoen, O.J., Storebakken, T., Baeverfjord, G., Lein, I. & Roem, A.J. (2000). Differing nutritional responses to dietary soybean meal in rainbow trout (*Oncorhynchus mykiss*) and Atlantic salmon (*Salmo salar*). *Aquaculture* 190, 49-63.

60. Regost, C., Arzel, J. & Kaushik, S.J. (1999). Partial or total replacement of fish meal by corn gluten meal in diet for turbot (*Psetta maxima*). *Aquaculture* 180, 99-117.

61. Robaina, L., Izquierdo, M.S., Moyano, F.J., Socorro, J., Vergara, J.M., Montero, D. & Fernandezpalacios, H. (1995). Soybean and lupin seed meals as protein-sources in diets for gilthead seabream (*Sparus-aurata*) - Nutritional and histological implications. *Aquaculture* 130, 219-233.

62. Rollin, X., Mambrini, M., Abboudi, T., Larondelle, Y. & Kaushik, S.J. (2003). The optimum dietary indispensable amino acid pattern for growing Atlantic salmon (*Salmo salar* L.) fry. *British Journal of Nutrition* 90, 865-876

63. Rowling, M.J., McMullen, M.H., Chipman, D.C. & Schaliniske, K.L. (2002). Hepatic glycine N-methyltransferase is up-regulated by excess dietary methionine in rats. *Journal of Nutrition* 132, 2545-2550.

64. Sugiura, S.H., Dong, F.M., Rathbone, C.K. & Hardy, R.W. (1998). Apparent protein digestibility and mineral availabilities in various feed ingredients for salmonid feeds. *Aquaculture* 159, 177-202

65. Silva, J.M.G., Espe, M., Conceção, L.E.C., Dias, J. & Valente, L.M.P. (2009). Senegalese sole juvenile (*Solea senegalensis* Kaup, 1858) grow equally well on diets devoid of fish meal provided the dietary amino acids are balanced. *Aquaculture* 296, 309-317.

66. Storebakken, T., Shearer, K.D., Refstie, S., Lagocki, S. & McCool, J. (1998). Interactions between salinity, dietary carbohydrate source and carbohydrate concentration on the digestibility of macronutrients and energy in rainbow trout (*Oncorhynchus mykiss*). *Aquaculture* 163, 3-4.

67. Storebakken, T., Baeverfjord, G., Skrede, A., Olli, J.J. & Berge, G.M. (2004). Bacterial protein grown on natural gas in diets for Atlantic salmon, *Salmo salar*, in freshwater. *Aquaculture*, 241, 413-425

68. Tacon, A.G.J. (1995). Feed ingredients for carnivorous fish species: Alternatives to fishmeal and other fishery resources. In: *Sustainable Fish Farming* (Reinertsen H. & Haaland H. eds.) pp. 89-114.)

69. Tacon, A.G.J. & Metian, M. (2008). Global overview on the use of fish meal and fish oil in industrially compounded aquafeeds: Trends and future prospects. *Aquaculture* 285, 146-158.

70. Torstensen, B.E., Frøyland, L., Ørnsrud, R. & Lie, Ø. (2004). Tailoring of a cardioprotective fillet fatty acid composition of Atlantic salmon (*Salmo salar*) fed vegetable oils. *Food Chemistry* 87, 567-580.

71. Torstensen, B.E., Bell, J.G., Rosenlund, G., Henderson, R.J., Graff, I.E., Tocher, D.R., Lie, Ø. & Sargent, J.R. (2005). Tailoring of Atlantic salmon (*Salmo salar* L.) flesh lipid composition and sensory quality by replacing fish oil with a vegetable oil blend. *Journal of Agricultural Food Chemi*stry 53, 10166-10178

72. Torstensen, B.E., Espe, M., Stubhaug, I., Waagbø, R., Hemre, G-I., Fontanillas, R., Nordgarden, U., Hevrøy, E.M., Olsvik, P. & Berntssen, B.H.G. (2008). Combined maximum replacement of fish meal and fish oil with plant meal and vegetable oil blends in diets for Atlantic salmon (*Salmo salar* L.) growing from 0.3 to 4 kilo. *Aquaculture* 285, 193-200.

73. Torstensen, B.E., Espe, M.., Stubhaug, I. & Lie, Ø. (2011). Combined dietary plant proteins and vegetable oil blends increase adiposity and plasma lipids in Atlantic salmon (*Salmo salar* L.). *British Journal of Nutrition* 106, 633-647. doi: 10.1017/ S0007114511000729

74. Treberg, J.R. & Driedze, W.R. (2007). The accumulation and synthesis of betaine in winter skate (*Leucoraja ocellata*). *Comparative Biochemistry and Physiology* 147, 475-483

75. Von derDecken, A. & Lied, E. (1993). Metabolic effects on growth and muscle of soybean protein feeding in cod (*Gadus morhua*). *British Journal of Nutrition* 69, 689-697.

76. Wang, T.C. & Fuller, M.F. (1987). An optimal dietary amino acid pattern for growing pigs. *Animal Production* 44, 486.

77. Wang, T.C. & Fuller, M.F. (1989). The optimum dietary amino acid pattern for growing pigs. 1. Experiment by amino acid deletion. *British Journal of Nutrition* 62, 77-89.

78. Wu G. (2009) Amino acids: metabolism, functions, and nutrition. *Amino Acids* 37, 1-17

79. Yanik, T., Dabrowski, K. & Bai, S.C. (2003). Replacing fish meal in rainbow trout (*Oncorhynchus mykiss*) diets. *Israeli Journal of Aquaculture* 55, 179-186

04 MICROALGAE'S NUTRITIONAL VALUE AND APPLICATIONS IN AQUACULTURE

1. INTRODUCTION: AN OVERVIEW

Microalgae, which are single-celled algae or phytoplankton, are the largest but least understood category of microorganisms on Earth. Microalgae serve as the principal source of essential nutrients in the aquatic food chain, similar to how plants do for terrestrial species. Microalgae are essential for the nourishment of marine animals in the open sea and in aquaculture. Many marine invertebrates rely on microalgae throughout their whole life cycle. Therefore, both commercial and experimental mollusc or fish hatcheries have integrated a microalga production system alongside their animal production. Microalgae are used as food for various growth stages of bivalve molluscs, such as oysters, scallops, clams, and mussels, as well as for the early stages of abalone, crustaceans, some fish species, and zooplankton in aquaculture food chains. The productivity of a hatchery depends directly on the quantity and quality of the food source used. Recently, the idea of aquaculture as a collection of constructed systems for wastewater treatment and recycling has gained momentum. Their design is tailored to meet precise treatment and wastewater requirements, potentially addressing environmental and sanitary issues while also being economically viable. High rate microalgal ponds have regained popularity for wastewater treatment. In these ponds, photosynthetic microalgae provide oxygen to heterotrophic bacteria, converting wastewater nutrients into biomass protein. Microalga culture is expected to become more significant in aquatic food production systems, particularly for producing feed for fish, converting CO2 to O2, and improving water quality.

2. CHARACTERISTICS OF MICROALGAE SPECIES IN AQUACULTURE

Aquatic creatures consumed by humans on a vast scale are often carnivorous or omnivorous after their larval stage, unlike land animals that breathe air. Only filtering molluscs and a few other creatures are reliant on plankton for their entire lifespan because of the extended nature of the linked food web. Microalgae are necessary for larval feeding for a short duration, either consumed directly by mollusks and penaeid shrimps or indirectly through live prey given to small larval fish. Post-larvae specimens are hatched, bred, and raised by specialized establishments known as hatcheries. These facilities are intricate to manage because to the extensive production of microalgae and small live prey like rotifers for small-larval fish. It is rare to find aquacultured species that do not face these limitations. Salmonids are an exception since their eggs contain enough nutrients for the larvae to hatch large and feed on dry particles directly. Over the past few decades, numerous microalgae species have been examined as feed, although only a small fraction, perhaps less than twenty, have been widely used in aquaculture. Microalgal species exhibit great variation in nutritional value, which is influenced by growing conditions. For enhanced animal growth and improved nutrition, a specific blend of microalgae should be provided to fish, either directly or indirectly through zooplankton enrichment, as it yields better outcomes compared to a diet consisting of a single microalga. Microalga production for feed is categorized into intensive monoculture for larval stages of bivalves, shrimp, and specific fish species, and extensive culture for the growth of bivalves, carp, and shrimp. Preferred genera for the former group are Chaetoceros, Thalassiosira, Tetraselmis, Isochrysis, Nannochloropsis, Pavlova, and Skeletonema [6,9,10]. The organisms are either directly or indirectly provided as food to the cultured larval creature. Indirect methods often involve using artemia, rotifers, and Daphnia, which are then fed to the target larval organisms. Microalgae are commonly believed to be consumed by shrimp larvae and are significant for nutrition at that life stage. However, it is unclear if juveniles and adults also actively consume microalgae. Microalgae are present in shrimp guts because shrimp inadvertently consume them along with trash, as indicated by certain reports [11,12]. The nutritional quality of a meal containing microalgae is crucially dependent on its capacity to provide necessary macro- and micronutrients to the intended animal consumer. A mixed microalgal diet, commonly employed during the hatchery and nursery stages of oyster farming, is expected to be more effective than diets consisting of a single kind of algae. Yet, the dietary needs of bivalves are not clearly established; studies using microalgae with known compositions have indicated that carbohydrate and polyunsaturated fatty acid (PUFA) levels significantly impact the growth of oysters [5]. Increasing the dietary carbohydrate supply was discovered to enhance oyster growth rate, as long as sufficient protein and critical fatty acids were also delivered simultaneously. Adding protein microcapsules to young mussel diets resulted in increased growth, showing that mussel growth rates are considerably limited when the protein concentration is below 40% (w/w) [15]. Diatoms and haptophytes are commonly utilized as a nutritious diet for oysters.

Isochrysis sp. and P. lutheri are high in docosahexaenoic acid (DHA, 22:6n-3), making up 8-10% of total fatty acids. Diatoms are abundant in eicosapentaenoic acid (EPA, 20:5n-3). Combining prymnesiophytes and diatoms in bivalve hatcheries is a common practice and is regarded very nutritious due to their important polyunsaturated fatty acids content. Microalgae suitable for aquaculture species should have specific characteristics: they should be easily digestible due to a digestible cell wall, grow rapidly, be able to be mass-cultured, tolerate fluctuations in environmental conditions, and have good nutritional qualities without toxins. Algologists face a significant difficulty in reducing production costs while ensuring the reliability of microalgal feed. Microalgae serve as food for zooplankton and can enhance the quality of the culture medium. Introducing phytoplankton to rearing ponds, known as the green-water approach, yields superior outcomes in terms of survival, growth, and transformation index for many freshwater and marine animal species compared to the traditional clear-water technique [19-21]. The reason for this observation is not fully understood, but it may involve factors such as oxygen production and pH stabilization improving water quality, the effects of excreted biochemical compounds, initiation of prey catching behavior, regulation of bacterial population, probiotic effects, and immune system stimulation.

3. NUTRITIONAL FEATURES OF MICROALGAE

Microalgal species can vary significantly in nutritional value, as a function of the prevailing culture conditions. Only a reduced number of species have been used, primarily for historical reasons and ease of cultivation – rather than supported by scientific evidence of any superior performance as nutritional or therapeutical supplements. Hence, formulations more carefully selected of microalgal origin may offer the opportunity for development of improved nutritional packages aimed at larval animals.

Several factors contribute to the nutritional value of a microalga – including its size and shape, and digestibility as related to cell wall structure and composition (as mentioned above), as well as biochemical composition (e.g. accumulation compounds, enzymes and toxins) and specific requirements of the target animal. For this reason, several studies have attempted to correlate the nutritional value of microalgae to their chemical profile. However, results from feeding experiments are often difficult to interpret because of the confounding effects of other formulation additives. An examination of literature data – including those pertaining to microalga-based, compounded diet emulsions, have meanwhile allowed a few general conclusions to be reached [25].

As primary producers in the aquatic food chain, microalgae provide many phytonutrients, including in particular PUFAs – e.g. EPA, arachidonic acid (AA) and DHA, which are known to be essential for various marine animals [25], as well as for growth and metamorphosis of many larvae [8,26]. However, the ratios of DHA, EPA and AA may actually be more important than their absolute levels [24,27]. Most microalgal species exhibit moderate to high percents of EPA (7 to 34%); and prymnesiophytes (e.g. *Pavlova* spp. and *Isochrysis* sp.) and cryptomonads are relatively rich in DHA (0.2 to 11%), whereas eustigmatophytes (e.g. *Nannochloropsis* spp.) and diatoms have the highest percentages of AA (up to 4%). Chlorophytes (*Dunaliella* spp. and *Chlorella* spp.) are deficient in both C20 and C22 PUFAs, although some species have small amounts of EPA (up to 3.2%); because of such a PUFA deficiency, chlorophytes are in general ascribed a poor nutritional value, so they are not suitable for use as single species-diet [6]. Prasinophyte species contain significant proportions of C20 (*Tetraselmis* spp.) or C22 (*Micromonas* spp.), but rarely of both. Therefore, the fatty acid contents of microalgae exhibit systematic differences according to taxonomic group – although there are examples of significant differences between microalgae, even within the same class.

The contents of antioxidants are also not uniform among microalgae; e.g. the concentrations of vitamins and carotenoids convey significant variations among species. Note that any mixed-algal diet should provide adequate concentrations of vitamins and carotenoids to be effective in aquaculture; unfortunately, the nutritional requirements of larval or juvenile animals that feed directly on microalgae are still poorly understood at present. In fact, artificial diets often lack natural pigments that allow such organisms as salmon or trout acquire their characteristic red color (muscle), which, in nature, is a result of eating microalgae containing red pigments; without such a color, a lower market value results. One way to alleviate this shortcoming is by adding astaxanthin to fish feed, with a consequently growing market for microalga-based sources, e.g. *Haematococcus pluvialis* [24,28].

On the other hand, the amino acid composition of microalgal proteins is rather similar between species [29], and relatively unaffected by their intrinsic growth phase and extrinsic light conditions [30,31]. Furthermore, the content in essential amino acids of microalgae is similar to that of oyster larvae. Overall, this indicates that protein quality is unlikely a factor that contributes to differences in nutritional value among microalgae. Finally, sterols [32], minerals [33] and pigments of microalgae also contribute to their nutritional performance in aquaculture.

Several studies have indicated that, in the late-logarithmic growth phase, microalgae contain typically 30-40 %(w/w) protein, 10-20 %(w/w) lipids and 5-15 %(w/w) carbohydrates [6,34]. When cultured through the stationary phase, the proximate composition of microalgae may significantly change; e.g. nitrate limitation leads carbohydrate levels to double at the expense of protein [31,35]. Hence, a strong correlation exists between composition of microalgae and their measurable nutritional value – even though diets containing high levels of carbohydrates have been reported to produce the best growth of juvenile oysters [9] and larval scallops [36], as long as PUFAs are also present to adequate proportions. Conversely, high dietary protein provides maximum growth for juvenile mussels [15] and oysters [18].

Another relevant issue is that marine environments are typically filled with bacteria and viruses that can attack fish and shellfish, and thus potentially devastate aquaculture farms. Bacteria and viruses can also attack single-celled microalgae, so these microorganisms have developed biochemical mechanisms for self-defense; such mechanisms involve secretion of compounds that inhibit bacterial growth or viral attachment. For instance, compounds synthesized by *Scenedesmus costatum*, and partially purified from its organic extract exhibited activity against aquacultured bacteria because of their fatty acids longer than 10 carbon atoms in chain length – which apparently induce lysis of bacterial protoplasts.

The ability of fatty acids at large to interfere with bacterial growth and survival has been known for some time, and recent structure-function relationship studies have proven that said ability depends on both their chain length and degree of unsaturation. Cholesterol and other compounds can antagonize antimicrobial features [37], so both composition and concentration of free lipids should be taken into account [38]. The activity of extracts of *Phaeodactylum tricornutum* against *Vibrio* spp. was attributed to EPA – a compound synthesized *de novo* by diatoms [39]; this PUFA is found chiefly as a polar lipid species in structural cell components (e.g. membranes) and is toxic to grazers [40], as well as a precursor of aldehydes with deleterious effects upon such consumers as copepods [41]. Similarly, unsaturated and saturated long chain fatty acids isolated from *S. costatum* [42] and organic extracts from *Euglena viridis* [43] display activity against that bacterial genus.

4. MICROALGAL BIOMASS PRODUCTION SYSTEMS

Commercial culture of microalgae targeted at their metabolites has been taking place for over 40 years, and the main microalgal species grown are *Chlorella* and *Spirulina* for healthy foods, *Dunaliella salina* for β-carotene, *H. pluvialis* for astaxanthin and several species for aquaculture [44].

There are several reactor configurations that met with success in mass cultivation of microalgae – chosen according to such factors as physiology of the microalga, cost of layout land, intensity of labour, cost of energy, availability of water, cost of nutrients, suitability of climate (if the culture is implemented outdoors) and specification of final product(s). Large- scale culture systems should be compared on the basis of such indicators as efficiency of light utilization, controllability of temperature, hydrodynamic stress allowable, ability to maintain unimicroalgal and/or axenic cultures and feasibility of scale-up.

A major decision to be made is whether to use closed photobioreactors (PBRs) or open ponds to cultivate a given microalga. The latter may entertain a large area, and are relatively cheap to build and easy to operate – but contamination is hard to control, stable environmental conditions (particularly temperature) are difficult to maintain, and the attainable cell density is relatively low because of mutual shading effects. On the other hand, extensive areas of land will be needed for commercial exploitation, besides substantial costs of harvesting afterwards [45,46]. The final choice of system is always a compromise between these parameters, aimed at achieving an economically acceptable outcome [44].

A common feature of most microalgal species produced commercially (i.e. *Chlorella*, *Spirulina* and *Dunaliella*) is that they grow in highly selective environments – which means that they can be grown in open air cultures and still remain relatively free of contamination by other microalgae and protozoa [47-49]. Species of microalgae that do not possess this selective advantage must be grown in closed systems; this includes most marine algae grown as aquaculture feeds (e.g. *Skeletonema*, *Chaetoceros*, *Thalassiosira*, *Tetraselmis* and *Isochrysis*), as well as the dinoflagellate *Crypthecodinium cohnii* [44].

Typical systems used indoors for microalgal mass culture include carboys (10 to 20 L), polythene bags (100 to 500 L) and tubs (1000 to 5000 L); these are usually operated batch- or continuouswise [44]. For larger volumes, outdoor tanks or ponds are preferred, which are operated semicontinuously; depending on their scale, hatcheries may produce between several hundreds to tens of thousands of liters of microalgal biomass per day. However, the culture systems employed at present are still fairly unsophisticated: e.g. *D. salina* is cultured in large (up to ca. 250 ha) shallow open-air ponds with no artificial mixing; and *Chlorella* and *Spirulina* are also grown outdoors, in either paddle-wheel mixed ponds or circular ponds (up to 1 ha each) with a rotating mixing arm. The production of microalgae for aquaculture occurs generally on a much lower scale. Other commercial large-scale systems include tanks used in aquaculture, the cascade system developed in the Czech Republic [50] and heterotrophic fermenter devices used for culture of *Chlorella* in Japan and Taiwan [47,51], and for culture of *C. cohnii* in USA [52,53].

The choice of which configuration is preferable depends obviously on the objective function; e.g. wastewater treatment might preclude open systems, owing to the unacceptably high costs arising from the large volumes to be processed and the low added value of the resulting products [54]. There has been a major effort directed at examining alternatives for the production of fresh microalgae, and also at more cost- efficient production systems.

4.1 Open cultivation systems

Microalgae cultivation in open ponds has been in current used since the 1950s [44]; these systems have been categorized as natural waters (lakes, lagoons and ponds), and artificial ponds or containers – with raceway ponds being the most frequently used artificial system [55]. The four major types of open-air systems currently in use (i.e. shallow big ponds, tanks, circular ponds and raceway ponds) have all advantages and disadvantages. This type of system usually consists of either circular ponds with a rotating arm to mix the culture, or long channels in a single or multiple loop configuration stirred by paddle wheels [56] – although simpler configurations have also met with sucess [54]. Raceway ponds are usually built in concrete, but compacted earth-lined ponds with (white) plastic have also been proposed. In a continuous production cycle, broth and nutrients required by microalgal growth are introduced in front of the paddlewheel, and circulated through the loop to the harvest extraction point; said paddlewheel undergoes a continuous motion to prevent sedimentation. The CO2 requirement is usually satisfied using the open atmosphere as source – yet submerged aerators may be installed to enhance CO2 supply, and thus absorption yield [57].

Compared to closed photobioreactors, open ponds represent a less expensive investment for large-scale production of microalgal biomass. On the other hand, open pond production does not necessarily compete with agricultural crops for land, since it can be implemented in areas with marginal crop production potential [58]. Open ponds also have low energy input requirements [59], and regular maintenance and cleaning are easier [60].

Open ponds and raceways were the first large-scale designs implemented, and are still the most widely applied in industrial processing. The main constraints related to their operation are the difficulty to control contamination and to keep the culture environment steady, and the cost associated with harvesting. Furthermore, the open character of the system makes it possible for naturally occurring microalgae or their predators to infiltrate, and thus compete with microalgae intended for cultivation. Therefore, a monoculture can only be maintained under extreme conditions of pH, salinity or temperature that guarantee dominance by the desired strain (e.g. *D. salina* dominance requires highly salted media, whereas *Spirulina platensis* demands high pH values). Unfortunately, high pH, temperature and salt concentration are not compatible with most microalgal species of interest.

Regarding biomass productivity, however, open pond systems are less efficient than closed photobioreactors [61]. This can be attributed to such parameters as evaporation losses, temperature variation, CO2 deficiency, inefficient mixing and light limitation. Although evaporation losses make a net contribution to cooling, they may also cause significant changes in the ionic composition of the medium – with detrimental effects upon microalga growth [62]. Although this type of reactor is extensively used in industrial microalgal production – e.g. to produce *Spirulina* and *Dunaliella* spp. up to worldwide totals of 5000 and 1200 ton/yr, respectively [24], open systems have apparently reached their upper limit – with little room for further technological improvement.

4.2 Closed cultivation systems

Despite the success of open systems, future advances in microalgal mass culture will require improved closed systems, as the most interesting microalgal species cannot grow in highly selective environments [44]. Hence, photobioreactor technology is on the rise, which is designed to overcome the major constraints associated with open pond production systems [63]; recall that both pollution and contamination risks preclude use of open ponds to prepare high-value products for eventual use as active ingredients in aquaculture feed formulation [60].

Closed systems include tubular, flat plate and fermenter types, among other possibilities. The former two are specifically designed for efficient recovery of sunlight, whereas the latter may require artificial illumination. Owing to the higher cell mass productivities attained, harvesting costs can be significantly reduced. Closed photobioreactors also provide reproducible cultivation conditions, good heat transfer, better biomass yield, higher product quality and opportunity for flexible technical design [44,60]. Note, however, that the costs of closed systems are substantially higher than their open pond counterparts [54]. A variety of closed photobioreactors have been tested (or at least proposed) for industrial microalgal biomass production [64,65], but engineering and economic analyses of such reactors still lag behind the open ponds [66-70].

A typical photobioreactor is essentially a four-phase system, consisting of solid microalgal cells, a liquid growth medium, a gaseous phase and a light radiation field [71]. Its productivity is limited by various design features – but, most importantly, the reactor is to be operated under favorable illumination conditions, with optimized surface-to-volume ratio and light/dark cycle, coupled with adequate mass transfer features [72].

The current consensus is that commercial (photoautotrophic) production of metabolites with interest for aquaculture by microalgae should resort to outdoor enclosed photobioreactors [56,62,65,73,74]. Tredici [65] reviewed the development of those type of reactors over the last decade; while many types of experimental PBRs have been considered, built and tested, very few have actually succeeded on a commercial level. Commercial application of photobioreactor technology remains indeed restricted to the production of two Chlorophyte microalgae: *Chlorella* and *Haematococcus* [62,75].

Scale up of photobioreactors from bench to commercial scale is not trivial – since it needs changes in illumination, gas transfer and temperature to be taken into account, all of which are severely affected by turbulence in the reactor, and consequently require a tight control. Therefore, scale up appears to be much more of an engineering problem than a biological one; and general recommendations as to possible maximum scales have accordingly been produced [75,76].

5. ALTERNATIVES TO FRESH MICROALGAE

Marine microalgae have been the traditional food component in finfish and shellfish aquaculture, e.g. for larval and juvenile animals [77]; they are indeed essential in hatchery and nursery of bivalves, shrimp and some finfish cultures. Microalgae are also used to produce zooplankton – typically rotifers, which are in turn fed to freshly hatched carnivorous fish [78]. As aquaculture industry expands [79] – and since microalgal biomass cultivation on-site may represent up to 30% of the operating costs [13], there is a demand for marine microalgae that cannot be met by the conventional methods used in hatcheries – thus forcing one to resort to substitutes with mediocre results that bring about several problems [44,80].

Despite the obvious advantages of alive microalgae in aquaculture, the current trend is to avoid using them because of their high cost and difficulty in producing, concentrating and storing them [8,81]. Alternatives that are potentially more cost-effective have been investigated – including nonliving food, viz. microalga pastes, dried microalgae, microencapsulates, cryopreservation, flocculation, bacteria or yeasts; they have been tested *in vitro* and in actual hatcheries, but met with variable degrees of success [82,83]. For instance, in Japan, where *Nannochloropsis oculata* is the most important cultured feed for the rotifer *Brachionus plicatilis*, concentrated suspensions and frozen biomass of this microalga are commercially available [84]; and partial replacement of alive microalgae by microencapsulated and yeast-based diets is indeed a routine practice in hatcheries for penaeid shrimp [24,85]. However, most these approaches have proven unsuitable as major dietary components, because of their lower nutritional value than mixtures of alive microalgae.

Several criteria should be addressed in attempts to find substitutes for alive microalgae as diet in aquaculture. From a nutrition standpoint, alive microalgae possess higher nutritional value and better digestibility than most substitutes; note that the nutritional quality depends critically on such biochemical constituents as PUFAs, vitamins, sterols and carbohydrates [86].

Useful bacteria can provide only a part of the metabolic requirements in aquaculture – by supplying a few organic molecules and vitamins. Under conditions close to those found in rearing facilities, the bacterial input should not represent more than 15% of the microalgal contribution for mollusk larvae and juveniles of many species [87,88]. Yeasts were as well investigated as an alternative food source – but poor results were observed [83,89]. Therefore, these two alternatives are not suitable to fully replace alive microalgae.

An alternative diet with an apparently better potential is microalgal pastes or concentrates [90-92]; these are prepared by centrifugation (up to 1:500 concentration) or flocculation (up to 1:100 concentration). Concentrates prepared from distinct microalgae vary in their suitability – with diatoms being the most promising; and they have a shelf life of between 2 and 8 weeks, when stored below 4°C. Commercially, microalgal concentrates can be prepared under two different scenarios: (a) by hatcheries on-site, which prepare concentrates as back-up or as a means to store overproduction; or (b) by remote production, centralized at a large facility – with a greater economy of scale, with the resulting concentrates dispatched to hatcheries upon request.

The advantage of such concentrates is that they can be used "off-the-shelf", thus contributing favorably to the cost-efficiency in hatcheries. On the other hand, the lower nutritional value of most dried microalgae compared to alive feed, and the limited availability of commercial dried products appear as main shortcomings. Globally speaking, concentrates have low levels (or even absence) of ω3-PUFA, and lead to a difficult digestion by bivalve larvae [93]. The genus *Tetraselmis* seems to be a good candidate for microalgal paste, but its nutritional quality deteriorates quite rapidly [94]; experiments have indicated that such substitutes should be used as supplement only when rations of live microalgae are insufficient. Furthermore, spray-dried microalgae and microalga paste may be useful to replace up to 50% of alive microalgae. Coutteau and Sorgeloos [13] reported that artificial or non-living diets are rarely applied in routine processing of bivalves, and are mostly considered as a backup food source only. Centrifuged concentrates of *P. lutheri*, in combination with *Chaetoceros calcitrans* or *S. costatum*, lead to 85-90% of the growth when a mixed diet of alive microalgae for oyster *Saccostrea glomerata* larvae is used [92].

Centrifugation has been successfully applied to prepare concentrates, but it has some limitations – i.e. the process involves exposing cells to high gravitational and shear forces that damage the cell structure. On the other hand, processing of large culture volumes is time-consuming and requires costly equipment, i.e. a specialized continuous centrifuge. Research on post-harvest preservation is required to extend shelf-life beyond 4 to 8 weeks, and also to prepare concentrates from flagellate species (e.g. *Isochrysis* sp. and *P. lutheri*).

Alternative processes have meanwhile been developed that are potentially less damaging to cells – including foam fractionation [95], flocculation [96,97] and filtration [98]. Sandbank [99] fed microalgae, grown in waste-water and flocculated with aluminium sulfate, to common carp (*Cyprinus carpia*); a diet containing 25% of microalgal meal led to a growth comparable to that by the control diet, with no harmful effects detected upon long term health of the fish. Millamena et al. [96] successfully fed *Penaeus monodon* larvae with dried, flocculated *C. calcitrans* and *Tetraselmis chuii* cells. However, a common disadvantage encountered was that the harvested cells are difficult to disaggregate back to single cells, which is a requirement to feed them to filter-feeding species such as bivalves [100].

A novel technique was developed for flocculation of marine microalgae that appears useful in aquaculture: it entails adjustment of pH of the culture using

NaOH, followed by addition of a non-ionic polymer, Magnafloc LT-25; the ensuing flocculate is then harvested and neutralized, thus leading to a final concentration of between 200- and 800-fold. This process was successfully applied to harvest cells of *C. calcitrans* and *C. muelleri, Thalassiosira pseudonana, Attheya septentrionalis, Nitzschia closterium, Skeletonema* sp., *Tetraselmis suecica* and *Rhodomonas salina*, with efficiencies above 80%; it proved rapid, simple and inexpensive, and relatively independent of processed volume (unlike concentration by centrifugation). The harvested material was readily disaggregated to single cell suspensions by dilution in seawater, coupled with mild agitation. Microscopic examination proved that the final cells are indistinguishable from the nonflocculated ones; and assay for chlorophyll of the concentrates prepared from cultures of up to 130 L showed marginal degradation by 2 weeks of storage [100].

Cryopreservation has been thoroughly adopted by culture collections to preserve strains, but may also find an application in aquaculture [80]. Viable cryostorage of biological specimens has followed various protocols of cooling/thawing rates and cryoprotectant addition, which have been developed and tuned more or less empirically [101]. Recall that temperatures used for cryostorage are well below freezing – down to even -196 ºC in liquid helium, when biological specimens are to be stored without limit [102]. While cryostorage is generally thought to be innocuous to the cell, the events occurring upon freezing or thawing can lead to severe damage, or even cell death. Moreover, cryoprotectants that enhance the cell viability at cryogenic temperatures are usually toxic at physiological temperatures [103] – an obstacle that is overcome by reducing the exposure time or the temperature of incubation prior to cryopreservation [104]. Knowledge of cryoprotectant tolerance levels for microalgae is still limited [105], as well as for early larval stages and for zooplankton that are cultivated and rely on the availability of microalgae for growth. In general, cryopreservation possesses a high potential for culture collections, and may also offer a solution for reliable supply of microalgae in aquaculture. For instance, marine microalgae used in aquaculture were successfully cryopreserved under 4, -20 and -80 ºC using common cryoprotectants (i.e. methanol, dimethylsulfoxide, propylene glycol and polyvinylpyrrolidone), with promising results at least for *Chlorella minutissima, Chlorella stigmatophora, Isochrysis galbana* and *Dunaliella tertiolecta* [80].

Several products based on thraustochytrids (i.e. microorganisms with a taxonomy related to certain microalgal classes), from the genus *Schizochytrium*, have been marketed through Aquafauna Biomarine and Sanders Brine Shrimp. These products have high concentrations of DHA [106], and have accordingly been applied as alternatives to commercial oil enrichment of zooplankton fed to larvae. As direct feeds, most such products have a lower nutritional value than mixtures of microalgae, yet some performed well as components of a mixed diet with alive microalgae [83,107].

In general, substitutes of alive microalgae should present an appropriate physical behavior – and this constitutes a significant challenge; in particular, they should not aggregate or easily break apart. Drying microalgae can cause, due to oxidation, a loss of PUFAs [108], which are essential components for larval growth [87]; the poor performance reported for dried microalgae was associated chiefly with the difficulty to keep cells in suspension without disintegrating them, so as to avoid said oxidation [13]. Moreover, when cell walls are broken, a high fraction of water-soluble components cannot be ingested by the organism, and may consequently interfere with the water quality of the aquaculture [109]. Therefore, pathogenic bacterial proliferation may occur, and cause costly production losses. Similar difficulties arise when using microalga paste, because the preparation procedures (i.e. centrifugation, flocculation or filtration) and/or preservation techniques (i.e. additives or freezing) must ensure that cell wall integrity is essentially preserved.

Products other than alive microalgae must obviously be free of bacterial contamination and devoid of toxicity. Consequently, the use of alive bacteria as a food source in hatcheries seems somehow inappropriate, since physical and chemical treatments are often used to limit bacterial contamination that would otherwise be responsible for drastic larval mortality [110]. Oyster larvae fed with alive microalga diets underwent improved growth via addition of some bacterial isolates [111,112], but this advantage may obviously not be possible in a treated microalgal product. However, in alive microalgae, the natural bacterial flora was proven to enhance the health of molluscs. Langdon and Bolton [88] showed that antibiotic suppression of the bacterial flora in artificial feed of juvenile oysters reduced their growth.

In conclusion, mitigated or unsuccessful results when using nonliving microalgae have turned alive microalgae into the first choice in aquaculture feeding. Only partial replacement thereof has been possible in studies encompassing preserved non-living algae [113], microencapsulated diets [88] or spray-dried algae [114]; but no whole replacement can be recommended, despite intensive research efforts in that direction [107]. Consequently, novel solutions to totally replace microalgae in aquaculture diets cannot at present be widely adopted [4,24,81].

6. USE OF MICROALGAE TO ENRICH ZOOPLANKTON

Microalgae have an important role in aquaculture, also as a means to enrich zooplankton for feeding fish and other larvae afterwards. In addition to providing proteins (that contain essential amino acids) and energy, they carry such other key nutrients as vitamins, essential PUFAs, pigments and sterols – which are transferred up through the food chain. For instance, PUFA-rich microalgae, such as *Pavlova* sp. and *Isochrysis* sp., have been successfully fed to zooplankton to enrich them in DHA [115]. However, when the level of enrichment attained is not sufficient, commercial oil-emulsions are often used. Recently, such products as dried preparations of *Schizochytrium* sp. (which contain 5-15% of their DW as DHA) have been utilized, which produce levels of DHA enrichment in zooplankton comparable to use of commercial oils [116] – and also produce DHA to EPA ratios of 1–2, which are considered favorable for fish larval nutrition [117].

Brown, Skabo & Wilkinson [118] described that rotifers fed with microalgae (e.g. *Isochrysis* sp. and *N. oculata*) become rapidly enriched with ascorbic acid (AsA), whereas rotifers fed on baker's yeast (which itself is deficient in AsA) contained only residual amounts of AsA.; after 16 h of starving, rotifers lost ca. 10% of their AsA, while retaining ca. 50% of the total AsA ingested. Similarly, the concentration of AsA in *Artemia* sp. may be increased by feeding with microalgae [119]. However, little information is available on the transfer of other vitamins from microalgae to fish larvae.

Rønnestad, Helland & Lie [120] demonstrated that microalgal pigments transferred to zooplankton may add to their nutritional value; recall that the dominant pigments in the copepod *Temora* sp. are lutein and astaxanthin, whereas in *Artemia* it is canthaxanthin. When these microalgae were fed to copepods and then to halibut larvae, adequate amounts of vitamin A were found, but not when halibut was fed on *Artemia*; this was attributed to the ability of larvae to convert lutein and/or astaxanthin, but not canthaxanthin to vitamin A. They accordingly recommended that *Artemia* should routinely be enriched with astaxanthin and lutein (the latter pigment is common in "green" microalgae, e.g. *Tetraselmis* sp.) to improve their nutritional value.

A common procedure during culture of both larval fish and prawns is to add microalgae (i.e. "green water") to intensive culture systems, together with the zooplankton prey [121]. The most popular microalga species used for this purpose are *N. oculata* and *T. suecica*. Addition of microalgae to larval tanks can also improve the production of larvae, but their exact mechanism of action remains unclear. Light attenuation (i.e. shading effects) may have a beneficial effect on larvae; however, maintenance of nutritional quality of the zooplankton, excretion of vitamins or other growth-promoting substances by the microalgae, and probiotic effects of the microalgae have also been hypothesized. Maintenance of NH3- and O2-balances has also been proposed, but this assumption failed to be supported by experimental evidence [121]. More research is still needed on the application of other microalgae – especially those species rich in DHA, to green water systems. Green water may also be applied to extensive outdoor production facilities, by fertilizing ponds in attempts to stimulate microalgal growth, and consequently zooplankton production.

7. POTENTIAL RESEARCH DIRECTIONS FOR MICROALGAE

The expensive production costs of microalgae continue to be a limitation for numerous hatcheries. Although there have been advancements in creating affordable artificial diets to substitute microalgae, the production of microalgae on-site is still essential for the functioning of many marine hatcheries. Advancements in alternate diets will certainly persist, but, the production costs of microalgae are predicted to decrease, indicating that microalgae will not be completely supplanted in the near future. There is a diverse range of microalgal species ready for use in aquaculture. Specific uses in industrial subsectors require new species with enhanced nutritional quality or growth characteristics that align with efforts to enhance hatchery efficiency and production. In addition to enhancing cost-effectiveness in on-site microalgal production, another option is to consolidate microalga production in specialized mass-culture facilities, utilizing heterotrophic techniques or unconventional photobioreactors. These technologies can be combined with post-harvest processing, such as spray-drying, or concentration methods like centrifugation or flocculation, to provide microalgal biomass that is readily available for delivery to hatcheries. Extracts from certain microalgae have demonstrated antifouling properties in microalga production tanks, proving to be more effective and less harmful than typical biocides. These natural compounds could be viewed as effective replacements for commercial biocides in antifouling paints. Paint coatings are still the most common method to prevent marine biofouling. Coatings designed for aquaculture equipment, which include a microalgae-based active ingredient to inhibit the main microorganisms that cause issues in cultivation, could be an effective solution to combat fouling. The specific molecules responsible for antifouling properties in microalgae have not been identified yet, thereby necessitating further research to isolate and characterize these active chemicals. Efforts to decrease water usage in aquaculture have been focused on developing recirculating systems. Regrettably, the focus of present research and development in aquaculture water re-use is mostly on bacteria-based systems, whereas the potential of employing microalga-based water re-use has been largely ignored. The bacteria in a water re-use system focus on removing excess nutrients, whereas a microalgae-based system produces microalgae that may be used to grow additional crops like bivalve seed or Artemia for additional income. The primary challenge in developing microalga-based water re-use systems is the difficulty in sustaining the specific microalgal species in an open system. An advancement in marine diatom production technology enables the development of water re-use systems where the effluent is transformed into a valuable resource. An integrated shrimp/microalga/oyster production system decreases water consumption and converts effluent waste into a profitable asset. This system also utilizes the antibacterial properties of marine diatoms to manage diseases and lower the shrimp's vulnerability to viral infections.

REFERENCES

1. G. Oron, L. R. Wildschut, D. Porath, "Wastewater recycling by duckweed for protein production and effluent renovation," *Water Science and Technology*, vol. 17, pp. 803- 818, 1985.
2. O. Hammouda, A. Gaber, N. Abdel-Raouf, "Microalgae and wastewater treatment," *Ecotoxicology and Environmental Safety*, vol. 31, pp. 205-210, 1995.
3. W. J. Oswald, "Algal Pond Systems–Habitats," in *Proceedings of a Seminar held at Murdoch University*, Western Australia, 1991, pp. 61-69.
4. A. Muller-Feuga, "The role of microalgae in aquaculture: situation and trends," *Journal of Applied Phycology*, vol. 12, pp. 527-534, 2000.
5. C. T. Enright, G. F. Newkirk, J. S. Craigie, J. D. Castell, "Evaluation of phytoplankton as diets for juvenile *Ostrea edulis* L.," *Journal of Experimental Marine Biology and Ecology*, vol. 96, pp. 1-13, 1986.
6. M. R. Brown, S. W. Jeffrey, J. K. Volkman, G. A. Dunstan, "Nutritional properties of microalgae for mariculture," *Aquaculture,* vol. 151, pp. 315–331, 1997.
7. K. Yamaguchi, "Recent advances in microalgal bioscience in Japan, with special reference to utilization of biomass and metabolites: a review," *Journal of Applied Phycology*, vol. 8, pp. 487–502, 1997.
8. W. Becker, "Microalgae for aquaculture. The nutritional value of microalgae for aquaculture." In Richmond, A. (ed.), *Handbook of Microalgal Culture*. Blackwell, Oxford, pp. 380–391, 2004.
9. C. T. Enright, G. F. Newkirk, J. S. Craigie, J. D. Castell, "Growth of juvenile *Ostrea edulis* L. fed *Chaetoceros gracilis* Schütt of varied chemical composition," *Journal of Experimental Marine Biology and Ecology,* vol. 96, pp. 15-26, 1986.
10. P. A. Thompson, M. Guo, P. J. Harrison, "The influence of irradiance on the biochemical composition of three phytoplankton species and their nutritional value for larvae of the Pacific oyster (*Crassostrea gigas*)," *Marine Biology*, vol. 117, pp. 259-268, 1993.
11. L. R. Marínez-Córdova, E. Peña-Messina, "Biotic communities and feeding habits of *Litopenaeus vannamei* (Boone 1931) and *Litopenaeus stylirostris* (Stimpson 1974) in monoculture and polyculture semi-intensive ponds," *Aquaculture Research*, vol. 36, pp. 1075-1084, 2005.
12. M. Kent, C. L. Browdy, J. W. Leffler, "Consumption and digestion of suspended microbes by juvenile Pacific white shrimp *Litopenaeus vannamei,*" *Aquaculture*, doi: 10.1016/j.aquaculture.2011.06.048, 2011.

13. P. Coutteau, P. Sorgeloos, "The use of algal substitutes and the requirement for live algae in the hatchery and nursery rearing of bivalve molluscs: an international survey," *Journal of Shellfish* Research, vol.11, pp. 467-476, 1992.

14. C. E. Epifanio, "Growth in bivalve molluscs: nutritional effects of two or more species of algae in diets fed to the American oyster *Crassostrea virginica* (Gmelin) and the hard clam *Mercenaria mercenaria* (L.)," *Aquaculture,* vol. 18, pp. 1-12, 1979.

15. D. A. Kreeger, C. J. Langdon, "Effect of dietary protein content on growth of juvenile mussels, *Mytilus trossulus* (Gould 1850)," *Biological Bulletin,* vol. 185, pp. 123-139, 1993.

16. J. K. Volkman, S. W. Jeffrey, P. D. Nichols, G. I. Rogers, C. D. Garland, "Fatty acid and lipid composition of 10 species of microalgae used in mariculture," *Journal of Experimental Marine Biology and Ecology*, vol. 128, pp. 219-240, 1989.

17. G. A. Dunstan, J. K. Volkman, S. M. Barrett, J.-M. Leroi, S. W. Jeffrey, "Essential polyunsaturated fatty acids from 14 species of diatom (Bacillariophyceae)," *Phytochemistry,* vol. 35, pp. 155-161, 1994.

18. R. M. Knuckey, M. R. Brown, S. M. Barrett, G. M. Hallegraeff, "Isolation of new nanoplanktonic diatom strains and their evaluation as diets for juvenile Pacific oysters (*Crassostrea gigas*)," *Aquaculture,* vol. 211, pp. 253-274, 2002.

19. D. Chuntapa, S. Powtongsook, P. Menasveta, "Water quality control using *Spirulina platensis* in shrimp culture tanks," *Aquaculture*, vol. 220, pp. 355-366, 2003.

20. G. D. Lio-Po, E. M. Leaño, M. M. D. Peñaranda, A. U. Villa- Franco, C. D. Sombito, N. G. Guanzon, "Antiluminous *Vibrio* factors associated with the 'green water' growout culture of the tiger shrimp *Penaeus monodon,*" *Aquaculture*, vol. 250, pp. 1-7, 2005.

21. L. Rodolfi, G. C. Zittelli, L. Barsanti, G. Rosati, M. R. Tredici, "Growth medium recycling in *Nannochloropsis* sp. mass cultivation," *Biomolecular Engineering*, vol. 20, pp. 243-248, 2003.

22. A. Muller-Feuga, "Microalgae for aquaculture. The current global situation and future trends." In Richmond, A. (ed.), *Handbook of Microalgal Culture*. Blackwell, Oxford, 2004, pp. 352-364.

23. A. Irianto, B. Austin, "Probiotics in aquaculture," *Journal of Fish Diseases*, vol. 25, pp. 633-642, 2002.

24. P. Spolaore, C. Joannis-Cassan, E. Duran, A. Isambert, "Review: commercial applications of microalgae," *Journal of Bioscience and Bioengineering,* vol. 101, pp. 87- 96, 2006.

25. D. S. Nichols, "Prokaryotes and the input of polyunsaturated fatty acids to the marine food web," *FEMS Microbiology Letters*, vol. 219, pp. 1-7, 2003.

26. C. Aragão, L. E. C. Conceição, M. T. Dinis, H.-J. Fyhn," Amino acid pools of rotifers and *Artemia* under different conditions: nutritional implications for fish larvae," *Aquaculture*, vol. 234, pp. 429-445, 2004.

27. K. E. Apt, P. W. Behrens, "Commercial developments in microalgal biotechnology," *Journal of Phycology*, vol. 35, pp. 215-226, 1999.

28. L. Waldenstedt, J. Inborr, I. Hansson, K. Elwinger, "Effects of astaxanthin-rich algal meal (*Haematococcus pluvalis*) on growth performance, faecal campylobacter and clostridial counts and tissue astaxanthin concentration of broiler chickens," *Animal Feed Science and Technology*, vol. 108, pp. 119-132, 2003.

29. M. R. Brown, "The amino acid and sugar composition of 16 species of microalgae used in mariculture," *Journal of Experimental Marine Biology and Ecology*, vol. 145, pp. 79- 99, 1991.

30. M. R. Brown, G. A. Dunstan, S. W. Jeffrey, J. K., Volkman, S. M. Barrett, J. M. Leroi, "The influence of irradiance on the biochemical composition of the prymnesiophyte *Isochrysis* sp. (clone T-ISO)," *Journal of Phycology*, vol. 29, pp. 601- 612, 1993.

31. M. R. Brown, C. D. Garland, S. W. Jeffrey, I. D. Jameson, J. M. Leroi, "The gross and amino acid compositions of batch and semi-continuous cultures of *Isochrysis* sp. (clone T.ISO), *Pavlova lutheri* and *Nannochloropsis oculata*," *Journal of Applied Phycology*, vol. 5, pp. 285-296, 1993.

32. J. Knauer, S. M. Barrett, J. K. Volkman, P. C. Southgate, "Assimilation of dietary phytosterols by Pacific oyster *Crassostrea gigas* spat," *Aquaculture Nutrition*, vol. 5, pp. 257-266, 1999.

33. J. Fabregas, C. Herrero, "Marine microalgae as a potential source of minerals in fish diets," *Aquaculture*, vol. 51, pp. 237-243, 1986.

34. S. M. Renaud, L. V. Thinh, D. L. Parry, "The gross composition and fatty acid composition of 18 species of tropical Australia microalgae for possible use in mariculture," *Aquaculture*, vol. 170, pp. 147-159, 1999.

35. P. J. Harrison, P. A. Thompson, G. S. Calderwood, "Effects of nutrient and light limitation on the biochemical composition of phytoplankton," *Journal of Applied Phycology*, vol. 2, pp. 45-56, 1990.

36. J. N. C. Whyte, N. Bourne, C. A. Hodgson, "Influence of algal diets on biochemical composition and energy reserves in *Patinopecten yessoensis* (Jay) larvae," *Aquaculture*, vol. 78, pp. 333-347, 1989.

37. J. A. Mendiola, C. F. Torres, P. J. Martín-Alvarez, S. Santoyo, A. Toré, B. O. Arredondo,

38. F. J. Señoráns, A. Cifuentes, E. Ibáñez, "Use of supercritical CO2 to obtain extracts with antimicrobial activity from *Chaetoceros muelleri* microalga. A correlation with their lipidic content," *European Food Research and Technology,* vol. 224, pp. 505-510, 2007.

39. K. Benkendorff, A. R. Davis, C. N. Rogers, J. B. Bremner, "Free fatty acids and sterols in the benthic spawn of aquatic molluscs, and their associated antimicrobial properties," *Journal of Experimental Marine Biology and Ecology,* vol. 316, pp. 29-44, 2005.

40. V. J. Smith, A. P. Desbois, E. A. Dyrynda, "Conventional and unconventional antimicrobials from fish, marine invertebrates and micro-algae," *Marine Drugs,* vol. 8, pp. 1213-1262, 2010.

41. F. Jüttner, "Liberation of 5,8,11,14,17-eicosapentaenoic acid and other polyunsaturated fatty acids from lipids as a grazer defense reaction in epiphithic diatom biofilms," *Journal of Phycology,* vol. 37, pp. 744-755, 2001.

42. G. d'Ippolito, S. Tucci, A. Cutignano, G. Romano, G. Cimino, A. Miralto, A. Fontana, "The role of complex lipids in the synthesis of bioactive aldehydes of the marine diatom *Skeletonema costatum,*" *Biochimica et Biophysica Acta,* vol. 1686, pp. 100-107, 2004.

43. M. Naviner, J.-P. Bergé, P. Durand, H. le Bris, "Antibacterial activity of the marine diatom *Skeletonema costatum* against aquacultural pathogens," *Aquaculture*, vol. 174, pp. 15-24, 1999.

44. K. Das, J. Pradhan, P. Pattnaik, B. R. Samantaray, S. K. Samal, "Production of antibacterials from the freshwater alga *Euglena viridis* (Ehren)," *World Journal of Microbiology and Biotechnology,* vol. 21, pp. 45-50, 2005.

45. M. A. Borowitzka, "Commercial production of microalgae: ponds, tanks, tubes and fermenters," *Journal of Biotechnology*, vol. 70, pp. 313–321, 1999.

46. H. M. Amaro, A. C. Guedes, F. X. Malcata, "Advances and perspectives in using microalgae to produce biodiesel," *Applied Energy*, vol. 88, pp. 3402-3410, 2011.

47. S. A. Scott, M. P. Davey, J. S. Dennis, I. Horst, C. J. Howe, D. J. Lea-Smith, "Biodiesel from algae: challenges and prospects," *Current Opinion in Biotechnology,* vol. 21, pp. 277-286, 2010.

48. P. Soong, "Production and development of *Chlorella* and *Spirulina* in Taiwan." In: Shelef, G., Soeder, C. J. (Eds.), *Algae Biomass.* Elsevier, Amsterdam, 1980, pp. 97-113.

49. L. J. Borowitzka, M. A. Borowitzka, "Industrial production: methods and economics." In: Cresswell, R. C., Rees, T. A. V., Shah, N. (Eds.), *Algal and Cyanobacterial Biotechnology*. Longman Scientific, London, 1989, pp. 294–316.

50. A. Belay, "Mass culture of *Spirulina* outdoors—the Earthrise Farms experience". In: Vonshak, A. (Ed.), *Spirulina platensis (Arthrospira): Physiology, Cell-biology and Biotechnology.* Taylor & Francis, London, 1997, pp. 131-158.

51. I. Setlík, S. Veladimir, I. Malek, "Dual purpose open circulation units for large scale culture of algae in temperate zones. I. Basic design considerations and scheme of pilot plant," *Algology Studies (Trebon)*, vol. 1, pp. 11, 1970.

52. K. Kawaguchi, "Microalgae production systems in Asia." In: Shelef, G., Soeder, C. J. (Eds.), *Algae Biomass Production and Use*. Elsevier, Amsterdam, 1980, pp. 25-33.

53. J. Kyle, R. M. Gladue, "Eicosapentaenoic acids and methods for their production," *World Patent,* vol. 9, pp. 114,427, 1991.

54. J. Kyle, S. E. Reeb, V. J. Sicotte, "Dinoflagellate biomass, methods for its production, and compositions containing the same," *USA Patent* 5,711,983, 1998.

55. A. P. Carvalho, L. A. Meireles, F. X. Malcata, "Microalgal reactors: a review of enclosed system designs and performances," *Biotechnology Progress,* vol. 22, pp. 1490-1506, 2006.

56. C. Jiménez, B. R. Cossío, D. Labella, N. F. Xavier, "The feasibility of industrial production of *Spirulina (Arthrospira)* in southern Spain," *Aquaculture,* vol. 217, pp. 179-190, 2003.

57. D. Chaumont, "Biotechnology of algal biomass production: a review of systems for outdoor mass culture," *Journal of Applied Phycology*, vol. 5, pp. 593-604, 1993.

58. L. Terry, L. P. Raymond, "System design for the autotrophic production of microalgae," *Enzyme and Microbial Technology,* vol. 7, pp. 474-487, 1985.

59. Y. Chisti, "Biodiesel from microalgae beats bioethanol," *Trends in Biotechnology,* vol. 26, pp. 126-131, 2008.

60. L. Rodolfi, G. C. Zittelli, N. Bassi, G. Padovani, N. Biondi, G. Bonini, "Microalgae for oil: strain selection, induction of lipid synthesis and outdoor mass cultivation in a low- cost photobioreactor," *Biotechnology and Bioengineering,* vol. 102, pp. 100-112, 2008.

61. U. Ugwu, H. Aoyagi, H. Uchiyama, "Photobioreactors for mass cultivation of algae," *Bioresource Technology*, vol. 99, pp. 4021-4028, 2008.

62. Y. Chisti, "Biodiesel from microalgae," *Biotechnology Advances*, vol. 25, pp. 294-306, 2007. 62 O. Pulz, "Photobioreactors: production systems for phototrophic microorganisms," *Applied Microbiology and Biotechnology,* vol. 57, pp. 287-293, 2001.

63. B. Metting, "Biodiversity and application of microalgae," *Journal of Industrial Microbiology,* vol. 17, pp. 477-489, 1996.

64. E. Molina-Grima, "Microalgae, mass culture methods." In M. C. Flickinger, & S. W. Drew (Eds.), *Encyclopedia of Bioprocess Technology: Fermentation, Biocatalysis and Bioseparation.* New York: Wiley, 1999, vol. 3, pp. 1753-1769.

65. R. Tredici, "Bioreactors, photo." In Flickinger, M. C. & Drew, S. W. (Eds.), *Encyclopedia of Bioprocess Technology: Fermentation, Biocatalysis and Bioseparation,* vol. 1. New York, Wiley, 1999, pp. 395-419.

66. G. Acién-Fernández, F. García-Camacho, J. A. Sánchez-Pérez, J. M. Fernández-Sevilla, E. Molina-Grima, "Modelling of biomass productivity in tubular photobioreactors for microalgal cultures: effects of dilution rate, tube diameter and solar irradiance," *Biotechnology and Bioengineering,* vol. 58, pp. 605-616, 1998.

67. S. Aiba, "Growth kinetics of photosynthetic microorganisms," *Advances in Biochemical Engineering,* vol. 23, pp. 85-156, 1982.

68. L. E. Erickson, H. Y. Lee, "Process analysis and design of algal growth system." In: Barclay, W. R. and McIntosh, R. P. (Eds), *Algal Biomass Technologies: an Interdisciplinary Perspective.* J. Cramer, Berlin-Stuttgart, 1986, pp. 197–206.

69. S. L. Pirt, Y. K. Lee, M. R. Walach, M. W. Pirt, H. H. Balyuzi, M. J. Bazin, "A tubular bioreactor for photosynthetic production of biomass from carbon dioxide: design and performance," *Journal of Chemical Technology and Biotechnology,* vol. 33B, pp. 35- 58, 1983.

70. L. Rorrer, R. K. Mullikin, "Modeling and simulation of a tubular recycle photobioreactor for macroalgal cell suspension cultures," *Chemical Engineering Science,* vol. 54, pp. 3153-3162, 1999.

71. C. Posten, "Design principles of photo-bioreactors for cultivation of microalgae," *Engineering in Life Sciences,* vol. 9, pp. 165-177, 2009.

72. A. A. Tsygankov, "Laboratory scale photobioreactors," *Applied Biochemistry and Microbiology,* vol. 37, pp. 333-341, 2001.

73. A. Borowitzka, "Microalgae as sources of pharmaceuticals and other biologically active compounds," *Journal* of *Applied Phycology,* vol. 7, pp. 3-15, 1995.

74. A. Vonshak, "Tubular photobioreactors for algal mass production; prospects and achievements," *Israeli Journal of Aquaculture (Bamidgeh),* vol. 44, pp. 151, 1992.

75. M. Olaizola, "Commercial development of microalgal biotechnology: from the test tube to the marketplace," *Biomolecular Engineering,* vol. 20, pp. 459-466, 2003.

76. E. Molina-Grima, F. G. Acién-Fernández, F. García-Camacho, F. Camacho-Rubio, Y. Chisti, "Scale-up of tubular photobioreactors," *Journal of Applied Phycology*, vol. 12, pp. 355-368, 2000.

77. N. de Pauw, G. Persoone, "Micro-algae for aquaculture." In: Borowitzka, M. A., Borowitzka, L. J. (Eds.), *Micro-Algal Biotechnology*. Cambridge University Press, Cambridge, 1988, pp. 197-221.

78. R. Benemann, W. J. Oswald, "Systems and economic analysis of microalgae ponds for conversion of CO2 to biomass," *Final Report, Subcontract XK 4-04136- 06*, Pittsburgh Energy Technology Center Grant No. DE-FG22-93PC93204, p. 260, 1996.

79. FAO, 2000. *Aquaculture Production 1992–1998*. FAO, Rome.

80. I. Tzovenis, G. Triantaphyllidis, X. Naihong, E. Chatzinikolaou, K. Papadopoulou, G. Xouri, T. Tafas, "Cryopreservation of marine microalgae and potential toxicity of cryoprotectants to the primary steps of the aquacultural food chain," *Aquaculture*, vol. 230, pp. 457-473, 2004.

81. M. A. Borowitzka, "Microalgae for aquaculture: opportunities and constraints," *Journal of Applied Phycology*, vol. 9, pp. 393-401, 1997.

82. J. Knauer, P. C. Southgate, "A review of the nutritional requirements of bivalves and the development of alternative and artificial diets for bivalve aquaculture," *Reviews in Fisheries Science*, vol. 7, pp. 241–280, 1999.

83. R. Robert, P. Trintignac, "Substitutes for live microalgae in mariculture: a review," *Aquatic Living Resources*, vol. 10, pp. 315–327, 1997.

84. G. Chini-Zittelli, F. Lavista, A. Bastianini, L. Rodolfi, M. Vincenzini, M. R. Tredici, "Production of eicosapentaenoic acid by *Nannochloropsis* sp. cultures in outdoor tubular photobioreactors," *Journal of Biotechnology*, vol. 70, pp. 299-312, 1999.

85. B. Spolaore-Robinson, T. M. Samocha, J. M. Fox, R. L. Gandy, D. A. McKee, "The use of inert artificial commercial food sources as replacements of traditional live food items in the culture of larval shrimp, *Farfantepenaeus aztecus*," *Aquaculture*, vol. 245, pp. 135-147, 2005.

86. C. Seguineau, A. Laschi-Loquerie, J. Moral, J.-F. Samain, "Vitamin requirements in great scallop larvae," *Aquaculture International*, vol. 4, pp. 315-324, 1996.

87. R. Brown, S. M. Barret, J. K. Volkman, S. P. Nearhos, J. Nell, G. L. Allan, "Biochemical composition of new yeasts and bacteria evaluated as food for bivalve aquaculture," *Aquaculture*, vol. 143, pp. 341-360, 1996.

88. J. Langdon, E. T. Bolton, "A microparticulate diet for suspension-feeding bivalve mollusc, *Crassostrea virginica* (Gmelin)," *Journal of Experimental Marine Biology and Ecology*, vol. 89, pp. 239-258, 1984.

89. P. Coutteau, M. Dravers, P. Leger, P. Sorgeloos, "Manipulated yeast diets and dried algae as a partial substitute for live algae in the juvenile rearing of the Manila clam *Tapes philippinarum* and the Pacific oyster *Crassostrea gigas*," *Special Publication of the European Aquatic Society Ghent*, Belgium, vol. 18, pp. 523-531, 1993.

90. J. A. Nell, W. A. O'Connor, "The evaluation of fresh algae and stored algal concentrates as a food source for Sydney rock oyster, *Saccostrea commercialis* (Iredale and Roughley) larvae," *Aquaculture*, vol. 99, pp. 277-284, 1991.

91. A. McCausland, M. R. Brown, S. M. Barrett, J. A. Diemar, M. P. Heasman, "Evaluation of live and pasted microalgae as supplementary food for juvenile Pacific oysters (*Crassostrea gigas*)," *Aquaculture and Research*, vol. 174, pp. 323-342, 1999.

92. M. Heasman, J. Diemar, W. O'Connor, T. Sushames, L. Foulkes, "Development of extended shelf-life micro-algae concentrate diets harvested by centrifugation for bivalve molluscs—a summary," *Aquaculture and Research*, vol. 31, pp. 637–659, 2000.

93. A. Muller-Feuga, R. Robert, C. Cahu, J. Robin, P. Divemach, "Use of microalgae in aquaculture." In: Støttrup, J. A., McEvoy, L. A., (Eds), *Live Feeds in Marine Aquaculture*. Blackwell, Oxford, 2003, pp. 253-299.

94. E. Montaini, G. C. Zittelli, M. R. Tredici, E. M. Grima, J. M. F. Sevilla, J. A. S. Perez, "Long-term preservation of *Tetraselmis suecica*: influence of storage on viability and fatty acid profile," *Aquaculture*, vol. 134, pp. 81-90, 1995.

95. A. Csordas, J.-K. Wang, "An integrated photobioreactor and foam fractionation unit for the growth and harvest of *Chaetoceros* spp. in open systems," *Aquacultural Engineering*, vol. 30, pp. 15-30, 2004.

96. M. Millamena, E. J. Aujero, I. G. Borlongan, "Techniques on algae harvesting and preservation for use in culture as larval food," *Aquaculture Engineering*, vol. 9, pp. 295-304, 1990.

97. E. Poelman, N. de Pauw, B. Jeurissen, "Potential of electrolytic flocculation for recovery of micro-algae," *Resources, Conservation* and *Recycling*, vol. 19, pp. 1-10, 1997.

98. N. Rossingol, L. Vandanjon, P. Jaouen, F. Quéméneur, "Membrane technology for the continuous separation microalgae/culture medium: compared performances of cross-flow microfiltration and ultrafiltration," *Aquaculture Engineering*, vol. 20, pp. 191-208, 1999.

99. E. Sandbank, "The utilization of microalgae as feed for fish," *Ergeb Limnology*, vol. 11, pp. 108-120, 1978.

100. R. M. Knuckey, M. R. Brown, R. Robert, D. M. F. Frampton, "Production of microalgal concentrates by flocculation and their assessment as aquaculture feeds," *Aquacultural Engineering*, vol. 35, pp. 300-313, 2006.

101. J. O. M. Karlson, M. Toner, "Long-term storage of tissues by cryopreservation: critical issues," *Biomaterials*, vol. 17, pp. 243-256, 1996.

102. P. Mazur, "Freezing of living cells: mechanisms and implications," *American Journal of Physiology*, vol. 247, pp. 125-142, 1984.

103. M. Fahy, "The relevance of cryoprotectant 'toxicity' to cryobiology," *Cryobiology*, vol. 23, pp. 1-13, 1986.

104. M. Fahy, T. H. Lilley, H. Lindsell, M. J. Douglas, H. T. Meryman, "Cryoprotectant toxicity and cryoprotectant toxicity reduction: in search of molecular mechanisms," *Cryobiology*, vol. 27, pp. 247- 268, 1990.

105. R. Taylor, R. L. Fletcher, "Cryopreservation of eukaryotic algae—a review of methodologies," *Journal of Applied Phycology*, vol. 10, pp. 481–501, 1999.

106. W. Barclay, S. Zeller, "Nutritional enhancement of n-3 and n-6 fatty acids in rotifers and *Artemia nauplii* by feeding spray-dried *Schizochytrium* sp.," *Journal of the World Aquaculture Society*, vol. 27, pp. 314-322, 1996.

107. C. Langdon, E. Önal, "Replacement of living microalgae with spray-dried diets for the marine mussel *Mytilus galloprovincialis*," *Aquaculture*, vol. 180, pp. 283-294, 1999.

108. A. Dunstan, J. K. Volkman, S. W. Jeffrey, S. M. Barret, "Biochemical composition of microalgae from the green algal classes Chlorophyceae and Prasinophyceae," *Journal of Experimental Marine Biology and Ecology,* vol. 161, pp. 115-134, 1992.

109. J. Dhont, G. van Stappen, *Live Feeds in Marine Aquaculture*, Blackwell Science, pp. 65- 121. 2003.

110. A. Elston, "Mollusc diseases, guide for the shellfish farmer," *Washington Sea Grant Program*, Seattle, WA, p. 73, 1990.

111. P. Douillet, C. J. Langdon, "Effects of marine bacteria on the culture of axenic oyster *Crassostrea gigas* (Thunberg) larvae," *Biological Bulletin,* vol. 184, pp. 36-51, 1993.

112. P. Douillet, C. J. Langdon, "Use of a probiotic for the culture of larvae of the Pacific oyster (*Crassostrea gigas* Thunberg)," *Aquaculture*, vol. 119, pp. 25-40, 1994.

113. J. Donaldson, *Proceedings of US-Asia Workshop*, Honolulu, HA, January 28-31, 1991, The Oceanic Institute, HA, pp. 229-236.

114. B. Zhou, W. Liu, W. Qu, C. K. Tseng, “Application of *Spirulina* mixed feed in the breeding of bay scallop,”*Bioresource Technology*, vol. 38, pp. 229-232, 1991.

115. P. D. Nichols, D. G. Holdsworth, J. K. Volkman, M. Daintith, S. Allanson, “High incorporation of essential fatty acids by the rotifer *Brachionus plicatilis* fed on the prymnesiophyte alga *Pavlova lutheri*,” *Australian Journal of Marine and Freshwater Research*, vol. 40, pp. 645-655, 1989.

116. B. Gara, R. J. Shields, L. McEvoy, “Feeding strategies to achieve correct metamorphosis of Atlantic halibut, *Hippoglossus hippoglossus* L., using enriched *Artemia*,” *Aquaculture Research*, vol. 29, pp. 935-948, 1998.

117. C. Rodríguez, J. A. Pérez, P. Badía, M. S. Izquierdo, H. Fernández-Palacios, A. Lorenzo Hernández, “The n-3 highly unsaturated fatty acid requirements of gilthead seabream (*Sparus aurata* L.) larvae when using an appropriate DHA/EPA ratio in the diet,” *Aquaculture*, vol. 169, pp. 9-23, 1998.

118. M. R. Brown, S. Skabo, B. Wilkinson, “The enrichment and retention of ascorbic acid in rotifers fed with microalgal diets,” *Aquaculture Nutrition*, vol. 4, pp. 151-156, 1998.

119. G. Merchie, P. Lavens, P. Dhert, M. Dehasque, H. Nelis, A. de Leenheer, P. Sorgeloos, “Variation of ascorbic acid content in different live food organisms,” *Aquaculture*, vol. 134, pp. 325-337, 1995.

120. I. Rønnestad, S. Helland, Ø. Lie, “Feeding *Artemia* to larvae of Atlantic halibut (*Hippoglossus hippoglossus* L.) results in lower larval vitamin A content compared with feeding copepods,” *Aquaculture*, vol. 165, pp. 159-164, 1998.

121. C. S. Tamaru, R. Murashige, C.-S. Lee, “The paradox of using background phytoplankton during the larval culture of striped mullet, *Mugil cephalus* L.,” *Aquaculture*, vol. 119, pp. 167-174, 1994.

05 VALIDATION OF INTERNAL REFERENCE GENES FOR QPCR QUANTIFICATION OF MUSCLE TRANSCRIPTS IN ATLANTIC COD EXPOSED TO VARIOUS PHOTOPERIOD CONDITIONS

1. INTRODUCTION: AN OVERVIEW

Atlantic cod (Gadus morhua, L.) is a globally significant species for commercial purposes, and over fishing has led to a decrease in wild populations to unsustainable levels. The aquaculture production of this species has significantly expanded over the last decade to over 20,000 tonnes in 2008, as reported by FAO. However, cod aquaculture encounters various obstacles in production including larval quality, nutrition, illnesses, and premature sexual development. Sexual maturation occurring at a young age of roughly two years under farmed environments significantly hinders the industry's profitability. Sexual development and spawning cause a decrease in appetite, worse feed conversion, and higher mortality rate, resulting in a longer period needed to attain the appropriate harvest size. Continuous light exposure, known as photoperiod manipulation, has been utilized to partially postpone sexual maturation in various aquaculture species such as Atlantic salmon (Salmo salar, L.), European sea bass (Dicentrachus labrax, L.), and Atlantic cod (Davie et al., 2003; Hansen et al., 2001; Norberg et al., 2004) as documented in the literature (Endal et al., 2000; Begtashi et al., 2004). Continuous exposure to light starting from the summer solstice before maturity is believed to interfere with the natural light signal that initiates gonadal development and spawning. Taranger et al. (2006) shown that the gonadal maturation of cod in sea cages can be postponed by three to five months by using continuous broad-spectrum light. Manipulating the amount of light exposure can affect both sexual maturation and body growth, especially in young stages of development (Davie et al., 2007; Taranger et al., 2006). Short-term exposure to continuous light led to a 5 to 9% increase in body weight compared to cod raised under regular light-dark circumstances. This difference was still noticeable at harvesting size over three years later (Imsland et al., 2007). The molecular mechanism behind the growth plasticity produced by light cues in the aquaculture industry is currently unknown. To improve control over the early sexual development of farmed cod, it is essential to pinpoint the transcriptional networks associated with this process and comprehend how they are impacted by photoperiod. Muscle is the primary component that facilitates the growth of fish. Teleost myogenesis is an intricate event controlled by various molecules that oversee different stages of the process. Muscle development can occur through hypertrophy, which involves the enlargement of muscle fibers by absorbing myoblast nuclei, or hyperplasia, which involves the production of new fibers on the surface of existing muscle fibers (Johnston, 1999). Muscle formation progresses through the sequential expression of key genes from the myogenic regulatory factors (MRFs) family, such as myoblast differentiation 1 (myoD), myogenic factor 5 (myf-5), myogenin (myoG), and myogenic factor 6/myogenic regulatory factor 4 (myf-6/MRF4) (Watabe, 1999). Myosin heavy chain (MyHC) is a molecule that plays a crucial role in muscle development and is used as a marker in several investigations (Johnston, 1999). Myhc genes encode a group of ATP-dependent motor proteins crucial for muscle contraction (Ikeda et al., 2007). Myosin heavy chain (Myhc) activity is a useful indicator for tracking fish growth due to their consistent hyperplasic growth and the significant role Myhc plays in muscle protein synthesis (Dhillon et al., 2009). Real-time PCR (qPCR) is the preferred approach for quantifying transcript levels due to its reliability and sensitivity in detecting low amounts of mRNA in tiny tissue samples (Bustin, 2002). The technique has been utilized in tiger pufferfish (Takifugu rubripes, Temminck & Schlegel) to validate suppression subtractive hybridization results, study the impact of temperature on the expression of growth-related genes myoG and forkhead box protein K1 (foxk1) during embryonic development, and investigate the differential regulation of splice variants of the master transcription factor myoD1. Although relative qPCR quantification has significant potential, it has certain problems that need to be properly evaluated (Bustin and Nolan, 2004). Choosing appropriate reference genes with consistent expression across all samples is crucial for normalizing qPCR results. Using reference genes that have not been verified can result in inaccurate findings that lack biological significance (Fernandes et al., 2008). There is a widespread agreement that a reference gene that is adaptable and consistent across different experimental circumstances is not available. Prior to assessing the expression of a target gene, it is essential to choose the most suitable reference genes for each species and tissue in a specific experimental arrangement. It is advisable to test many genes in each experiment and utilize statistical methods to determine the optimal combination of the two or three most reliable genes for normalizing qPCR data (Andersen et al., 2004; Vandesompele et al., 2002). The current research paper aimed to find appropriate reference genes for relative quantification using qPCR of growth- and maturation-related genes influenced by photoperiod alteration. Five frequently utilized reference genes were assessed: β-actin (actb), acidic ribosomal protein (arp), eukaryotic elongation factor 1α (eef1a), glyceraldehyde-3-phosphate dehydrogenase (gapdh), and ubiquitin (ubi). The transcript levels in the fast muscle samples were measured using qPCR using SYBR chemistry. GeNorm and NormFinder were utilized to assess the expression stability of the candidate genes mentioned above. Furthermore, the significance of utilizing verified reference genes in qPCR analysis was demonstrated by examining myhc expression. We demonstrated the influence of normalization strategies on myhc expression levels by comparing individual candidate genes to the normalization factor from the two most validated reference genes. We also emphasized the importance of validation in choosing the most stable reference gene for each experimental condition.

2. MATERIALS AND METHODS

2.1 Photoperiod experiment and sampling

2.1.1 Fish husbandry

Atlantic cod juveniles, *Gadus morhua* L. with an initial size of 2.7 ± 0.8 g (mean ± standard deviation [SD], n=123) were divided into six tanks (250 m3) with an open flow system at a density of approximately 130 individuals per tank. Sea water was continuously supplied to each tank at 7.4 ± 0.4 °C. A commercial diet (Amber Neptun, Skretting AS, Stavanger, Norway) was provided daily by automatic belt feeders, at 5% (w/w) body weight of the fish and adjusted on a weekly basis.

2.1.2 Photoperiod experiment

Each group of fish in three tanks was either kept under continuous light (LL) or reared under a normal light regime (NL) that corresponded to natural environmental photoperiod conditions in Bodø (67°N), Norway during 6 months from January to July 2010. Day light time was recreated indoors using white light fluorescent tubes (Aura Light International AB, Karlskrona, Sweden) and controlled by a scheduled timer according to local sunrise and sunset times in Bodø. Light intensity on the central surface of each tank was 120 Lux.

2.1.3 Sample collection

The fish were killed by immersion in seawater containing 0.2 g·L-1 tricaine methanesulfonate (MS222; Sigma, Oslo, Norway). Fast muscle samples were carefully dissected from the trunk area below the second dorsal fin from six fish at the start of the experiment and 0.5, 1, 7, 30, 60, 120 and 180 days thereafter. Tissues were snap-frozen in liquid nitrogen and stored at – 80 °C until RNA extraction.

2.1.4 Ethics statement

All procedures of fish rearing and tissue sampling were in accordance with the guidelines set by the National Animal Research Authority (Forsøksdyrutvalget, Norway).

2.2 Real-time PCR (qPCR)

2.2.1 Primer design

To validate expression stability of reference genes, five candidate genes (*β-actin* (*actb*), *acidic ribosomal protein* (*arp*), *elongation factor 1 alpha* (*eef1a*), *glyceraldehyde-3-phosphate dehydrogenase* (*gapdh*), *ubiquitin* (*ubi*) and *myosin heavy chain* (*myhc*)) were selected and analysed for qPCR validation.

2.2.2 RNA extraction and cDNA synthesis

Total RNA was extracted from the fast muscle samples above and used to synthesized cDNA as detailed elsewhere (Campos et al., 2010). Two micrograms of total RNA were used for cDNA synthesis by reverse transcription, following treatment with gDNA wipe out buffer (Qiagen, Nydalen, Sweden) to remove genomic DNA contamination.

2.2.3 qPCR amplification

Table 1. Primers used in this study. Primer sequences, Genbank accession numbers, amplicon sizes and PCR efficiencies are indicated.

Gene	GenBank	Sequence	Size (bp)	E (%)	Reference
actb	AJ555463	Fw: TGACCCTGAAGTACCCCATC Rv: TCTTCTCCCTGTTGGCTTTG	162	77	(Lilleeng et al.
arp	EX741373	Fw: TGATCCTCCACGACGATGAG Rv: CAGGGCCTTGGCGAAGA	113	86	(Olsvik et al.
eef1a	CO541820	Fw: CACTGAGGTGAAGTCCGTTG Rv: GGGGTCGTTCTTGCTGTCT	142	84	(Lilleeng et al.
gapdh	AY635584	Fw: GGTCGCAACCGCAAGGT Rv:TGACCGTTGAGCATTTCCTTCT	83	88	(Hall et al.
ubi	EX735613	Fw: GGCCGCAAAGATGCAGAT Rv: CTGGGCTCGACCTCAAGAGT	69	87	(Olsvik et al.
myhc	AY093703	Fw:CAGAAGCTATAAAAGGTGTCCG Rv:GCAGCCATTCTTCTTATCCTCCTC	86	81	(Koedijk et al.

qPCR reactions were conducted with the primer sets indicated on table 1. Quantification of transcripts were analysed by qPCR with SYBR Green chemistry (SYBR Green I Master, Roche) on a LightCycler® 480 (Roche) as previously described (Campos et al., 2010). Fifty- fold diluted muscle cDNA were run in duplicate, and minus reverse transcriptase and no template controls were included in the reactions. Thermocycling parameters were as follows: 95°C for 15 min, followed by 45 cycles of 15 s at 94°C, 20 s at 60°C and 20 s at 72°C. Five-point standard curves of a 2-fold dilution series (1:1, 1:2, 1:4, 1:8 and 1:16) were prepared from pooled RNA that was reverse transcribed as above. These dilution curves were used to calculate amplification efficiencies of the PCR reactions (Fernandes et al., 2006). Cycle threshold (Ct) values were determined by the LightCycler® 480 software with a fluorescence level arbitrarily set to 1.

2.3 Data analyses

2.3.1 Expression stability analyses

Raw qPCR data were converted to expression level using the above dilution curves. These were then analysed for expression stability using the statistical applications GeNorm (Vandesompele et al., 2002) and NormFinder (Andersen et al., 2004).

2.3.2 Statistical analyses

Differences in expression levels of *gapdh, actb, eef1a, arp, ubi* and *myhc* during the photoperiod manipulation experiment were examined by one-way ANOVA with Holm- Sidak post-hoc tests. ANOVA assumptions were checked prior to carrying out the analyses and when the data did not follow the Gaussian distribution or did not meet the equal variance requirements, a Kruskal-Wallis one-way ANOVA on ranks with Dunn's test for post-hoc comparisons was used instead. Statistical analyses were performed with the SigmaStat statistical package (Systat software, London, UK). In all cases, significance was set at $P < 0.05$.

3. RESULTS AND DISCUSSION

3.1 Validation of reference genes

Specificity of qPCR reactions was confirmed by melting curve analysis, which revealed a single dissociation peak for each gene. Global variation on expression profiles of the candidate reference genes can be observed in Fig.1. Mean Ct values of the candidate reference genes were 19.99, 18.36, 18.03, 17.76 for *actb, ubi, eef1a* and *arp*, respectively, whereas the median Ct value for *gapdh* (25.96) was above the range of values for the other four genes. Expression of *ubi* showed the least variation across samples, in contrast to *gapdh, actb* and *eef1a.*

Detailed expression of the individual reference genes during the 6-month photoperiod manipulation experiment is presented in Fig. 2, showing a differential expression trend between sampling points and photoperiod regimes. There were no significant differences ($P < 0.05$) on the overall expression amongst any of the candidate reference genes (*arp*: $P = 0.997$; *eef1*: $P = 0.735$; *ubi*: $P = 0.124$; *gapdh*: $P = 0.386$; *actb*: $P = 0.554$). It is also important to note that the transcript levels of *actb*, *ubi*, *eef1a* and *arp* were almost the same based on the close range of their Ct values and all showed a perceptible increase on their transcript levels at the last 3 sampling points. In contrast, transcript levels of *gapdh* were relatively lower compared with the other reference genes as characterized by having higher Ct values. There was an apparent difference on the expression of *actb* and *gapdh* between treatments.

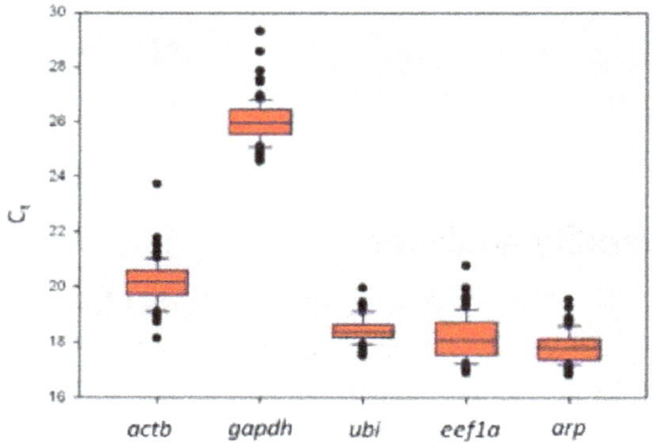

Fig. 1. Overall expression patterns of candidate reference genes in the muscle of Atlantic cod reared under different photoperiod regimes. Raw cycle threshold (Ct) qPCR data of individual reference genes in all samples (n=96) are represented as box-and-whisker plots. Median values are indicated by a solid line inside the boxes.

Expression stability indices of the candidate reference genes as assessed by geNorm varied with time and photoperiod regime (Table 2). *Arp* and *ubi* were identified as the most stable reference genes. Specifically, *arp* was the most stable in fish group under normal photoperiod (0 h, 6 h, 1 d, 1 w, 1 m and 6 m), while *ubi* was the most stable in the group reared under continuous light (6 h, 1 d, 1 w, 1 m and 2 m). *Actb* and *gapdh* were two of the least stable reference genes, with *gapdh* as the least stable in 3 sampling points (0 h, 6 h and 1 m) regardless of photoperiod regime. Collating all the stability indices of each reference gene, the order of stability from the most to the least was as follows: *arp* > *ubi* > *eef1a* > *actb* > *gapdh*. Pairwise comparisons revealed that *arp* and *ubi* were the best pair for two-gene normalisation with a joint stability value of 0.138 (Fig. 3A).

The validation software, NormFinder identified *arp* (0.084) and *gapdh* (0.259) as the most and least stable reference genes, respectively. The overall ranking reference genes from the most to the least stable was as follows: *arp* > *ubi* > *actb* > *eef1a* > *gapdh* (Fig. 3B). It was also determined that the best pair of candidate reference genes was *arp* and *ubi* with a joint stability value of 0.084. This pairwise result is similar to the result in geNorm that the best pair for two-gene normalisation was *arp* and *ubi*. It was also found that regardless of photoperiod regimes, *gapdh* was the least stable gene at 0 h, 6 h, 1 m and 4 m, corresponding broadly to the results obtained in geNorm.

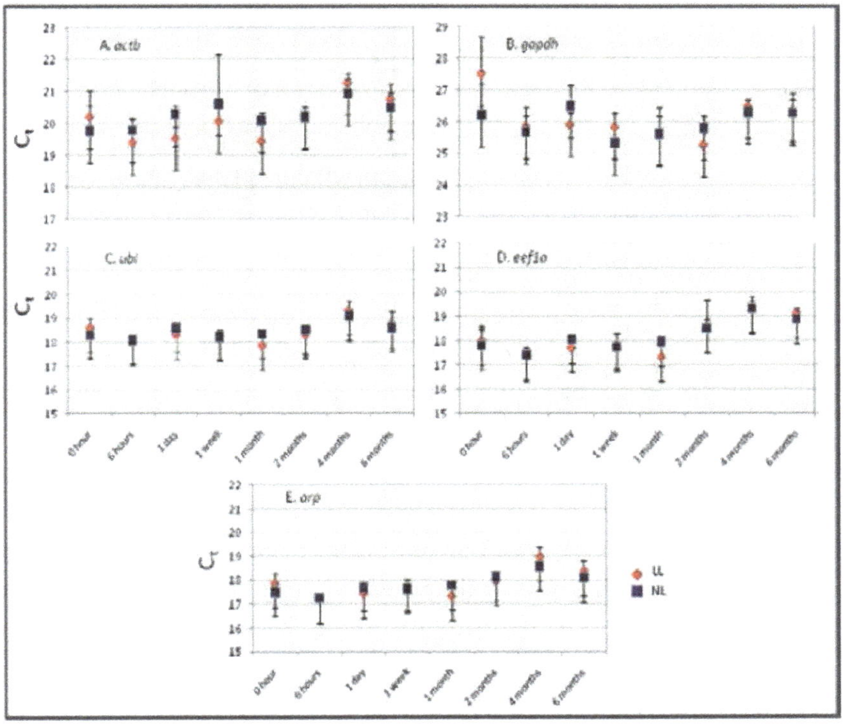

Fig. 2. Expression profiles of potential reference genes in fast muscle of cod kept under two photoperiod regimes: continuous illumination (LL) or normal photoperiod (NL). Data are presented as mean ± SD of the raw cycle threshold (Ct) values of A) *actb*, B) *gapdh*, C) *ubi*, D) *eef1a* and E) *arp* as determined by real-time qPCR (n=6).

Table 2. Expression stability indices of the five reference genes, as determined by geNorm and NormFinder applications. Relative mRNA level were determined in fast muscle of cod kept under two photoperiod regimes: continuous illumination (LL) and normal photoperiod (NL). The most and the least stable reference genes are shaded in blue and red, respectively.

		geNorm					NormFinder				
Time	Photoperiod	arp	eef1a	ubi	actb	gapdh	arp	eef1a	ubi	actb	gapdh
0 h	LL	0.389	0.425	0.401	0.445	0.862	0.023	0.031	0.025	0.001	0.342
	NL	0.372	0.389	0.381	0.464	0.602	0.027	0.016	0.034	0.049	0.145
6 h	LL	0.363	0.349	0.329	0.640	0.665	0.001	0.002	0.001	0.174	0.190
	NL	0.159	0.171	0.206	0.166	0.326	0	0.003	0.013	0.001	0.047

		geNorm					NormFinder				
1 d	LL	0.182	0.189	0.153	0.258	0.195	0.007	0.007	0.001	0.026	0.010
	NL	0.298	0.357	0.313	0.356	0.602	0.010	0.025	0.013	0.013	0.160
1 w	LL	0.264	0.347	0.250	0.430	0.388	0.004	0.032	0.001	0.069	0.051
	NL	0.540	0.623	0.544	1.390	0.676	0.001	0.072	0.001	0.904	0.094
1 m	LL	0.287	0.277	0.275	0.367	0.605	0.014	0.009	0.009	0.020	0.164
	NL	0.230	0.298	0.231	0.295	0.542	0	0.022	0	0.013	0.134
2 m	LL	0.389	0.946	0.367	0.382	0.429	0.021	0.419	0.009	0.007	0.010
	NL	0.173	0.193	0.179	0.166	0.184	0.009	0.012	0.011	0.004	0.008
4 m	LL	0.210	0.225	0.239	0.224	0.321	0.008	0.010	0.016	0.006	0.044
	NL	0.188	0.239	0.186	0.316	0.313	0	0.011	0	0.039	0.039
6 m	LL	0.329	0.332	0.387	0.469	0.390	0.020	0.006	0.053	0.085	0.033
	NL	0.180	0.208	0.192	0.252	0.350	0	0.004	0.004	0.016	0.052

In order to obtain more robust conclusions it is advisable to do a parallel validation of reference gene stability using different alternative software applications, since there is no each method uses different mathematical models that can lead to different outputs. For example, NormFinder takes all candidate reference genes into account and ranks them with the intragroup and intergroup variation, whereas geNorm sequentially excludes the worst gene ending with two and ranks genes with the degree of similarity of expression. The overall results from NormFinder and geNorm applied to our data revealed that the most stable genes were *arp* and *ubi* and these two is the most suitable pair for two-gene normalisation. In previous qPCR studies in cod, it was also shown that *arp* and *ubi* were the most stable genes and could be used in studying wild populations of cod living in contaminated areas (Olsvik et al., 2008) and ontogeny in cod larvae (Sæle et al., 2009). To optimise the results in selecting the most suitable reference gene, candidate genes should belong to different biological pathways, so as to minimise errors associated with co-regulation. Co-regulation is still possible between *ubi* and *arp*, since they fall on the same biological pathway as important molecules in protein degradation and elongation step of protein synthesis, respectively. However, since these genes were validated by two applications generating similar results, it is fair to consider them as stable in this experimental setup. Both applications identified *gapdh* as the most unstable reference gene. It has been observed that *gapdh* is regulated under varying physiological conditions, which could render this gene inappropriate as reference gene (Olsvik et al., 2005). In a study in cod where candidate reference genes were evaluated during ontogeny with emphasis on the development of gastrointestinal tract, *gapdh* was also rated as one of the least suitable normalisation genes regardless of categorisation and analysis (Sæle et al.,

2009). It is also mentioned that in cod, there are two isoforms of *gapdh*, one that is muscle-specific and the other as a brain-specific. Validation of reference genes in cod exposed to thermal stress revealed that *gapdh* was also the least favourable gene for normalisation (Aursnes et al., 2011).

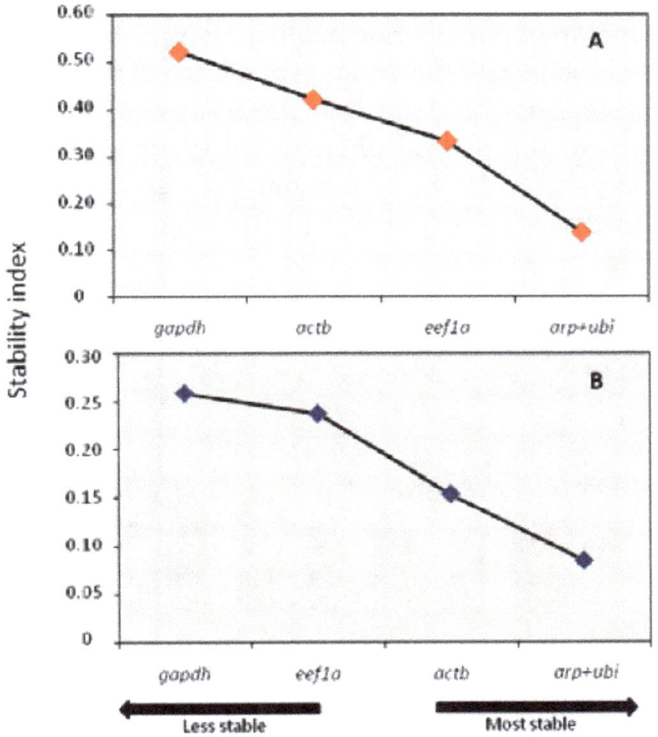

Fig. 3. Ranking of reference genes according to their expression stability in fast muscle of Atlantic cod reared under different photoperiod regimes. The average expression stability values were calculated with (A) geNorm and (B) NormFinder.

3.2 Influence of normalisation on photic-induced expression of *myhc* in cod muscle

Myosin is a ubiquitous eukaryotic motor that interacts with actin to generate the force for cellular movements as diverse as cytokinesis and muscle contraction (Cheney et al., 1993). This motor protein accounts for the majority of myofibrils, which themselves make up to two-thirds of muscle protein synthesis (Mommsen, 2001). For this reason, *myhc* has been used to study muscle growth and development in teleosts (Johnston, 2001). Environmental stimuli such as light influence most of the physiological processes in fish and muscle development is not an exception. The influence of photoperiod manipulation on the muscle physiology of Atlantic cod was assessed in this study by profiling the expression of this gene during a photoperiod manipulation experiment.

For comparison, raw expression data of *myhc* were normalised in two different ways: i) with the use of a two-gene normalisation factor from the most stable genes (*arp* and *ubi*) and, ii) with the least stable reference genes (Fig. 4). Using the best two-gene normalisation factor from geNorm, it was observed that from 0 h to 1 week the expression of *myhc* did not change significantly in either photoperiod regime. However, after a month of photoperiod manipulation, a significant difference was noted between treatments and the group exposed to normal photoperiod showed a significantly higher *myhc* expression than the group exposed to continuous light. Expression of *myhc* in the natural photoperiod group was approximately 40% higher than the expression in the continuous light group. From 2 to 6 months, *myhc* expression increased equally in both photoperiod regimes.

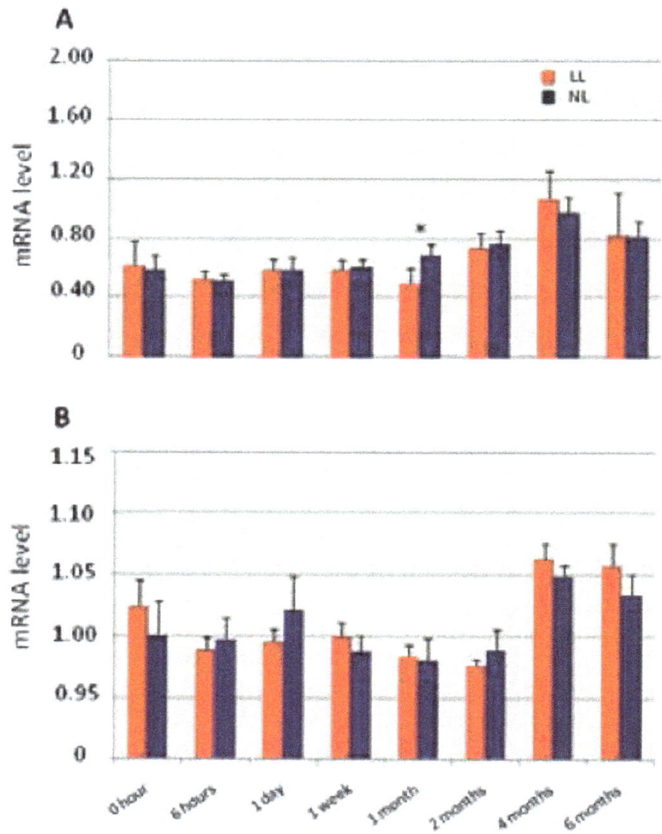

Fig. 4. Expression of *myhc* in the fast muscle of Atlantic cod subjected to different photoperiod regimes. A) Transcript levels of *myhc* gene normalised using *arp* and *ubi*, the best combination for a two-gene normalisation. B) Transcript levels of *myhc* gene normalised by the least stable reference gene, *gapdh* using ΔΔCt method, Data are shown as mean ± SD of the normalised values (n=6). Asterisk (*) indicates that a significant difference was detected between photoperiod treatments ($P < 0.001$).

No significant difference was noted between light regimes throughout whole photoperiod manipulation experiment when *myhc* expression data were normalised using *gapdh*, the least stable reference gene. This stresses the importance of identifying suitable reference genes for a particular biological system, not only to draw robust conclusions but also to identify subtle and important differences in mRNA levels.

4. AT A GLANCE

Morphometric analysis remains a valid method for examining muscle formation in fish, but molecular techniques have introduced new opportunities to investigate this process and comprehend the function of essential regulatory molecules in myogenesis. Quantitative polymerase chain reaction (qPCR) is the most dependable technique for measuring gene expression, as long as appropriate reference genes are utilized for data normalization. This work is the first to validate reference genes for qPCR quantification of muscle transcripts in Atlantic cod raised under various photoperiod conditions, based on our current understanding. NormFinder and geNorm determined that ubi and arp are the optimal gene pair for normalizing our expression data. When normalized with a two-gene factor, a 40% variation in myhc transcript levels was seen between photoperiod conditions, which was not evident when gapdh was used for normalization. Using inappropriate reference genes for normalizing qPCR data can result in scientifically invalid results. This study is a great resource for future gene expression investigations focusing on understanding the molecular mechanisms of muscle growth in Atlantic cod related to light sensitivity. Furthermore, it is relevant to broader subjects within aquaculture research, such as growth and nutrition.

REFERENCES

1. Andersen, C.L., Jensen, J.L. & Orntoft, T.F. (2004). Normalisation of real-time quantitative reverse transcription-PCR data: A model-based variance estimation approach to identify genes suited for normalisation, applied to bladder and colon cancer data sets. *Cancer Research*, 64, 15, 5245-5250, ISSN 0008-5472

2. Aursnes, I.A., Rishovd, A.L., Karlsen, H.E. & Gjøen, T. (2011). Validation of reference genes for quantitative RT-qPCR studies of gene expression in Atlantic cod (*Gadus morhua l.*) during temperature stress. *BMC Research Notes*, 4, ISSN 1756-0500

3. Begtashi, I., Rodriguez, L., Moles, G., Zanuy, S. & Carrillo, M. (2004). Long-term exposure to continuous light inhibits precocity in juvenile male European sea bass (*Dicentrarchus labrax, L.*). I. Morphological aspects. *Aquaculture*, 241, 1-4, 539-559, ISSN 0044-8486

4. Bustin, S.A. (2002). Quantification of mRNA using real-time reverse transcription PCR (RT- PCR): trends and problems. *Journal of Molecular Endocrinology*, 29, 1, 23-39, ISSN 0952-5041
5. Bustin, S.A. & Nolan, T. (2004). Pitfalls of quantitative real-time reverse-transcription polymerase chain reaction. *Journal of Biomolecular Techniques*, 15, 3, 155-166, ISSN 1524-0215
6. Campos, C., Valente, L.M., Borges, P., Bizuayehu, T. & Fernandes, J.M. (2010). Dietary lipid levels have a remarkable impact on the expression of growth-related genes in Senegalese sole (*Solea senegalensis* Kaup). *The Journal of Experimental Biology*, 213, 2, 200-209, ISSN 1477-9145
7. Cheney, R.E., Riley, M.A. & Mooseker, M.S. (1993). Phylogenetic analysis of the myosin superfamily. *Cell Motility and the Cytoskeleton*, 24, 4, 215-223, ISSN 0886-1544
8. Davie, A., Porter, M.J.R. & Bromage, N.R. (2003). Photoperiod manipulation of maturation and growth of Atlantic cod (*Gadus morhua*). *Fish Physiology and Biochemistry*, 28, 1-4, 399-401, ISSN 0920-1742
9. Davie, A., Porter, M.J.R., Bromage, N.R. & Migaud, H. (2007). The role of seasonally altering photoperiod in regulating physiology in Atlantic cod (*Gadus morhua*). Part I. Sexual maturation. *Canadian Journal of Fisheries and Aquatic Sciences*, 64, 1, 84-97, ISSN 0706- 652X
10. Dhillon, R.S., Esbaugh, A.J., Wang, Y.S. & Tufts, B.L. (2009). Characterization and expression of a myosin heavy-chain isoform in juvenile walleye *Sander vitreus*. *Journal of Fish Biology*, 75, 5, 1048-1062, ISSN 0022-1112
11. Endal, H.P., Taranger, G.L., Stefansson, S.O. & Hansen, T. (2000). Effects of continuous additional light on growth and sexual maturity in Atlantic salmon, *Salmo salar*, reared in sea cages. *Aquaculture*, 191, 4, 337-349, ISSN 0044-8486
12. FAO, http://www.fao.org/fishery/species/2218/en
13. Fernandes, J.M., Mackenzie, M.G., Elgar, G., Suzuki, Y., Watabe, S., Kinghorn, J.R. & Johnston, I.A. (2005). A genomic approach to reveal novel genes associated with myotube formation in the model teleost, *Takifugu rubripes*. *Physiological Genomics*, 22, 3, 327-338, ISSN 1531-2267
14. Fernandes, J.M., Mackenzie, M.G., Wright, P.A., Steele, S.L., Suzuki, Y., Kinghorn, J.R. & Johnston, I.A. (2006). Myogenin in model pufferfish species: Comparative genomic analysis and thermal plasticity of expression during early development. *Comparative Biochemistry and Physiology Part D: Genomics and Proteomics*, 1, 1, 35-45, ISSN 1878-0407
15. Fernandes, J.M., MacKenzie, M.G., Kinghorn, J.R. & Johnston, I.A. (2007a). FoxK1 splice variants show developmental stage-specific plasticity of expression with

temperature in the tiger pufferfish. *Journal of Experimental Biology*, 210, 19, 3461-3472, ISSN 1477-9145.

16. Fernandes, J.M., Kinghorn, J.R. & Johnston, I.A. (2007b). Differential regulation of multiple alternatively spliced transcripts of MyoD. *Gene*, 391, 1-2, 178-185, ISSN 0378-1119
17. Fernandes, J.M., Mommens, M., Hagen, O., Babiak, I. & Solberg, C. (2008). Selection of suitable reference genes for real-time PCR studies of Atlantic halibut development. *Comparative Biochemistry and Physiology Part B: Biochemistry and Molecular Biology*, 150, 1, 23-32, ISSN 1096-4959
18. Hall, J.R., Short, C.E. & Driedzic, W.R. (2006). Sequence of Atlantic cod (*Gadus morhua*) GLUT4, GLUT2 and GPDH: Developmental stage expression, tissue expression and relationship to starvation-induced changes in blood glucose. *The Journal of Experimental Biology*, 209, Pt 22, 4490-4502, ISSN 0022-0949
19. Hansen, T., Karlsen, O., Taranger, G.L., Hemre, G.I., Holm, J.C. & Kjesbu, O.S. (2001). Growth, gonadal development and spawning time of Atlantic cod (*Gadus morhua*) reared under different photoperiods. *Aquaculture*, 203, 1-2, 51-67, ISSN 0044-8486
20. Ikeda, D., Ono, Y., Snell, P., Edwards, Y.J., Elgar, G. & Watabe, S. (2007). Divergent evolution of the myosin heavy chain gene family in fish and tetrapods: evidence from comparative genomic analysis. *Physiological Genomics*, 32, 1, 1-15, ISSN 1531-2267
21. Imsland, A.K., Foss, A., Koedijk, R., Folkvord, A., Stefansson, S.O. & Jonassen, T.M. (2007). Persistent growth effects of temperature and photoperiod in Atlantic cod *Gadus morhua*. *Journal of Fish Biology*, 71, 5, 1371-1382, ISSN 0022-1112
22. Johnston, I.A. (1999). Muscle development and growth: potential implications for flesh quality in fish. *Aquaculture*, 177, 1-4, 99-115, ISSN 0044-8486
23. Johnston, I.A. (2001). *Muscle development and growth*, Academic Press, ISBN 0123504422, San Diego, CA, USA
24. Karlsen, O., Norberg, B., Kjesbu, O.S. & Taranger, G.L. (2006). Effects of photoperiod and exercise on growth, liver size, and age at puberty in farmed Atlantic cod (*Gadus morhua* L.). *Ices Journal of Marine Science*, 63, 2, 355-364, ISSN 1054-3139
25. Koedijk, R.M., Le Francois, N.R., Blier, P.U., Foss, A., Folkvord, A., Ditlecadet, D., Lamarre, S.G., Stefansson, S.O. & Imsland, A.K. (2010). Ontogenetic effects of diet during early development on growth performance, myosin mRNA expression and metabolic enzyme activity in Atlantic cod juveniles reared at different salinities. *Comparative Biochemistry and Physiology Part A: Molecular & Integrative Physiology*, 156, 1, 102-109, ISSN 1531-4332

26. Lilleeng, E., Froystad, M.K., Vekterud, K., Valen, E.C. & Krogdahl, A. (2007). Comparison of intestinal gene expression in Atlantic cod (*Gadus morhua*) fed standard fish meal or soybean meal by means of suppression subtractive hybridization and real-time PCR. *Aquaculture*, 267, 1-4, 269-283, ISSN 0044-8486

27. Mommsen, T.P. (2001). Paradigms of growth in fish. *Comparative Biochemistry and Physiology Part B: Biochemistry & Molecular Biology*, 129, 2-3, 207-219, ISSN 10964959

28. Norberg, B., Brown, C.L., Halldorsson, O., Stensland, K. & Bjornsson, B.T. (2004). Photoperiod regulates the timing of sexual maturation, spawning, sex steroid and thyroid hormone profiles in the Atlantic cod (*Gadus morhua*). *Aquaculture*, 229, 1, 451-467, ISSN 0044-8486

29. Olsvik, P.A., Lie, K.K., Jordal, A.E.O., Nilsen, T.O. & Hordvik, I. (2005). Evaluation of potential reference genes in real-time RT-PCR studies of Atlantic salmon. *BMC Molecular Biology*, 6, ISSN 1471-2199

30. Olsvik, P.A., Softeland, L. & Lie, K.K. (2008). Selection of reference genes for qRT-PCR examination of wild populations of Atlantic cod *Gadus morhua*. *BMC Research Notes*, 1, 47, ISSN 1756-0500

31. Sæle, Ø., Nordgreen, A., Hamre, K. & Olsvik, P.A. (2009). Evaluation of candidate reference genes in Q-PCR studies of Atlantic cod (*Gadus morhua*) ontogeny, with emphasis on the gastrointestinal tract. *Comparative Biochemistry and Physiology Part B: Biochemistry & Molecular Biology*, 152, 1, 94-101, ISSN 1096-4959

32. Taranger, G.L., Aardal, L., Hansen, T. & Kjesbu, O.S. (2006). Continuous light delays sexual maturation and increases growth of Atlantic cod (*Gadus morhua* L.) in sea cages. *Ices Journal of Marine Science*, 63, 2, 365-375, ISSN 1054-3139

33. Vandesompele, J., De Preter, K., Pattyn, F., Poppe, B., Van Roy, N., De Paepe, A. & Speleman, F. (2002). Accurate normalisation of real-time quantitative RT-PCR data by geometric averaging of multiple internal control genes. *Genome Biology*, 3, 7, RESEARCH0034, ISSN 1465-6914

34. Watabe, S. (1999). Myogenic regulatory factors and muscle differentiation during ontogeny in fish. *Journal of Fish Biology*, 55, SUPPL. A, 1-18, ISSN 0022-1112

06 IDENTIFYING FISH ORIGIN VIA 16S RDNA FINGERPRINTING OF MICROBIAL COMMUNITIES USING PCR-DGGE: AN APPLICATION ON FISH FROM DIFFERENT TROPICAL ORIGINS

1. INTRODUCTION: AN OVERVIEW

Food safety is now a mandatory requirement for food imported into the European Union. European customers remember Bovine Spongiform Encephalitis, pathogens, and avian influenza. Vigilance and thorough monitoring are essential due to the prevalence of comparable shocks worldwide. EU rule 178/2002 mandated traceability for any food imported into the EU. Historically, the food sector utilized basic traceability systems, but with the growing adoption of current Good Manufacturing Practice, traceability systems have gained significance in the production process. Only a limited number of analytical techniques allow for the tracing of food. To overcome the challenges of implementing documentary systems in developing countries and to monitor the product throughout processing, we suggest identifying and validating specific biological markers from the fish's environment to ensure traceability of aquaculture products during international trade. We suggested investigating the source of fish by globally analyzing the bacterial populations present on the fish samples. The dominating bacterial flora can help identify the capture location, production process, or sanitary and hygienic conditions during post-harvest operations (Montet et al., 2004; Le Nguyen et al., 2007, Montet, 2008).

Aquatic microorganisms are directly linked to the physiological condition of fish. Several research have been conducted on the microbiota in fish from different geographical regions (Grisez et al., 1997; Spanggaard et al., 2000; Al-Harbi and Uddin, 2003; Leesing, 2005). Water composition, temperature, meteorological conditions, and agricultural activities can impact the bacterial populations in fish. PCR product separation in DGGE relies on the reduction of the electrophoretic mobility of partially melted double-stranded DNA molecules in polyacrylamide gels with a linear gradient of DNA denaturants such as formamide and urea at 60°C. Molecules with distinct sequences exhibit varying melting behavior and halt migration at different positions in the gel (Muyzer et al., 1993; Leesing, 2005). PCR-DGGE has been utilized to examine various fish bacterial assemblage distributions by different researchers (Murray et al., 1996; Øvreas et al., 1997; Moeseneder et al., 1999; Riemann et al., 1999; Maiworé et al., 2009a, 2009b; Tatsadjieu et al., 2010). Our team also employed this technique to investigate freshwater fish bacteria for traceability purposes (Le Nguyen et al., 2007; Montet et al., 2008). This methodology allows for the examination of both cultivable and non-cultivable, anaerobic and aerobic bacteria. It also offers a quick way to examine changes in community structure in response to various environmental conditions (Yang et al., 2001). Our work aims to utilize the PCR-DGGE method to analyze the bacteria present in fish. This will help establish a way to correlate bacterial communities with their geographical origin, eliminating the need for studying each bacterial strain individually. The band patterns of bacterial species from various fish in Vietnam were compared and statistically analyzed to identify the fish's origin. We provide an example of studying the ecology of bacteria in a tropical and traditional fish fermentation process in Ivory Coast, specifically focusing on Adjuevan.

2. MATERIALS AND METHODS

2.1 Fish sampling

Pangasius hypophthalmus were gathered from a distinct pond located in five aquaculture farms throughout five different districts in South Vietnam, specifically Chau Phu, An Phu, Phu Tan, Chau Doc, and Tan Chau of An Giang Province. The samples were gathered in Vietnam during two distinct seasons: the rainy season in October 2005 and the dry season in February 2006. From every farm in each area, samples were collected from the identical pond and then transported to storage bags using aseptic techniques. The samples were kept on crushed ice and taken to the laboratory. The skin, gills, and intestines of each fish specimen were aseptically removed and placed in individual sealed plastic bags. They were then stored frozen at -20°C until examination. Tilapia fish (Oreochromis niloticus) were obtained from freshwater tanks maintained at a temperature of 27°C at Cemagref-Cirad in Montpellier, France, in

April and June 2010. The sample weights were 500 ± 10 g. After sampling using aseptic gloves, each fish was packed in separate sterile plastic bags and carried in chilled containers cooled by crushed ice to the laboratory located 500 meters away. Fermentation commenced right after the fish arrived.

2.2 Production of fermented fish "Adjuevan"

Adjuevan is a salted and fermented fish traditionally produced in the west coast of Ivory Coast at ambient temperature (28-30°C) following two traditional methods. First method of production took place in jars covered with plastics and stones for 5 days and second method followed the same fermentation process and then fish were dried on racks or nets for at least 10 days. We studied the changes in physicochemical characteristics and the dynamic of bacterial flora by PCR-DGGE on artificial Adjuevan made in our laboratory at 30°C, with different percentages of salts of 10%, 15% 20%, 25%, and 30% following the both methods.

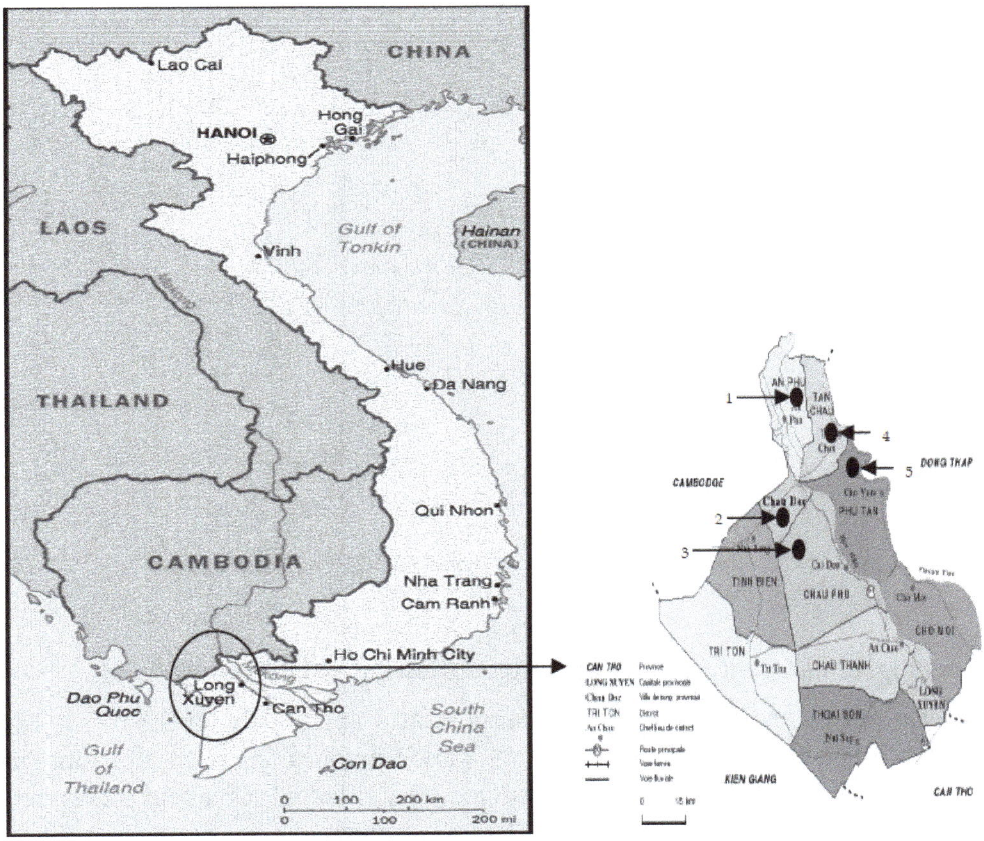

(a) (b)

Fig. 1. (a). The map of Viet nam (b) The expansion of An giang province with five different sampling locations: 1) An Phu; 2) Chau Doc; 3) Chau Phu; 4) Tan Chau; 5) Phu Tan.

Fermentation methods are described in Figure 2. About 1.5 kg of fish was gutted, washed and salt was added at the following concentrations 10%, 15%, 20%, 25%, 30% (w/w) after 24 h of maturation (stored at room temperature). For method 1, fish was salted, then wrapped in sterile plastic containers and arranged in different coolers. They are left to ferment for 5 days at ambient temperature (30°C) followed by drying in a ventilated dryer (at 30°C with minimal ventilation) for 24 h. For method 2, fish after been salted, was deposited on a sterile plastic surface then placed in sterile trays. Fermentation was done simultaneously as well as drying for 5 days at 30°C followed by ventilation for 24 h (Fig. 2). Fermentation experiments were done in triplicate.

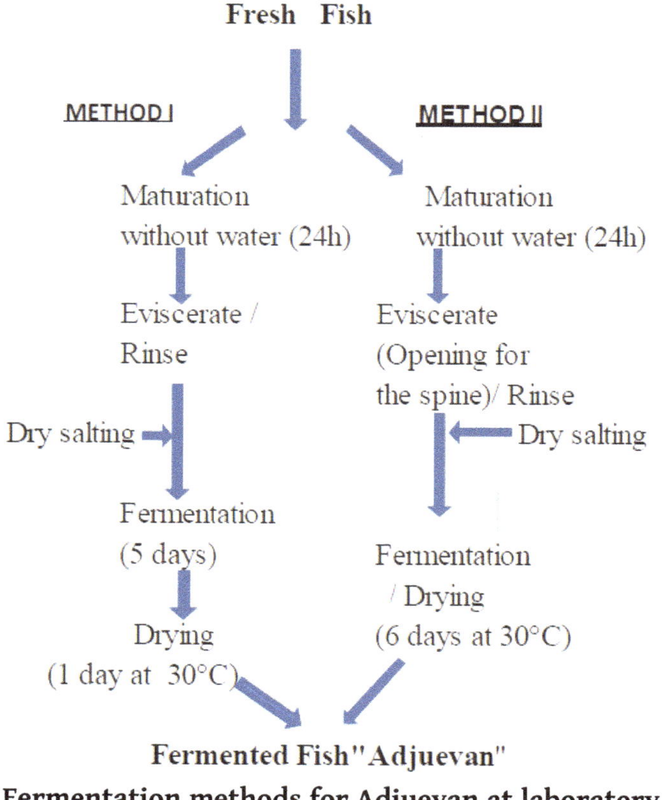

Fig. 2. Fermentation methods for Adjuevan at laboratory level

Fish samples were collected during fermentation primarily on fresh fish, then after 24 h followed by maturation and 1, 2, 3, 4 and 5 days during fermentation. They were put in aseptic sterile tubes, stored in a cooler filled with ice and transported immediately to laboratory for physicochemical analysis.

2.3 Total DNA extraction

DNA extraction was based on the methods of Ampe et al. (1999) and Leesing (2005) but modified and optimised. Around 2g each of gills, skin and intestine were homogenized for 3 min with vortexing after addition of 6 mL sterile peptone water (pH 7.0, Dickinson, France). Four 1.5-mL tubes containing the resulting suspension were then centrifuged at 10,000g for 10 min. 100 µL of lysis buffer TE (10 mM Tris-HCl; 1 mM EDTA; pH 8.0, Promega, France) and 100 µL of lysozyme solution (25 µg.µL-1, Eurobio, France) and 50 µL of proteinase K solution (10 µg.µL-1, Eurobio, France) were added to each pellet. Samples were vortexed for 1 min and incubated at 42°C for 30 min. Then 50µL of 20% SDS (Sodium Dodecyl Sulphate, Sigma, France) were added to each tube, and the tubes were incubated at 42°C for 10 min. 300 µL of MATAB (Mixed Alkyltrimethyl Ammonium Bromide, Sigma France) were added to each tube, and the tubes were incubated at 65°C for 10 min. The lysates were then purified by repeated extraction with 700 µL of phenol- chloroform-isoamyl alcohol (25:24:1, Carlo Erba, France), and the residual phenol was removed by extraction with an equal volume of chloroform-isoamyl alcohol (24:1). The DNA was precipitated with isopropanol, washed with 70% ethanol and then air dried at room temperature. Finally, the DNA was resuspended in 100 µL of ultra pure water and stored at - 20°C until analysis.

2.4. PCR-denaturing gradient gel electrophoresis (DGGE) analysis

The V3 variable region of bacterial 16S rDNA from fish was amplified using primers gc338f (5'CGCCCGCCGCGCGCGGCGGGCGGGGCGGGGGCACGGGGGGACTCCTACGGGAG GCAGCAG, Sigma, France) and 518r (5'-ATTACCGCGGCTGCTGG, Sigma, France) (Øvreas et al., 1997; Ampe et al., 1999; Leesing, 2005). A 40-bp GC-clamp (Sigma, France) was added to the forward primer in order to insure that the fragment of DNA will remain partially double-stranded and that the region screened is in the lowest melting domain (Sheffield et al., 1989). Each mixture (final volume 50 µL) contained about 100ng of template DNA, all the primers at 0.2µM, all the deoxyribonucleotide triphosphate (dNTPs) at 200µM, 1.5mM MgCl2, 5µL of 10x of reaction *Tag* buffer (MgCl2 free) (Promega, France) and 5U of Taq polymerase (Promega, France). In order to increase the specificity of amplification and to reduce the formation of spurious by-products, a "touchdown" PCR was performed according to the protocol of Díez et al. (2001). An initial denaturation at 94°C for 1 min and 10 touchdown cycles of denaturation at 94°C for 1 min, then annealing at 65°C (with the temperature decreasing 1°C per cycle) for 1 min, and extension at 72°C for 3 min, followed 20 cycles of 94°C for 1 min, 55°C for 1 min and 72°C for 3 min. During the last cycle, the extension step was increased to 10 min. Aliquots (5µL) of PCR products were analysed first by conventional electrophoresis in 2% (w/v) agarose gel with TAE 1X buffer (40 mM Tris- HCl pH 7.4, 20 mM sodium acetate, 1.0 mM Na2-EDTA), stained with ethidium bromide (Sigma, France) 0.5 µg/mL in TAE 1X and quantified by using a standard (DNA mass ladder 100 bp, Promega, France).

The PCR products were analyzed by Denaturing Gradient Gel Electrophoresis (DGGE) by using a Bio-Rad DCodeTM universal mutation detection system (Bio-Rad Laboratories, Hercules, USA) and the procedure first described by Muyzer et al. (1993) and improved by Leesing (2005, 2011). Samples containing approximately equal amounts of PCR amplicons were loaded into 8% (wt/v) polyacrylamide gels (acrylamide/NN'-methylene bisacrylamide, 37.5:1, Promega, France) in 1X TAE buffer (40 mM Tris-HCl pH 7.4, 20 mM sodium acetate, 1.0mM Na2-EDTA). All electrophoresis experiments were performed at 60°C using a denaturing gradient ranging from 30 to 60% (100% corresponded to 7M urea and 40% [v/v] formamide, Promega, France). The gels were electrophoresed at 20 V for 10 min and then at 180 V for 12h.

After electrophoresis, the gels were stained for 30 min with ethidium bromide and rinsed for 20 min in distilled water and then photographed on a UV transilluminator with the Gel Smart 7.3 system (Clara Vision, Les Ulis, France).

2.5 Image and statistical analysis

Individual lanes of the gel images were straightened and aligned using ImageQuant TL software version 2003 (Amesham Biosciences, USA). Banding patterns were standardized with the two reference patterns included in all gels which are the patterns of *Escherichia coli* DNA and *Lactobacillus plantarum* DNA. This software permitted to identify the bands and their relative position compared with the standard patterns.

In DGGE analysis, the generated banding pattern is considered as an "image" of all of the major bacterial species in the population. An individual discrete band refers to a unique "sequence type" or phylotype (Muyzer et al., 1995; van Hannen et al., 1999), which is treated as a discrete bacterial population. It is expected that PCR fragments generated from a single population will display an identical electrophoretic mobility in the analysis. This was confirmed by Kowalchuk et al. (1997) who showed that co-migrating bands generally corresponded to identical sequence.

The DGGE fingerprints were manually scored by the presence and absence of co-migrating bands, independent of intensity. Pairwise community similarities were quantified using the Dice similarity coefficient (SD) (Heyndrickx et al., 1996).

$$S_D = 2\,N_c / N_a + N_b$$

Where Na represented the number of bands detected in the sample A, Nb represented the number of bands detected in the sample B, and Nc represented the numbers of bands common to both sample. Similarity index were expressed within a range of 0 (completely dissimilar) to 1.0 (perfect similarity). Dendograms were constructed using the Statistica version 6 software (StatSoft, France). Similarities in community structure were determined using the cluster analysis by the single linkage method with the Euclidean distance measure. Significant differences of bacterial communities of fish between seasons were determined by factorial correspondence analysis using the first 2 variances which described most of the variation in the data set.

3. RESULTS

3.1 DGGE pattern of different locations in Viet Nam within the same sampling period

Fish were collected during the rainy season (October 2005) in 5 different districts of An Giang province, Viet Nam. PCR-DGGE fingerprinting of 5 replicates for each location revealed the presence of 8 to 12 bands of bacteria in the fish (Fig. 3). Some of the bands were common to all the different regions. The pattern obtained for the bacterial community for 5 replicates of the same pond of a unique farm in each district was totally similar among the same season (Fig. 4). We observed also high similarities on bacteria patterns for the samples from the same districts, as well as the neighbouring district where the water was supplied by the same branch of the Mekong River. Statistical analysis of the DGGE gel patterns for the 5 replicates of fish samples from 5 different districts of An Giang province harvested in the rainy season (25 samples), showed the community similarity among the different geographical locations where the fish samples were collected (Fig. 3). Two main clusters were observed at 70% similarities level (Fig. 3): 1. The first cluster included the samples from Chau Doc and Chau Phu districts; 2. The second cluster comprised the samples from An Phu, Tan Chau and Phu Tan districts. The bacterial communities of Chau Doc and Chau Phu districts were closely related, as well as the bacterial communities from An Phu, Tan Chau and Phu Tan districts.

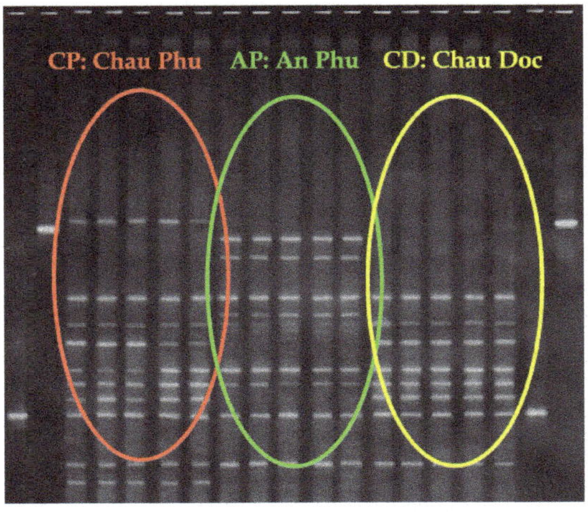

Fig. 3. PCR-DGGE 16S rDNA banding profiles of fish bacteria from 3 districts of An Giang province (five fish from the same pond in the same farm in each district), Viet Nam in rainy season : CP: Chau Phu district; AP: An Phu district; CD: Chau Doc district;

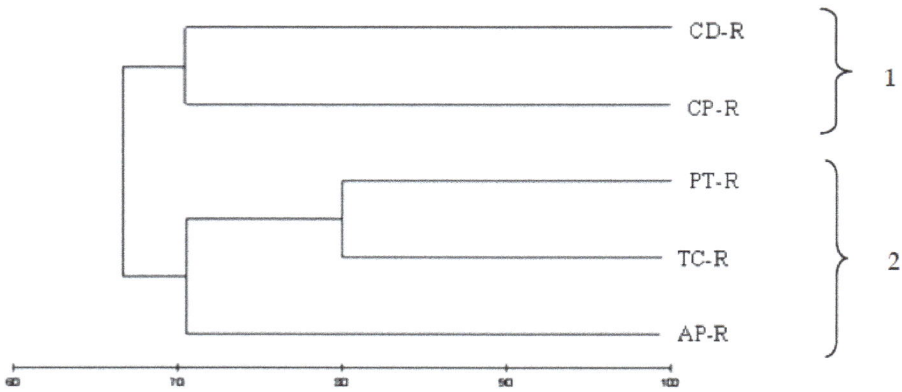

Fig. 4. Cluster analysis of 16S rDNA banding profiles for fish bacterial communities from 5 districts of An Giang province, Viet Nam in rainy season (R) 2006: CP: Chau Phu district; AP: An Phu district; CD: Chau Doc district; TC: Tan Chau district; PT: Phu Tan district.

3.2 DGGE pattern of fish within the same location in Viet Nam at two sampling seasons

Samples were taken at 8 months interval from the same ponds as previous experiments in the dry (February 2006) and rainy season (October 2005). Only four districts were studied for the dry season due to the closing down of Phu Tan pond. Differences on the DGGE band patterns of the same ponds at two different seasons can be clearly noted (Fig. 5). However, some dominant DNA bands could be observed that remained the same during the two seasons. Cluster analysis by Statistica software also showed that the bacterial communities of the same ponds were quite similar for the two seasons. Two main clusters were also observed at 75% similarities levels (Fig. 5): 1. The first cluster included samples from An Phu and Tan Chau districts; 2. The second cluster comprised all the samples from Chau Doc and Chau Phu districts. It was also shown that throughout the 2 seasons for the same location, the number of main bands observed on the gel was slightly different. For example, for the farm in An Phu district, 10 bands were observed in the dry season and 9 bands in the rainy season but only 7 bands are common and the calculation by Statistica showed that there was 75 % similarity between the two seasons. The similarities were higher for the other farms (88% for Than Chau, 84% for Chau Phu and 86% for Chau Doc).

In addition, similarities of the bacterial communities from the 4 different geographical locations for two seasons were compared at 8 months interval by Factorial Correspondence Analysis (FCA). The first two variances described 76 % of the variation in the data set (Fig. 6). Four different groups of 4 different districts were observed regardless the seasons.

3.3 Microbiological changes during "Adjuevan" fermentation in Ivory Coast

Fish samples were collected between April and June 2010 at the pilot plant of Cemagref- Cirad. There were used to produce fermented fish following the two methods of Adjuevan fermentation used in Ivory Coast. PCR-DGGE analysis was done on samples taken successively from the first day to the fifth day of fermentation.

3.3.1 Microbiological changes following method 1

DGGE analyses conducted on fish collected after 24 h of maturation and before salting, gave 3 to 5 common bands of bacteria that were identified by sequencing as *Pseudomonas putida*, *Pseudomonas fluorescens*, *Aeromonas* spp., *Staphylococcus* spp. A similar DGGE profile was obtained at the second day of fermentation. But more bands appeared, from 9 to 21 bands composed of 94 % of positive gram bacteria dominated by *Staphylococcus* spp. (55%), *Bacillus* spp. (24 %), *Miccrococcus* spp. (13%) and 5% of aerobic mesophilic bacteria. Three percent of unidentified bacteria were revealed by PCR-DGGE in the third day of fermentation. *Staphylococcus xylosus, Staphylococcus auricularis, Staphylococcus piscifermentans, Staphylococcus saprophyticus, Micrococcus luteus, Bacillus subtilis, Bacillus licheniformis and Streptococcus* spp. were the genera identified (Fig.7). *Staphylococcus saprophyticus, Bacillus licheniformis, and Micrococcus luteus* were present in all samples. Some bacteria were common to some samples. However a loss of about 40% of bands of aerobic mesophilic bacteria and some genera of *Micrococcus* was observed at the fifth day of fermentation.

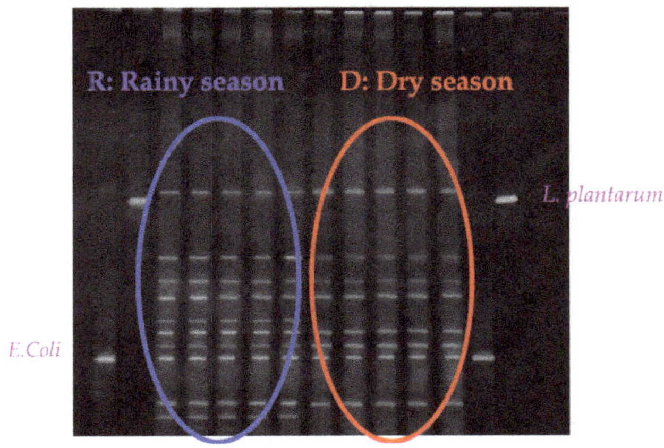

Fig. 5. PCR-DGGE 16S rDNA banding profiles of fish bacteria from one district of An Giang province, Viet Nam in the dry season (D) and rainy season (R)

Factorial analysis of correspondences (FCA) was used to compare the levels of similarities between bacteria communities in function of salt content. Three different groups of samples were observed regardless the initial percentage of salt. Two variances described 75.1% of the variation (Fig. 8). More the percentage of salt was great, more differences between profiles were observed. Up to 25% of salt,

DGGE profiles were almost identical and the same bacterial species were identified. This showed the influence of salt on the strains present on meat samples from method 1.

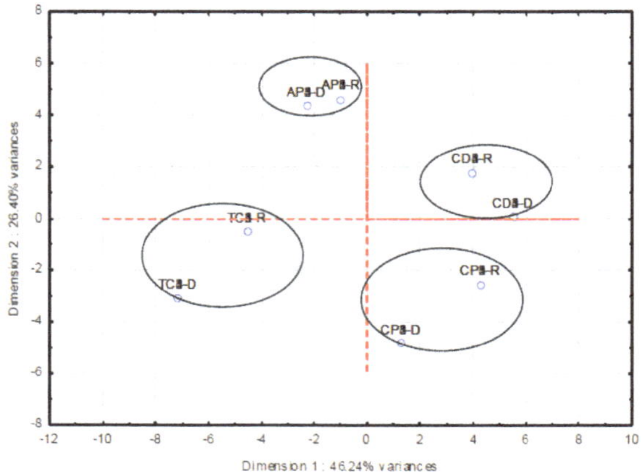

Fig. 6. Factorial variance analysis of 16S rDNA banding profiles of fish bacteria from 4 districts of An Giang province, Viet Nam in the dry season (D) and rainy season (R) 2006: CP: Chau Phu district; AP: An Phu district; CD: Chau Doc district; TC: Tan Chau district.

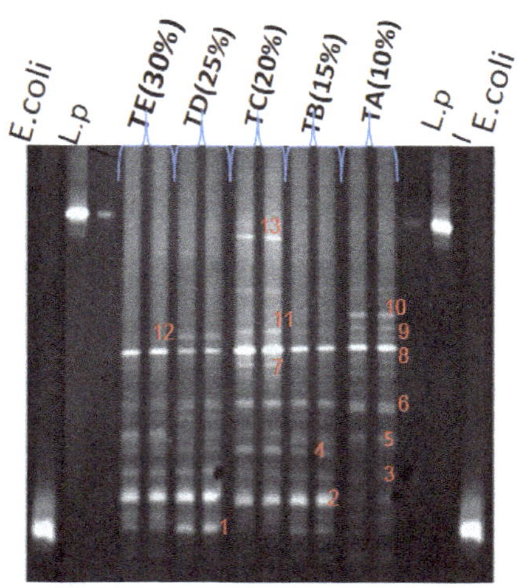

1: P. putida; 2:B. licheformis; 3, 5: Bacillus spp.; 4: B. subtilis; 6: Micrococcus luteus; 7: S. Piscifermentans; 8:S. saprophyticus; 9, 11: S. xylosus; 12: Staphylococcus spp.; 10: S. auricularis; 13: Steptococcus spp.. (Abbreviations: B.; Bacillus; S, Staphylococcus. P, Pseudomonas).

TA (10%): sample with 10% (w/w) of salt; TB (15%): sample with 15% (w/w) of salt; TC (20%): sample with 20% (w/w) of salt; TD (25%): sample with 25% (w/w) of salt; TE (30%): sample with 30% (w/w) of salt.

Fig. 7. PCR- DGGE of 16s rDNA profiles of bacteria from five samples at the third day of fermentation following the method 1 in Ivory Coast.

3.3.2 Microbiological changes following method 2

After 24 hours of maturation, DGGE tests of fish revealed two bands corresponding to citrobacter freundii and Pseudomonas putida, which were present in all samples. On the third day of fermentation, PCR-DGGE analysis detected the presence of five to 12 bands, predominantly constituted of 89.2% gram-positive bacteria, including Bacillus spp. (49.8%), Staphylococcus spp. (17.3%), and Miccrococcus spp. (22.1%). Additionally, 6.7% of the bands were classified as aerobic mesophilic bacteria, and 4.1% were unidentified bacteria (Fig.9). The genera identified were Bacillus subtilis, Bacillus licheniformis, Staphylococcus auricularis, Staphylococcus piscifermentans, and Micrococcus spp. Pseudomonas putida, Bacillus licheniformis, Bacillus subtilis, and Micrococcus spp. were the main microorganisms present in all samples during the fermentation process and remained until the end of fermentation. Both samples D (25% salt) and E (30% salt) have similar bands. Only a 10% reduction in bands was noticed on the fifth day of fermentation.

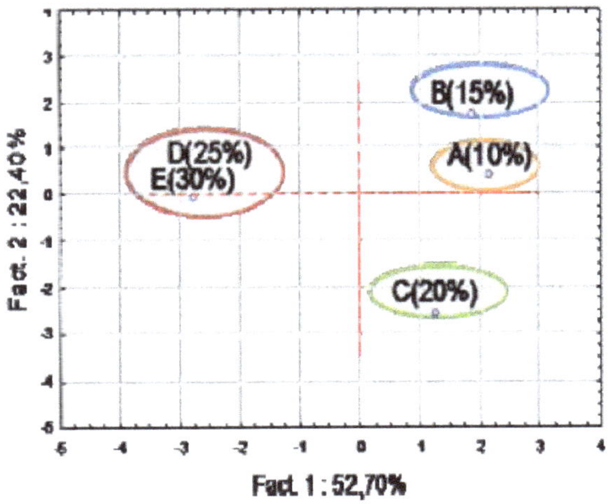

A (10%); B (15%); C (20%); D (25%); E (30%) five samples of fermented fish prepared with different percentages of salt.

Fig. 8. Factorial variance analysis of 16s rDNA banding profiles of bacteria from five samples at the third day of fermentation following method 1 in April 2010 in Ivory Coast.

An FCA was conducted to compare the similarities of bacterial communities on the third day of fermentation. The analysis revealed that 67.8% of the variation in the data set was accounted for, showing minimal differences across bacterial communities in all samples (Fig. 10). Overall, the plant life showed minimal variation across different samples. However, due to a 20% salt content, the DGGE profiles exhibited nearly similar bacterial species. The results indicated that salt had a lesser impact on the bacterial flora in method 2 compared to method 1.

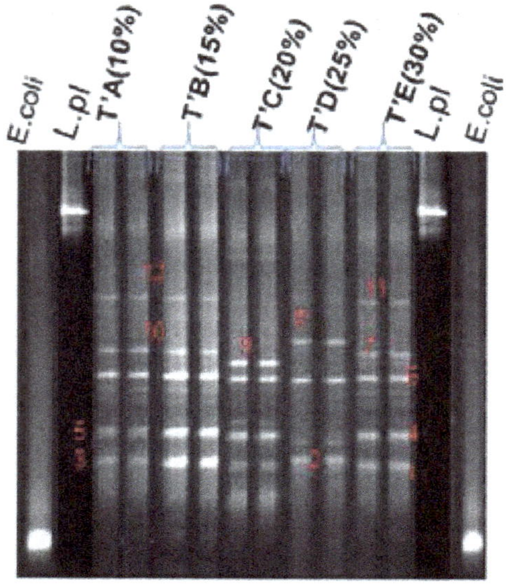

1: Pseudomonas spp.; .2: P. fluorescens ; 3: P. putida; 4,5: B. licheformis; 6: B. subtilis; 7,10: S. piscifermentans; 8: Micrococcus spp.; 9 : Bacillus spp.; 11,12: S. auricularis. (Abbreviations: B.; Bacillus; S, Staphylococcus; P, Pseudomonas).

T'A (10%): sample with 10% (w/w) of salt; T'B (15%): sample with 15% (w/w) of salt; T'C (20%): sample with 20% (w/w) of salt; T'D (25%): sample with 25% (w/w) of salt; T'E (30%): sample with 30% (w/w) of salt.

Fig. 9. PCR- DGGE 16s rDNA profiles of bacteria from five samples at the third day of fermentation following method 2 in Ivory Coast.

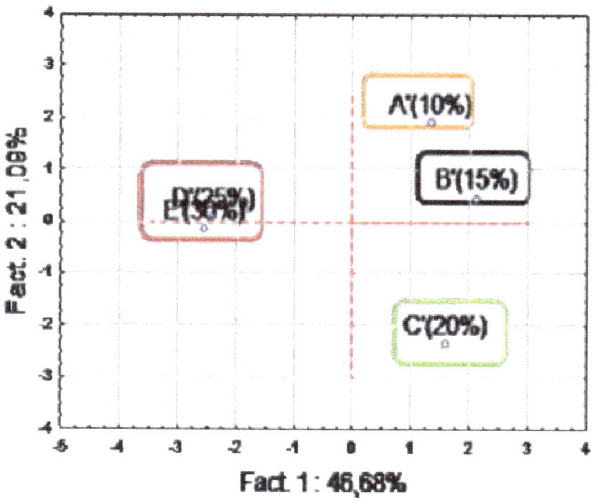

A' (10%); B' (15%); C' (20%); D' (25%); E' (30%) five samples of fermented fish prepared following different percentages of salt.

Fig. 10. Factorial variance analysis of 16s rDNA banding profiles of bacteria from five samples at the third day of fermentation following method 2 in June 2010 in Ivory Coast.

4. DISCUSSION

Studies on bacterial populations in fish samples have been often conducted using both culture-dependent and culture-independent approaches such as random amplified polymorphic DNA (RAPD) (Spanggaard et al., 2000). Only a limited number of studies have examined the bacterial communities in fish samples using PCR-DGGE methods, as documented by Spanggaard et al. (2000), Huber et al. (2004), Maiworé et al. (2009), and Tatsadjieu et al. (2010). The band pattern of bacterial communities isolated from fish using PCR-DGGE was closely associated with the fish's microbial environment in our investigation. The skin of fish directly contacts water, and the gills, which filter air from water entering the fish's lungs, are effective at accumulating environmental germs. The intestine harbors a significant population of bacteria that is influenced by the fish's dietary patterns. Examining fish samples from several places during the rainy season revealed notable variations in migration patterns on the DGGE gel. The five replicates for each sampling point exhibited statistically identical DGGE patterns consistently during the research. The variations in the band profiles are due to variances in feeding strategies among farms and the type of aquaculture system used. The changes may also be attributed to the water supply, which can be influenced by pollution from urban areas. Additionally, medications used to treat diseases and stressors can impact the fish's microbial populations (Sarter et al., 2007). Nevertheless, many typical bands identified using DGGE were present in all profiles from the same sampling dates and locations. When looking at the map of An Giang province in South Vietnam, you can observe the Mekong River splitting into two main branches, which then further divide into numerous minor canals. Chau Phu and Chau Doc are located on the western branch of the Mekong river, whereas An Phu, Tan Chau, and Phu Tan are situated on the eastern branch of the river. Differences in water quality and fish habitat have a significant impact on bacterial ecology. Examining fish samples from identical locations in Vietnam during two distinct seasons revealed notable statistical differences in the DGGE bands profiles. The significant fluctuations in salinity and pH of the pond water may be attributed to the high rainfall throughout the rainy season. The microbial communities of the fish are heavily influenced by these elements, as they rely on the external environment. The results indicated that the DGGE band profile of the fish's bacterial community is distinctive and characteristic of a specific farm and season. Samples from the same location throughout several sampling seasons showed more similarity compared to samples from different locations. The DGGE profiles displayed consistent dominant bands at specific sites that were present consistently throughout all sample times. Our study on fermented fish from Ivory Coast found that the salt content in the meat was affected by the salt concentration in the brine and the fermentation technique employed. After analyzing the DGGE bands, it was discovered that several kinds of halophilic bacteria were strongly dominant based on the fermentation method used. The bacterial flora in fermented fish varied mostly due to the fermentation procedure and the usage of high salt content, reaching at least 7% in the meat. PCR-DGGE study of fermented fish bacteria communities can be used to distinguish between different fermentation processes. PCR-DGGE technology can be utilized to distinguish the geographical origin of fish production based on their bacterial community. This technique could potentially be used to track the impact of a process on the bacterial flora. Biological indicators remained consistent for particular areas over many seasons and demonstrated enough statistical specificity for each farm. This worldwide technique is faster (less than 24 hours) than traditional microbiological techniques and eliminates the need for detailed bacterial examination through biochemistry or molecular biology. Sequencing can be performed straight from the DGGE gel extracts to identify microorganisms. This approach is a fast analytical tool for tracing fish products and assigns a distinct biological bar code to each fish.

REFERENCES

1. Al Harbi, A.H., Uddin, N., 2003. Quantitative and qualitative studies on bacterial flora of hybrid tilapia (*Oreochromis niloticus x O. aureus*) cultured in earthen pond in Saudi Arabia. Aquaculture 34, 43-48.
2. Ampe, F., Omar, N.B., Moizan, C., Wacher, C., Guyot, J.P., 1999. Polyphasic study of the spatial distribution of microorganisms in Mexican pozol, a fermented maize dough, demonstrates the need for cultivation-independent methods to investigate traditional fermentations. Appl. and Environ. Microbiol. 65, 5464-5473.
3. Díez, B., Pedrós–Alió, C., Marsh, T.L., Massana, R., 2001. Application of denaturing gradient gel electrophoresis (DGGE) to study the diversity of marine picoeukaryotic assemblage and comparison of DGGE with other molecular techniques. Appl. and Environ. Microbiol. 67, 2942-2951.
4. Grisez, L., Reyniers, J., Verdonck, L., Swings, J., Ollevier, F., 1997. Dominant intestinal microbiota of sea breeam and sea bass larvae, from two hatcheries, during larval development. Aquaculture 155, 387-399.
5. Heyndrickx, M., Vauterin, L., Vandamme, P., Kersters, K., De Vos, P., 1996. Applicability of combined amplified ribosomal DNA restriction analysis (ARDRA) patterns in bacterial phylogeny and taxonomy. Journal of Microbiol. Methods 26, 247-259.
6. Huber, I., Spanggaard, B., Appel, K.F., Rossen, L., Nielsen, T., Gram, L., 2004. Phylogenetic analysis and in situ identification of the intestinal microbial community of rainbow trout (*Oncorhynchus mykiss*, Walbaum). Journal of Appl. Microbiol. 96, 117-132.

7. Kowalchuk, G.A., Stephen, J.R., de Boer, W., Prosser, J.I., Embley, T.M., Woldendorp, J.W., 1997. Analysis of ammonia-oxydizing bacteria of the beta subdivision of the class *Proteobacteria* in coastal sand dunes by denaturing gradient gel electrophoresis and sequencing of PCR amplified 16S ribosomal DNA fragments. Applied and Environ. Microbiol. 63, 1489-1497.

8. Leesing, R., 2005. Identification and validation of specific markers for traceability of aquaculture fish for import/export. PhD dissertation. University of Montpellier 2, France.

9. Leesing Ratanaporn, Dijoux Daniel, Le Nguyen Doan Duy, Loiseau Gérard, Ray Ramesh C, Montet Didier. Improvement of DNA Extraction and Electrophoresis Conditions for the PCR-DGGE Analysis of Bacterial Communities Associated to Two Aquaculture Fish Species. Dynamic Biochemistry, Process Biotechnology and Molecular Biology, In press July 2011.

10. Le Nguyen D.D., Ha Ngoc H., Dijoux D., Montet D., Loiseau G., 2008. Determination of fish origin by using 16S rDNA fingerprinting of bacterial communities by PCR-DGGE: application to Pangasius fish from Viet Nam. Food control. 19, 454-460.

11. Moeseneder, M.M., Arrieta, J.M., Muyzer, G., Winter, C., Herndl, G.J., 1999. Optimization of terminal-restriction length polymorphism analysis for complex marine bacterioplankton communities and comparison with denaturing gradient gel electrophoresis. Appl. and Environ. Microbiol. 65, 3518-3525.

12. Maiworé J., Tatsadjieu Ngouné L., Montet D., Loiseau G., Mbofung C.M., 2009a. Comparison of bacterial communities of Tilapia fish from Cameroon and Vietnam using PCR-DGGE. African Journal of Biotech. 8 (24), 7156-7163.

13. Maiworé J., Tatsadjieu Ngouné L., Montet D., Loiseau G., Mbofung C.M., 2009b. Comparison of bacterial communities of tilapia fish from Cameroon and Vietnam using PCR-DGGE (polymerase chain reaction-denaturing gradient gel electrophoresis). African journal of Biotechnol 8 (24), 7156-7163.

14. Montet, D., Leesing, R., Gemrot, F., Loiseau, G., 2004. Development of an efficient method for bacterial diversity analysis: Denaturing Gradient Gel Electrophoresis (DGGE). In: Seminar on Food Safety and International Trade, Bangkok, Thailand.

15. Montet, D., Le Nguyen D.D., El Sheikha A.F., Condur A., Métayer I., Loiseau G., 2008. Application of PCR-DGGE in determining food origin: Cases studies of fish and fruits. Greening the Food Chain 3 and 4; 3: Traceability: Tracking and tracing in the food chain. Aspects of Applied Biol 87, 11-22.

16. Murray, A.E., Hollibaugh, J.T, Orrego, C., 1996. Phylogenetic composition of bacterioplankton from two California estuaries compared by denaturing gradient gel electrophoresis of 16S rDNA fragments. Appl. Environ. Microbiol. 6, 2676-2680.

17. Muyzer, G., De Waal, E.C., Uitterlinden, A.G., 1993. Profiling of complex microbial populations by denaturing gradient gel electrophoresis analysis of polymerase chain reaction-amplified genes coding for 16S rRNA. Appl. Environ. Microbiol. 59, 695-700.

18. Muyzer, G., Teske, A., Wirsen, C.O., Jannasch, H.W., 1995. Phylogenetic relationships of Thiomicrospira species and their identification in deep-sea hydrothermal vent sample by denaturing gradient gel electrophoresis of 16S rDNA fragment. Archives of Microbiol. 164, 165-172.

19. Øvreas, L., Forney, L. Dae, F.L., Torsvik., V., 1997. Distribution of Bacterioplankton in Meromictic Lake Sælenvannet, as determined by Denaturing Gradient Gel Electrophoresis of PCR-Amplified gene fragment coding for 16S rRNA. Appl. and Environ. Microbiol. 63, 3367-3373.

20. Riemann, L., Steward, G.F., Fandino, L.B., Campbell, L., Landry, M.R., Azam, F., 1999. Bacterial community composition during two consecutive NE Monsoon periods in the Arabian Sea studied by denaturing gradient gel electrophoresis (DGGE) of rRNA genes. Deep-Sea Res. 46, 1791-1811.

21. Sheffield, V.C., Beck, J.S., Stone, E.M., Myers, R.M., 1989. Attachment of a 40 bp G+C rich sequence (GC-clamp) to genomic DNA fragments by polymerase chain reaction results in improved detection of single-base changes. Proceeding of the National Academy of Sciences of the United States of American 86, 232-236.

22. Sarter S., Hoang Nam Nguyen K., Le Thanh H., Lazard J., Montet D., 2007 Antibiotic resistance in Gram-negative bacteria isolated from farmed catfish. Food Control 18, 1391–1396.

23. Spanggaard, B., Huber, I., Nielsen, T.J., Nielsen, T., Appel, K., Gram, L., 2000. The microbiota of rainbow trout intestine: a comparison of traditional and molecular identification. Aquaculture 182, 1-15.

24. Tatsadjieu Ngouné L., Maiworé J., Hadjia B.M., Loiseau G., Montet D., Mbofung C.M., 2010. Study of the microbial diversity of *Oreochromis niloticus* of three lakes of Cameroon by PCR-DGGE: Application to the determination of the geographical origin. Food control 21, 673-678.

25. Van Hannen, E.J., Zwart, G., van Agterveld, M.P., Gons, H.J., Ebert, J., Laanbroek, H.J., 1999. Changes in bacterial and eukaryotic community structure after mass lysis of filamentous cyanobacteria associated with viruses. Appl. and Environ. Microbiol. 65, 795-801.

26. Yang, C.H., Crowley, D.E., Menge, J.A., 2001. 16S rDNA fingerprinting of rhizosphere bacterial communities associated with healthy and Phytophthora infected avocado roots. FEMS Microbiol. Ecol. 35, 129-136.

07 USING MITOCHONDRIAL DNA VARIATION TO CLARIFY THE SYSTEMATIC STATUS OF COMMERCIAL SPECIES – INVESTIGATING TWO DISTINCT COMMERCIAL FLEXOPECTEN VARIETIES IN THE AEGEAN SEA

1. INTRODUCTION: AN OVERVIEW

Identifying and determining species are fundamental initial stages in biodiversity monitoring and protection, as stated by Dayrat (2005). Species identification is typically conducted by taxonomists, however it is often limited by the absence of physical characteristics. Two Flexopecten species are being discussed: Flexopecten glaber L. and F. proteus Dillwyn ex Solander ms. F. glaber is a consumable bivalve mollusk that has a smooth shell and may grow up to 8.66 cm in length (Pisor & Poppe, 2008). It is an epibenthic species that inhabits soft and/or hard ground in muddy and sandy environments with organic debris. The species has rapid growth. Reproduction primarily occurs in late summer, however gonad activity is observable throughout the year. The species can be found at depths ranging from 5 to 900 meters, and occasionally as deep as 1600 meters, in the Mediterranean and Black Seas as shown by Poppe & Goto (1993). Its distribution is illustrated in Figure 1. There is controversy around the growth of F. proteus, as some writers have reported that the species is limited to the Adriatic Sea (Pountiers, 1987). The form has been identified as a commercial species in the Thessaloniki Gulf area of the North Western Aegean Sea, where it lives with F. glaber (Zenetos, 1996;

Galinou-Mitsoudi & Sinis, 2000). The authors reported that the ratio of F. glaber to F. proteus in three random fishery samples of 99, 130, and 120 individuals was 0.74, 1.28, and 6.50, respectively. F. proteus is present in the Thessaloniki Gulf, Ionian Sea (Amvrakikos and Korinthiakos Gulfs), and Aegean Sea (Pagasitikos Gulf, N. Evoikos Gulf, Saronikos Gulf, Lesvos Island, and Lemnos Island) according to Koutsoubas et al. (2007) and Zenetos (1996) (Fig. 1).

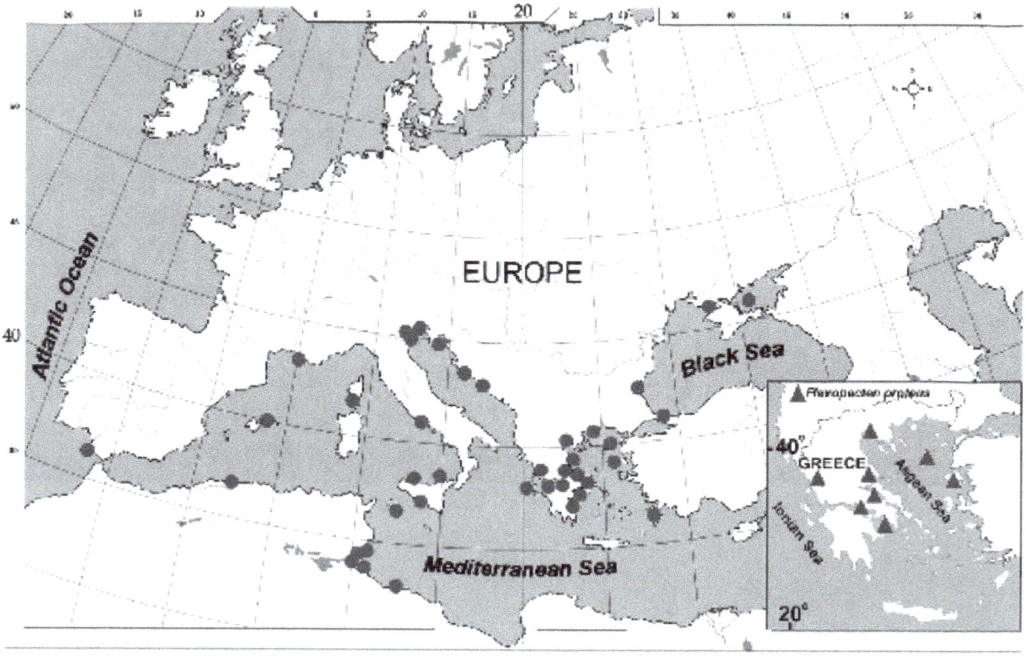

Fig. 1. *F. glaber* distribution in Mediterranean and Black Sea. In inlaid map of Greece the *F. proteus* distribution is denoted, according to Zenetos (1996).

Commercial scallops are a primary focus due to their strong demand in both the national and European markets. However, the species appears to be excessively harvested, as the yearly Hellenic output decreased from over 20 t between the mid-1970s and late 1980s to 5 t from the late 1980s to the 1990s (Koutsoubas et al., 2007). Flexopecten scallops fishing has been prohibited in the Thessaloniki Gulf since 2002 because of elevated levels of heavy metals, particularly cadmium. Additionally, the scallop stocks in the central Aegean Sea experienced a significant decline in 2003 (Koutsoubas et al., 2007). F. glaber and F. proteus can also be utilized in the Adriatic Sea, as mentioned by Pujolar et al. (2010). In addition to being a popular choice for fishing, Flexopecten scallops are also considered a promising species for aquaculture due to their rapid development rate, strong demand, and favorable pricing per kilogram. Mitochondrial DNA (mtDNA) is valuable for studying relationships among bivalve populations because of its maternal inheritance, haploid nature, and effective population size being one fourth that of nuclear DNA. The invertebrate mitochondrial 16S rDNA gene has been extensively utilized to investigate population structure and genetic variation levels in several Pectinidae species (Kong et al., 2003; Mahidol et al., 2007; Yuan et al., 2009; Pujolar et al., 2010). Our work aimed to examine the genetic resources of the Aegean populations of F. glaber and F. proteus in Greece by analyzing mitochondrial 16S rDNA nucleotide sequences, as these are the primary commercial stocks of these bivalves in the region. Chlamys proteus, also known as F. proteus Dillwyn, has long been regarded as a member of the bivalve family Pectinidae according to various sources (Cossignani et al., 1992; Delamotte & Vardala-Theodorou, 1994; Doneddu & Trainito, 2005; Repetto et al., 2005; Zenetos et al., 2005). Most sources indicate that F. proteus is another species closely resembling F. glaber. The primary systematic feature distinguishing the two pectinids is the variation in the number of major and dominant ribs. F. glaber has 9-12 ribs, whereas F. proteus has only five ribs. Poppe & Goto (1993) state that CLEMAM & ERMS consider F. proteus as a subspecies or a variant of F. glaber. Raines & Poppe (2006) emphasize that F. proteus is considered just a synonym of F. glaber.

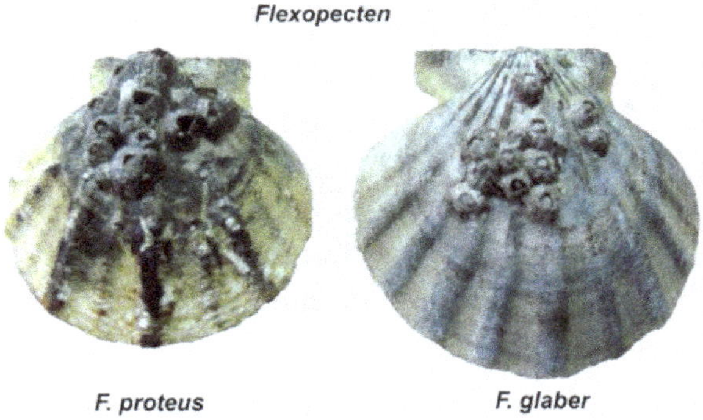

Fig. 2. Fished scallops in the study area of the two *Flexopecten* forms *proteus* and *glaber*, with epibiots on their left valves

Several scientists believe that the future of descriptive taxonomy relies on effectively adopting new methodologies. Various ideas have been suggested and significant advancements have been made by utilizing molecular data to distinguish species that are differentiated by subtle morphological traits (Knowlton, 1993; Avise, 1994; Chan & Chu, 1996; Sarver et al., 1998; Mathews et al., 2002; Goetze, 2003). The invertebrate mtDNA 16S rDNA gene has been extensively researched using universal primers (Kocher et al., 1989; Palumbi, 1996). It has been applied to address taxonomic issues in the families Mytilidae (Rawson & Hilbish, 1995), Veneridae (Canapa et al., 1996), and Pectinidae (Canapa et al., 2000a; Saavedra & Pena, 2004; Pujolar et al., 2010). Considering all relevant factors and the ambiguous systematic classification of these two typical Flexopecten "forms," our study aimed to address the identification issue of the two Aegean taxa by analyzing the mtDNA 16S rDNA gene sequences. A study conducted by Pujolar et al. (2010) focused on the Adriatic populations of the Flexopecten complex in a parallel and similar manner.

2. METHODS AND TOOLS

2.1 Sampling

Specimens were collected from three different regions of the Gulf of Thessaloniki, northern Aegean Sea (Fig. 3). In total, 78 adult individuals of *F. glaber* and 57 adult individuals of *F. proteus* from the three different locations, were collected (Table 1). Specimens were classified according to the number of dominant ribs.

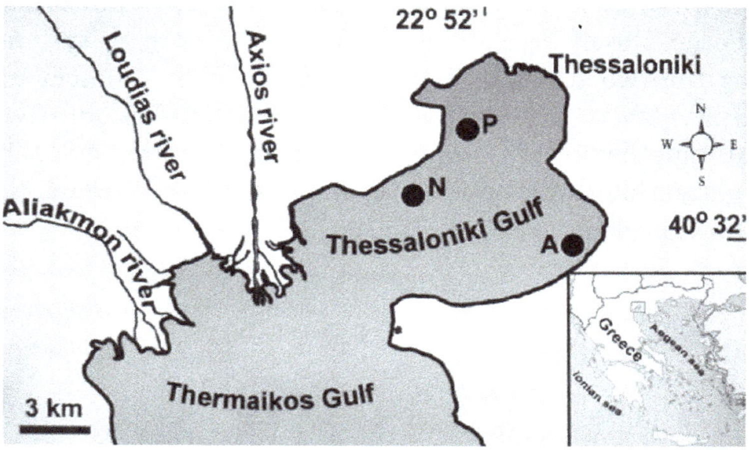

Fig. 3. Sampling sites of the analyzed population for both *F. glaber* and *F. proteus*. A: Airport; P: Paliomana; N: Naziki.

2.2 DNA extraction and PCR amplification

Total DNA was extracted from the anterior adductor muscle according to Hillis et al. (1996). A universal primer set (Palumbi, 1996) was used for the amplification of the 16S rDNA gene in both *F. glaber* and *F. proteus*. The reaction mixture contained template DNA (approximately 100 ng), 1X PCR buffer, 2.2 mM MgCl2, 20 pmol of each primer, 0.25 mM of each dNTP and 0.5 U of Promega polymerase. Amplification was started at 94°C for 3 min, followed by 31 cycles at 94°C for 50 s, 50°C for 50 s, 72°C for 50 s and a final extension at 72oC for 5 min.

Electrophoresis of 3 µl of the PCR product was performed in 1XTBE buffer for 1 h at 150 V, in 1.5% agarose gel containing 0.5 µg/ml ethidium bromide. The size of the PCR products was checked against a 100 bp DNA ladder and was approximately 500 bp for both taxa. The resulting DNA fragments were visualized by UV transilumination and photographed.

2.3 DNA sequencing

A sequencing analysis on a 3730Xl DNA Analyzer (Applied Biosystems) was followed using both forward and reverse primers for crosschecking. DNA sequences were deposited to GenBank (accession numbers GU320272 – GU320288; HM 627014 – HM627051).

2.4 Data analysis

The nucleotide sequences of all individuals were aligned using the Clustal X software (Thompson et al., 1997) and the BioEdit software (Hall, 1999), set to default parameters and corrected by eye. Mt DNA 16S rDNA diversity was estimated by the number of haplotypes, haplotype frequency, haplotype diversity (h) and nucleotide diversity (π), via Arlequin version 3.5 (Excoffier & Lischer, 2010). Pairwise genetic distances among haplotypes were computed based on the 2-parameter Kimura distance model (Kimura, 1980) using MEGA software (Tamura et al., 2007; Kumar et al., 2008). The obtained pairwise genetic distances were used to construct a Neighbor-Joining tree with the same software. The species *Aequipecten opercularis* (GenBank: AM494412) was used as outgroup and two relevant ingroups (*Argopecten irradians*: GU119971; *Chlamys multistriata*: FN667665) were also included in the analysis. Pairwise sample differentiation was assessed using the exact test described by Raymond & Rousset (1995), included in the Arlequin package. Analysis of molecular variance was conducted also with the same package (Arlequin version 3.5; Excoffier & Lischer, 2010), to determine the genetic differentiation of populations.

To test whether the populations underwent recent demographic population growth we calculated the mismatch distribution in Arlequin package, which assumes an infinite sites model of selectively neutral nucleotide substitutions and assesses significance via coalescent simulations of a large, neutrally evolving population of constant size (Slatkin & Hudson, 1991). Finally, Tajima's *D* (Tajima, 1989) was used to examine the selective neutrality of mitochondrial fragment and it was also used in mismatch distribution, because significant negative *D*-values are what the hypothesis of population growth predicts (Bertorelle & Slatkin, 1995).

3. RESULTS

A total of 436 base pairs of the 16S rRNA gene fragment were successfully sequenced for 135 individuals from both taxa, and 50 polymorphic sites were identified (Appendix 1). The transition/transversion rate ratios were 2.312 for purines and 6.283 for pyrimidines while the overall transition/transversion bias was 1.7. No species specific positions were detected that could discriminate the two species. Fifty one haplotypes were detected among all samples (37 for *F. glaber* and 32 for *F. proteus*) and their frequencies are given in Appendix 1. Haplotype 5 (20% in total) was shared among all samples of both taxa. Haplotype 1 revealed a higher percentage of appearance among samples (25.2%) and it was dominant in all samples apart from Paliomana population of *F. proteus* (Appendix 1). Haplotypes 3, 10 and 15 were present in three samples in total of the same and different taxa, whereas haplotypes 6, 19 and 21 were shared between two samples of different taxa. Haplotypes 8, 45 and 50 were observed twice in the same population and all the other haplotypes were unique.

Haplotype diversity, nucleotide diversity, number of polymorphic sites and number of haplotypes are given also in Appendix 1. The highest values of haplotype and nucleotide diversity were revealed in the population of *F. proteus* from Paliomana (0.9744 and 0.0078 respectively) while the lowest values of these indices were observed in the sample of *F. glaber* from Airport (0.7793 and 0.0027). Mean haplotype and nucleotide diversity for *F. glaber* populations were h=0.854 and π=0.004 respectively, whereas for *F. proteus* were h=0.892 and π=0.005.

Values of genetic distances between pairs of populations are given in Table 1 and ranged from 0.003 (between pA and gA) to 0.008 (between pP and pN). Genetic distance between the two taxa was estimated in a value of D=0.005. Pairwise genetic distances among haplotypes were used to construct a neighbour joining tree. As it can be seen by Figure 4, no clustering of haplotypes corresponded to specific taxon and/or sampling site was detected. All internal nodes are supported by relatively low bootstrap values (≤67%).

Table 1. Genetic distances among the studied population samples, based on 16S rDNA sequences and the 2-parameter Kimura model. SE values are shown in brackets. gA: *F. glaber* from Airport; gP: *F. glaber* from Paliomana; gN: *F. glaber* from Naziki; pA: *F. proteus* from Airport; pP: *F. proteus* from Paliomana; pN: *F. proteus* from Naziki.

	gA	gP	gN	pA	pP	pN
gA		[0.001]	[0.002]	[0.001]	[0.002]	[0.001]
gP	0.005		[0.002]	[0.001]	[0.002]	[0.002]
gN	0.005	0.006		[0.002]	[0.002]	[0.002]
pA	0.003	0.005	0.005		[0.002]	[0.001]
pP	0.006	0.008	0.006	0.007		[0.002]
pN	0.004	0.006	0.007	0.004	0.008	

The analysis of the partitioning of the haplotype diversity indicated that the majority of the genetic variation (96.98%) was distributed within populations (Table 2) and only a percentage of 4.53% could be attributed to variation among populations within groups. AMOVA with two groups (i.e. *F. glaber* versus *F. proteus*) revealed a low FST value of 0.0295 and showed that only a 1.52% of the genetic variation occurred among groups. Pairwise exact test (Raymond & Rousset, 1995) for samples of both taxa showed no population differentiation with all *P* values>0.05 (*P*=1.000 for all the estimates). When pooling together the three sampling sites of *F. glaber* and *F. proteus* respectively, the exact test of population differentiation was still not statistically significant (*P* values=0.20).

Table 2. Analysis of molecular variance, in populations of both taxa. Df: degrees of freedom. The fixation indices are: *FCT*=- 0.01518; *FSC*=0.03017; and *FST*=0.04467.

Source of variation	Df	Variance components	Percentage of variation
Among groups	1	-0.006	-1.52
Among populations within groups	4	0.020	4.53
Within populations	129	0.430	96.98
Total	134	0.444	100

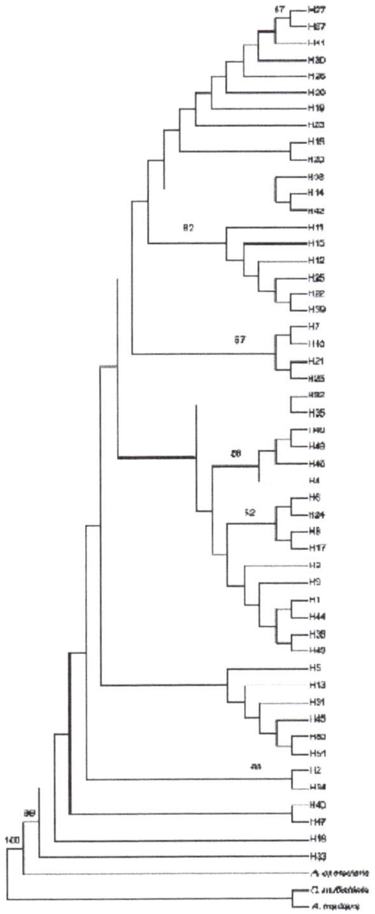

Fig. 4. Phylogenetic tree of the 51 haplotypes of *F. glaber* and *F. proteus* taxa recovered from 16S rRNA sequences, estimated by the Neighbor-joining method. Only bootstrap values based on 100 replications higher than 50% are displayed.

Mismatch distribution was calculated for all studied populations of both taxa (Fig. 5). All populations followed unimodal distribution (i.e a bell-shaped distribution) that it is assumed to be the signature of population expansion occurred probably after a bottleneck event. However, Tajima's *D*-values were not significantly negative for any of the studied populations (p=0.05–0.1), thus rejecting the hypothesis for population growth.

A

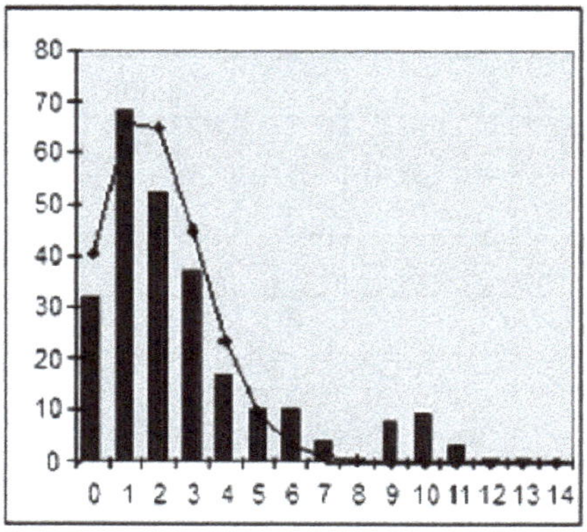

B

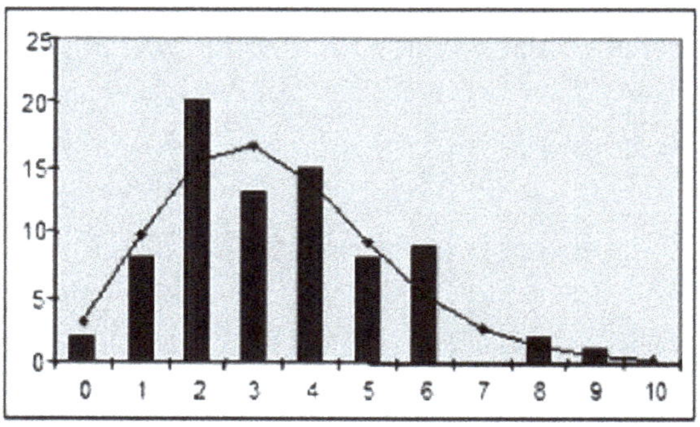

Fig. 5. Mismatch distributions for A) *F. glaber* and B) *F. proteus*. All studied samples followed the same unimodal distribution and thus graphs depict frequencies of pairwise differences for one of the populations of each taxon. The observed frequency of pairwise nucleotide differences among sequences is represented by black bars, and expected frequencies under a model of sudden population expansion are represented by continuous line.

4. DISCUSSION

4.1 Levels of population variation

Several scientists (Palumbi, 1996; Stepien et al., 1999; Saavedra & Pena, 2004; Pujolar et al., 2010) have noted low levels of genetic variation among species, suggesting that the mt 16S rDNA gene is largely conserved across taxa, perhaps because of its importance in protein assembly.

However, it was not the case in the present study. Values of nucleotide diversity found for *F. glaber* (π=0.004) and *F. proteus* (π=0.005) are higher than the values reported for Adriatic populations of the same taxa (π=0.001 and π=0.000 respectively) (Pujolar et al., 2010). Also our values are higher than those estimated for other Pectinidae taxa as *Chlamys nobilis* (π =0.0012; Yuan et al., 2009), *C. farreri* (π=0.0035; Kong et al., 2003), *Pecten jacobaeus* (π=0.0015), *P. novaezelandiae* (π=0.0020), *P. fumatus* (π=0.000) and *P. maximus* (π=0.0039) (Saavedra & Pena, 2004).

The levels of intrapopulation variation for *F. glaber* and *F. proteus*, revealed by sequences of the mt DNA 16S rDNA region in the present study are also relatively high. Values of haplotype diversity ranged from 0.7793 to 0.9744 for populations of both taxa, are much higher than the values of 0.000 and 0.400 reported for the Adriatic populations (Pujolar et al., 2010) and higher than the values of 0.5356 and 0.7328 reported for two wild populations of *C. nobilis*, using sequences of the same region (Yuan et al., 2009). Also the highest value of π revealed in our study (0.0078) exceeds much the value of π=0.0012 observed in *C. nobilis* populations (Yuan et al., 2009). Kong et al. (2003) reported lower to similar values of π (0.00175 to 0.00591) and similar values of h (0.750 to 1.000) for different populations of Zhikong scallop *C. farreri*, based on the sequencing analysis of 16S ribosomal gene. Also, values of nucleotide and haplotype diversity observed in our samples are considerably higher than the values observed for different populations of the Asian moon scallop *Amusium pleuronectes* (0.000 to 0.0017 for π; 0.000 to 0.0511 for h) (Mahidol et al., 2007b).

The demographic analysis revealed that all the studied samples of both taxa were probably stable over time. Pujolar et al. (2010) who have studied the demographic history of the Adriatic *Flexopecten* complex using three mitochondrial and one nuclear gene proposed that the lack of large demogarphic expansion since the last glaciations, suggests that colonization must have been conducted by a substantial number of individuals that occupied the new habitat relatively fast. Demographic expansion does not appear either as a correct explanation of the observed genetic variation in our samples. A possible explanation for high values of intrapopulation indices for *F. glaber* and *F. proteus* could be the large effective population sizes as genetic variability is abundant. It is common knowledge that large *Ne* results to large amounts of neutral variability in populations.

An alternative hypothesis to explain this finding is that there is more than one species in the data set, but the character used to identify it is not valid. In other words, there may be more than one evolutionary entity, so we are artificially lumping more than one species in the analysis. One way to sort this out was to include other taxa in the analysis and make a phylogenetic analysis of the haplotypes, to see if they form unique phylogenetic species. There may simply be cryptic species to be considered or perhaps the morphotype doesn't correspond to the phylotype, which would certainly not be the first time in molluscs (Knowlton, 2000). The phylogenetic analysis of haplotypes revealed no clustering corresponding to each putative species (*F. glaber* and *F. proteus*). Also, the taxa *A. irradians* and *C. multistriata* formed a separate group in the Neighbour – Joining tree.

The decreased intrapopopulation genetic diversity can have serious consequences on the survival and reproduction of the species (Soule, 1980). Nevertheless, in the marine invertebrates reduction of genetic variation is expressed mainly with the loss of rare haplotypes despite with the total reduction of heterozygosity in a population (Gosling, 2003). Our samples are characterized from high DNA variability and the existence of some rare haplotypes, a result which indicates the good population status for both taxa. As the fishing of the Aegean *F. glaber* and *F. proteus* stocks is forbidden since 2002 because of heavy metals (Cd) presence in high concentrations (Koutsoubas et al., 2007), our data are fully justified.

Although there was significant genetic heterogeneity within populations, the statistical analysis of the 16S rDNA haplotypes showed minimal spatial structuring within populations. Marine organisms with pelagic larval development typically exhibit little genetic variation within populations due to their strong dispersion abilities and extensive oceanic mixing. Following spawning and external fertilization, larvae have a variable duration in the plankton, where they can be carried by water currents passively (Seed, 1969; Cho et al., 2007). Gene flow among the communities in the research area in the Gulf of Thessaloniki does not appear to be particularly restricted. Our data confirm that marine invertebrates with a planktonic larval stage exhibit significant genetic relatedness, likely due to enhanced larval dispersal potential (Crisp, 1978).

4.2 Systematic status of the two taxa

F.proteusfoundintheAegeanSeacontradictsthelimitedgeneticexchangeortherecent emergenceofthistypeasstatedbyPulojaretal.(2010).Thewidespreaddissemination of the larval stage has led to the extension of this morph outside the Adriatic Sea, indicating the potential appearance of F. proteus in other Mediterranean regions. The taxonomic classification of Chlamys glabra, now known as F. glaber, is now being debated. Earlier phylogenetic studies (Canappa et al., 2000a, 2000b; Barucca et al., 2004; Liu et al., 2007) suggest that C. glabra may be classified in a separate genus from Chlamys, aligning with certain experts who have placed it in the genus Flexopecten (Vaught, 1989; Rombouts, 1991). These investigations show that C. glabra is closely related to Aequipecten opercularis rather than other species in the Chlamys genus such as C. islandica, C. varia, and C. farreri. Later on Saavedra & Pena (2006) used the updated latin name F. glaber instead of C. glabra in a phylogenetic analysis for American scallops, and they also found that taxa F. glaber and A. opercularis clustered together. An attempt was made to distinguish C. glabra larvae from larvae of other bivalve species using a method that combines PCR-SSCP with sequencing of a partial 18S rDNA region, however it was unsuccessful (Livi et al., 2006). Pujolar et al. (2010) propose that F. glaber and F. proteus are the same species based on their study of the two Adriatic Flexopecten forms using multiple genes. The genetic trees showed both species mixed together without forming distinct clusters. The current study aimed to elucidate the relationship between two Aegean taxa by analyzing a mitochondrial gene. Analysis of 16S rDNA sequencing could not identify any unique sites that could differentiate between various species. Moreover, two common haplotypes and six uncommon haplotypes are shared between the two groups, with no distinct grouping observed among different groups. This indicates interbreeding between F. glaber and F. proteus, implying they belong to the same taxonomic group. Wilding et al. (1999) discovered that there were no common unique genetic variations in P. maximus and P. jacobaeus after analyzing their mitochondrial DNA using PCR-RFLP. This, together with the minimal genetic differences observed, indicates that the two species can interbreed. Saavedra & Pena (2004) observed in a phylogenetic analysis of P. maximus and P. jacobaeus that the two most prevalent haplotypes were present in both species, indicating a shared ancestry between the two European taxa. Another indication of conspecific taxa was the genetic distance levels. The genetic distance of 0.005 between the two taxa does not indicate significant genetic variations between the two species. It is comparable in scale to values found within a species rather than between different species. This number is far below the anticipated range for species that are closely related (Liu et al., 1998; Stepien et al., 1999; Saavedra & Pena 2004; Yu et al., 2004; Mahidol et al., 2007a). The data we have is insufficient to support classifying F. proteus as a subspecies of F. glaber. Therriault et al. (2004) proposed revising the classification of Dreissena bugensis as a subspecies of D. rostriformis due to a low genetic distance between them (0.003–0.004), indicating they may be a single species. The presence of only one species in our research is further supported by the precise examination of population differentiation. Combining the data from the three sampling sites of F. glaber and F. proteus did not reveal any statistically significant differences in population. Our research confirms that F. proteus is either a variant or merely a synonym of F. glaber. Shell morphological changes in Bivalve are frequently due to phenotypic flexibility, leading to many previous species classifications being deemed unwarranted (Knowlton, 2000). Therefore, bivalves likely have a higher number of genetic investigations leading to the synonimization of species compared to most other marine invertebrate phyla. Our research is one instance where differences between sympatric taxa identified through shell structure were contradicted by genetic information (Wilding et al., 1999; as discussed by Knowlton, 2000). The current data further support the reliability of mtDNA markers in resolving ambiguous taxonomic relationships. A universal DNA-based barcode identification system for all animal species serves as a global tool for species identification. The barcode system relies on genetic variation within a specific gene area, the mitochondrial DNA cytochrome c oxidase I gene (COI). Once the reference sequence library is established, fresh specimens can be identified by comparing their DNA barcode sequences to this library. Herbert and colleagues (2004a, 2004b) showed that the COI region is suitable for distinguishing closely related species, leading to global initiatives to speed up the identification of cryptic species (Herbert et al., 2003). In the future, a survey using the barcode identification system might be conducted for the two Flexopecten forms to verify if they belong to the same species.

5. APPENDIX

Variable nucleotide sites													
	1	1111112222	2222222223	3333333333	3344444444								
	555556790	1559990111	2335666781	2223566889	9900013333								
	8567890628	7244785678	9450035343	2699836693	7912603456	gA(30)		gP(23)	gN(25)	pA(18)	pP(13)	pN(26)	Total
H1	TGTCTTTTCT	TGCAAATGGG	GGTTGAATCT	TTTTGTCTTT	GCGTTTATTG	12	7	3	7	-	5	34	
H2	....	....G.....	....	....	.T........	1	-	-	-	-	-	1	
H3	....	....	....	....	........A.	1	1	-	-	-	2	4	
H4	....	....	....	......c...	....	1	-	-	-	-	-	1	
H5	....	....	....	....	.T........	8	5	1	4	2	8	28	
H6	C	....	....	....	....	1	-	-	-	-	1	2	
H7	....	....	....	C.........	.A........	1	-	-	-	-	-	1	
H8	....	....	.....G....	....	....	2	-	-	-	-	-	2	
H9	....	......C...	....	....	....	1	-	-	-	-	-	1	
H10	....	....	....	C.........	.T........	2	2	-	1	-	-	5	
H11	....	...G......	....	.......C..	.T........	-	1	-	-	-	-	1	
H12	C.........	...G......	....	....	.T......A.	-	1	-	-	-	-	1	
H13	....	....	....	....	.T......A.	-	1	-	-	-	-	1	
H14	..CA......	...G......	C	....	.T...A....	-	1	-	-	-	-	1	
H15	....	...G......	....	....	.T........	-	1	3	-	1	-	5	
H16	....	...G...TTC	....C.....	.CA.A.....	..C.......	-	1	-	-	-	-	1	
H17	C	....	.....G....	.CA.A.....	..C.......	-	1	-	-	-	-	1	
H18	....	....	......C...	....	.T......G.	-	1	-	-	-	-	1	
H19	....	...G......	....	....	....	-	-	7	-	2	-	9	

Variable nucleotide sites														
H20	....	...G......	....	....	.TC.......	-	-	1	-	-	-	1	
H21	....	...G......	....	C.........	.T........	-	-	1	-	1	-	2	
H22	....	...G......	....	...A......	.T.....G..	-	-	1	-	-	-	1	
H23	....	...G......	.A........	....	....	-	-	1	-	-	-	1	
H24	C	...G......	C.........	....	....	-	-	1	-	-	-	1	
H25	....	...G......	........T.	....	.T........	-	-	1	-	-	-	1	
H26	....	C..G......	....	C.........	....	-	-	1	-	-	-	1	
H27	.C........	...G......	....	....	....	-	-	1	-	-	-	1	
H28	....	...G......	......C...	....	....	-	-	1	-	-	-	1	
H29	....	...G......	....	....	......G...	-	-	1	-	-	-	1	
H30	....	...G......	....	......T...	....	-	-	1	-	-	-	1	
H31	....	....	....	....	.T.......T	-	-	-	1	-	-	1	
H32	....	....	...C......	....	.T........	-	-	-	1	-	-	1	
H33	....	....	......C...	....	.T........	-	-	-	1	-	-	1	
H34	....	....	....	....	.T......G.	-	-	-	1	-	-	1	
H35	....	....	...C......	....	....	-	-	-	1	-	-	1	
H36	..C.......	...G......	....	....	....	-	-	-	1	-	-	1	
H37	.T........	...G......	..C.......	....	....	-	-	-	-	1	-	1	
H38	.....A..T.	....	....	....	....	-	-	-	-	1	-	1	
H39	....A.....	...G......	....	....	.T........	-	-	-	-	1	-	1	
H40	....	.AT.......	....	....	.T........	-	-	-	-	1	-	1	
H41	....	...G......	....	.....G....	.......G..	-	-	-	-	1	-	1	
H42	..CT..A...	...G......	...C......	....	.T.....G..	-	-	-	-	1	-	1	

Variable nucleotide sites													
H43	....	....	....	....	........G.	-	-	-	-	1	-	1	
H44	....	.....G....	....	....	....	-	-	-	-	-	1	1	
H45	....	....	....	....	.T....T.G.	-	-	-	-	-	2	2	
H46	.......A..	....	....	....A....G	....	-	-	-	-	-	1	1	
H47	....	..T.......	....	....	.T........	-	-	-	-	-	1	1	
H48	.......A..	....	....	....A..AC.	A..CC.....	-	-	-	-	-	1	1	
H49	.......A..	....	...C......	....	...CC.....	-	-	-	-	-	1	1	
H50	....	....	....	....	.T......G.	-	-	-	-	-	2	2	
H51	....	....	....	....	.T........	-	-	-	-	-	1	1	
No. polymorphic sites						9	21	15	8	16	15		
No. haplotypes						10	12	15	9	11	0.8769		
Haplotype diversity (h)						0.7793	0.8735	0.9100	0.8235	0.9744	0.8769		
Nucleotide diversity (π)						0.0027	0.0065	0.0042	0.0032	0.0078	0.0053		

Appendix 1. Polymorphic sites, haplotype frequencies and genetic diversity indices, in all three studied populations of both taxa. The nucleotides in each position are given in comparison to haplotype H1. For all haplotypes variable sites are indicated, while identity is given by dots. In brackets it is the sample size. gA: *F. glaber* from Airport; gP: *F. glaber* from Paliomana; gN: *F. glaber* from Naziki; pA: *F. proteus* from Airport; pP: *F. proteus* from Paliomana; pN: *F. proteus* from Naziki.

6. AT A GLANCE

We created a DNA technique using PCR amplification and sequencing of the mitochondrial 16S rDNA gene to differentiate between two species of the Pectinidae family - Flexopecten glaber and Flexopecten proteus - and to conduct a population study of both species. A total of fifty polymorphic nucleotide locations and fifty-one distinct haplotypes were identified. 96.98% of the genetic variance was found within populations. Both taxa exhibited high levels of haplotype and nucleotide diversity, likely attributed to their large effective population sizes and favorable population status. No distinct location unique to a particular species was identified, and the two most common genetic variations are present in multiple taxa, indicating potential interbreeding between them. The genetic distance between F. glaber and F. proteus was D = 0.005, and there was no genetic differentiation detected among taxa when combining samples from various locations. The results indicate that F. glaber and F. proteus are of the same species, consistent with the most recent classification.

REFERENCES

1. Avise, J.C. (1994). *Molecular markers, Natural history and Evolution.* Chapman & Hall, ISBN 9780412037818, New York
2. Barucca, M.; Olmo, E; Schiaparelli, S. & Canapa, A. (2004). Molecular phylogeny of the family Pectinidae (Mollusca: Bivalvia) based on mitochondrial 16S and 12S rRNA genes. *Molecular Phylogenetics and Evolution*, Vol.31, No.1 (April 2004), pp. 89-95
3. Bertorelle, G. & Slatkin, M. (1995). The number of segregating sites in expanding human populations, with implications for estimates of demographic parameters. *Molecular Biology and Evolution*, Vol.12, No.5 (July 1995), pp. 887-892
4. Canapa, A.; Marota, I.; Rollo, F. & Olmo, E. (1996). Phylogenetic analysis of Veneridae (Bivalvia): comparison of molecular and palaeontological data. *Journal of Molecular Evolution*, Vol.43, No.5 (November 1996), pp. 517–522
5. Canapa, A.; Barucca, M.; Marinelli, A. & Olmo, E. (2000a). Molecular data from the 16S rRNA gene for the phylogeny of Pectinidae (Mollusca: Bivalvia). *Journal of Molecular Evolution*, Vol.50, No.1 (January 2000), pp. 93–97
6. Canapa, A.; Barucca, M.; Caputo, V.; Marinelli, A.; Cerioni, P.N.; Olmo, E.; Capriglione, T. & Odierna, G. (2000b). A molecular analysis of the systematics of three Antarctic bivalves. *Italian Journal of Zoology*, Vol.67, S1, pp. 127-132
7. Chan, T.Y. & Chu, K.H. 1996. On the different forms of *Panulirus longipes femoristriga* (von Martens, 1872) (Crustacea: Decapoda: Palinuridae), with description of a new species. *Journal of Natural History*, Vol.30, No.3, pp. 367-387

8. Cho, E.S.; Joung, C.G.; Sohn, S.G.; Kim, C.D. & Han, S.J. (2007). Population genetic structure of the ark shell Scapharca broughtonii Schrenck from Korea, China, and Russia based on COI gene sequences. *Marine Biotechnology*, Vol.9, No.2 (March 2007), pp. 203-216
9. CLEMAM, Taxonomic Database on European Marine Molusca. (http://www.somali.asso.fr/clemam/index.php) (5/2010)
10. Cossignani, T.; Cossignani, V.; Di Nisio, A. & Passamonti, M. (1992). *Atlante delle conchiglie del medio Adriatico*. L' Informatore Piceno, ISBN 8886070004, Ancona
11. Crisp, D.J. (1978). Genetic consequences of different reproductive strategies in marine invertebrates, In: *Marine Organisms: Genetics, Ecology and Evolution*, Battaglia, B. & Beardmore, J.A., (Eds.), 257–273, Plenum Press, ISBN 978-0306400209, New York
12. Dayrat, B. (2005). Towards integrative taxonomy. *Biological Journal of the Linnean Society of London*, Vol.85, No.3, (July 2005), pp. 407–415
13. Delamotte, M. & Vardala–Theodorou, E. (1994). *Shells from the Greek Seas*. The Goulandris National History Museum, ISBN 9608519381, Athens
14. Doneddu, M. & Trainito, E. (2005). *Conchiglie del Mediterraneo*. Il Castello, ISBN 9788865200582, Trezzano sul Naviglio, Italy
15. ERMS. European Register of Marine Species (http://www.marbef.org) (5/2010)
16. Excoffier, L. & Lischer, H.E.L. (2010). Arlequin suite ver 3.5: A new series of programs to perform population genetics analyses under Linux and Windows. *Molecular Ecology Resources*, Vol.10, No.3, (May 2010), pp. 564-567
17. Galinou-Mitsoudi, S. & Sinis, A.I. (2000). Influence of exploitation on to natural bivalve populations in Thessaloniki gulf, *Proceedings of the 9th Panhellenic Ichtyhologists Conference*, pp. 21-24, Mesologi, Greece, January 20-23, 2000
18. Goetze, E. (2003). Cryptic speciation on the high seas; global phylogenetics of the copepod family Eucalanidae. *Proceedings of the Royal Society of London B*, Vol.270, No.1531, (November 2003), pp. 2321-2331
19. Gosling, E. (2003). Genetics in Aquaculture, In: *Bivalve, Molluscs. Biology, Ecology and Culture*, Gosling, E. (Ed.), 333-369, Fishing New Books, Blackwell Science, ISBN 978-0-85238- 234-9, USA
20. Hall, T.A. (1999). BioEdit: a user-friendly biological sequence alignment editor and analysis program for Windows 95/98/NT. *Nucleic Acids Symposium Series*, Vol.41, No.41, pp. 95-98, ISSN 02613166
21. Hare, M.P. & Avise, J.C. (1996). Molecular genetic analysis of a stepped multilocus cline in the American oyster (*Crassostrea virginica*). *Evolution*, Vol.50, No.6, (December 1996), pp. 2305-2315

22. Hebert, P.D.N.; Cywinska, A.; Ball, S.L. & Ward, J.R. (2003). Biological identifications through DNA barcodes. *Proceedings of the Royal Society of London, Series B*, Vol.270, No.1512, (February 2003), pp. 313–321
23. Hebert, P.D.N.; Penton, E.H.; Burns, J.M.; Janzen, D.H. & Hallwachs, W. (2004a). Ten species in one: DNA barcoding reveals cryptic species in the neotropical skipper butterfly *Astraptes fulgerator*. *Proceedings of the National Academy of Sciences USA*, Vol.101, No.41, (October 2004), pp. 14812–14817
24. Hebert, P.D.N.; Stoeckle, M.Y.; Zemlak, T.S. & Francis, C.M. (2004b) Identification of birds through DNA barcodes. *Public Library of Science*, Vol.2, No.10, (September 2004), 1657–1663
25. Hillis, D.M.; Moritz, C. & Mable, B.K. (1996). *Molecular Systematics*. Sinauer Associates, Inc., Sunderland, ISBN 0878932828 9780878932825, MA, USA
26. Katsares, V.; Tsiora, A.; Galinou – Mitsoudi, S. & Imsiridou, A. (2008). Genetic structure of the endangered species *Pinna nobilis* Linnaeus, 1758 (Mollusca: Bivalvia) inferred from mtDNA sequences. *Biologia*, Vol.63, No.3, (June 2008), pp. 412 – 417
27. Kimura, M. (1980). A simple method for estimating evolutionary rate of base substitution through comparative studies of nucleotide sequences. *Journal of Molecular Evolution*, Vol.16, No.2, (June 1980), pp. 111-120
28. Knowlton, N. (1993). Sibling species in the sea. *Annual Review of Ecology, Evolution and Systematics*, Vol.24, (November 1993), pp. 189-216
29. Knowlton, N. (2000). Molecular genetic analyses of species boundaries in the sea. *Hydrobiologia*, Vol.420, No.1, (February 2000), pp. 73-90
30. Kocher, T.D.; Thomas, W.K.; Meyer, A.; Edwards, S.V.; Paabo, S.; Villablanca, F.X. & Wilson, A.C. (1989). Dynamics of mitochondrial DNA evolution in animals: Amplification and sequencing with conserved primers. *Proceedings of the National Academy of Sciences of USA*, Vol.86, No.16, (August 1989), pp. 6196-6200
31. Kong, X.; Yu, Z.; Liu, Y. & Chen, L. (2003). Intraspecific genetic variation in mitochondrial 16S ribosomal gene of Zhikong scallop *Chlamys farreri*. *Journal of Shellfish Research*, Vol.22, No.3, (December 2003), pp. 655-660
32. Koutsoubas, D.; Galinou-Mitsoudi, S.; Katsanevakis, S.; Leontarakis, P.; Metaxatos, A. & Zenetos, A. (2007). Bivalve and gastropod mollusks of commercial interest for human consumption in the Hellenic Seas, In: *State of Hellenic Fisheries*, Papaconstantinou, C.; Zenetos, A.; Vassilopoulou, V. & Tserpes, G., (Eds.), 70-84, HCMR publications, ISBN 9789609805414, Athens, Greece
33. Kumar, S.; Dudley, J.; Nei, M. & Tamura, K. (2008). MEGA: A biologist-centric software for evolutionary analysis of DNA and protein sequences. *Briefings in Bioinformatics*, Vol.9, No.4, (July 2008), pp. 299-306

34. Liu, B.; Dai, J. & Yu, Z. (1998). The investigation of populations by RAPD markers on oyster, Crassostrea talienwhanesis. *Journal of Ocean University of Qingdao*, Vol.28, pp. 82-88

35. Liu, B.; Dong, B.; Xiang, J. & Wang, Z. (2007). The phylogeny of native and exotic scallops cultured in China based on 16S rDNA sequences. *Journal of Ocean University of Qingdao*, Vol.25, pp. 85-90

36. Livi, S.; Cordisco, C.; Damiani, C.; Romanelli, M. & Crosetti, D. (2006). Identification of bivalve species at an early developmental stage through PCR-SSCP and sequence analysis of partial 18S rDNA. *Marine Biology*, Vol.149, No.5, (August 2006), pp. 1149-1161

37. Mahidol, C.; Na-Nakorn, U.; Sukmanomon, S.; Yoosuk, W.; Taniguchi, N. & Nguyen, T.T.T. 2007a. Phylogenetic relationships among nine scallop species (Bivalvia: Pectinidae) inferred from nucleotide sequences of one mitochondrial and three nuclear regions. *Journal of Shellfish Research*, Vol.26, No.1, (April 2007), pp. 25-32

38. Mahidol, C.; Na-Nakorn, U.; Sukmanomon, S.; Taniguchi, N. & Nguyen, T.T.T. 2007b. Mitochondrial DNA diversity of the Asian moon scallop, *Amusium pleuronectes* (Pectinidae) in Thailand. *Marine Biotechnology*, Vol.9, No.3, (May 2007), pp. 352-359

39. Mathews, L.M.; Schubart, C.D.; Neigel, J.E. & Felder, D.L. (2002). Genetic, ecological and behavioural divergence between two sibling snapping shrimp species (Crustacea: Decapoda: *Alpheus*). *Molecular Ecology*, Vol.11, No.8, (August 2002), pp. 1427-1437

40. Palumbi, S.R. (1996). Nucleic acids II: The polymerase chain reaction, In: *Mclecular Systematics*, Hillis, D.M.; Moritz, C. & Mable, B.K. (Eds.), 205-247, Sinauer Associates, Inc., Sunderland, ISBN 0878932828 9780878932825, MA, USA

41. Pisor, D. & Poppe, G. (2008). *Registry of World Record Size Shells*. Conchbooks & Conchology, ISBN 9783939767121, Germany

42. Poppe, G.T. & Goto, Y. (1993). *European seashells, Vol. II*. ConchBooks Hachenheim, ISBN 3925919104, Germany

43. Poutiers, J.M. (1987). Bivalves, In: *Fiches FAO d'identi⇒cation des espèces pour les besoins de la pêche. Méditerranée et Mer Noire. Zone de pêche* 37, Fischer, W.; Schneider, M. & Bauchot, M.L., (Eds.), 369–514, FAO, Rome, Italy

44. Pujolar, J.M.; Marčeta, T.; Saavedra, C.; Bressan, M. & Zane, L. (2010). Inferring the demographic history of the Adriatic *Flexopecten* complex. *Molecular Phylogenetics and Evolution*, Vol.57, No.2, (November 2010), pp. 942–947

45. Repetto, G.; Orlando, F. & Arduino, G. (2005). *Conchiglie del Mediterraneo*. Amici del Museo "Federico Eusebio", Alba, Italy

46. Raines, B.K. & Poppe, G.T. (2006). *A conchological Iconography: the family Pectinidae.* ConchBooks, Hackenheim, Germany

47. Rawson, P.D. & Hilbish, T.J. (1995). Evolutionary relationships among the male and female mitochondrial DNA lineages in the *Mytilus edulis* species complex. *Molecular Biology and Evolution*, Vol.12, No.5, (July 2005), pp. 893-901

48. Rawson, P.D. & Hilbish, T.J. (1998). Asymmetric introgression of mitochondrial DNA among European populations of blue mussels (*Mytilus* spp.). *Evolution*, Vol.52, No.1, (February 2008), pp. 100-108

49. Raymond, M. & Rousset, F. (1995). An exact test for population differentiation. *Evolution*, Vol.49, No.6, (December 1995), pp. 1280-1283

50. Reichow, D. & Smith, M.J. (2001). Microsatellites reveal high levels of gene flow among populations of the California squid *Loligo opalescens*. *Molecular Ecology*, Vol.10, No.5, (May 2001), pp. 1101–1109

51. Rivera, M.A.J.; Kelley, C.D. & Roderick, G.K. (2004). Subtle population genetic structure in the Hawaiian grouper, *Epinephelus quernus* (Serranidae) as revealed by mitochondrial DNA analysis. *Biological Journal of the Linnean Society. Linnean Society of London*, Vol.81, No.3, (March 2004), pp. 449–468

52. Rombouts, A. (1991). *Guidebook to Pecten shells: recent Pectinidae and Propeamussiidae of the world*. Oegstgeest Universal Book Service/Dr. W. Backhuys, ISBN 1863330267, The Netherlands

53. Saavedra, C. & Pena, J.B. (2004). Phylogenetic relationships of European and Australasian king scallops (*Pecten* spp.) based on partial 16S ribosomal RNA gene sequences. *Aquaculture*, Vol.235, No.1-4, (June 2004), pp. 153-166

54. Saavedra, C. & Pena, J.B. (2006). Phylogenetics of American scallops (Bivalvia: Pectinidae) based on partial 16S and 12S ribosomal RNA gene sequences. *Marine Biology*, Vol.150, No.1, (October 2006), pp. 111-119

55. Sarver, S.K.; Silberman, J.D. & Walsh, P.J. (1998). Mitochondrial DNA sequence evidence supporting the recognition of two subspecies or species of the Florida spiny lobster *Panilurus argus*. *Journal of Crustacean Biology*, Vol.18, No.1, (February 1998), pp. 177- 186

56. Seed, R. (1969). The ecology of *Mytilus edulis* L. (Lamelibrabchiata) on exposed rocky shores. 1. Breeding and settlement. *Oecologia*, Vol.3, No.3-4, (September 1969), pp. 277–316 Slatkin, M. & Hudson, R.R. (1991). Pairwise comparisons of mitochondrial DNA sequences in stable and exponential growth populations. *Genetics*, Vol.129, No.2, (October 1991), pp. 555–562

57. Soule, M.E. (1980). Thresholds for survival: maintaining fitness and evolutionary potential, In: *Conservation biology: an evolutionary-ecological perspective*, Soule, M.E. & Wilcox, B.A., (Eds.), 151-170, Sinauer Association, ISBN 978-0878938001, Sunderland, Massachusetts

58. Stepien, C.A.; Hubers, A.N. & Skidmore, J.L. (1999). Diagnostic genetic markers and evolutionary relationships among invasive Dreissenoid and Corbiculoid Bivalves in North America: phylogenetic signal from mitochondrial 16S rDNA. *Molecular Phylogenetics and Evolution*, Vol.13, No.1, (October 1999), pp. 31-49

59. Tajima, F. (1989). Statistical method for testing the neutral mutation hypothesis by DNA polymorphism. *Genetics*, Vol.123, No.3, (November 1989), pp. 585–595

60. Tamura, K.; Dudley, J.; Nei, M. & Kumar, S. (2007). MEGA4: Molecular Evolutionary Genetics Analysis (MEGA) software version 4.0. *Molecular Biology and Evolution*, Vol.24, No.8, (August 2007), pp. 1596-1599

61. Therriault, T.W.; Docker, M.F.; Orlova, M.I.; Heath, D.D. & MacIsaac, H.J. (2004). Molecular resolution of the family Dreissenidae (Mollusca: Bivalvia) with emphasis on Ponto- Caspian species, including first report of Mytilopsis leucophaeata in the Black Sea basin. *Molecular Phylogenetics and Evolution*, Vol.30, No.3, (March 2004), pp. 479-489

62. Thompson, J.D.; Gibson, T.J.; Plewniak, F.; Jeanmougin, F. & Higgins, D.G. (1997). The CLUSTAL X windows interface: Flexible strategies for multiple alignments aided by quality analysis tool. *Nucleic Acids Research*, Vol.25, No.24, (December 2007), pp. 4876-4882

63. Vaught, K.C. (1989). *A classification of the living mollusca*. American Malacologists, ISBN 9780915826223, Melbourne, FL

64. Wilding, C.S.; Beaumont, A.R. & Latchford, J.W. (1999). Are *Pecten maximus* and *Pecten jacobaeus* different species? *Journal of the Marine Biological Association of the United Kingdom*, Vol. 79, No.5, (September 2000), pp. 949-952

65. Yu, X.Y.; Mao, Y.; Wang, M.F.; Zhou, L. & Gui, J.F. (2004). Genetic heterogeneity analysis and RAPD marker detection among four forms of *Atrina pectinata* Linnaeus. *Journal of Shellfish Research*, Vol.23, No.1, (April 2004), pp. 165-171

66. Yuan, T.; He, M. & Huang, L. (2009). Intraspecific genetic variation in mitochondrial 16S rRNA and COI genes in domestic and wild populations of Huaguizhikong scallop *Chlamys nobilis* Reeve. *Aquaculture*, Vol.289, No.1-2, (April 2009), pp. 19-25

67. Zenetos, A. (1996). *Fauna Graeciae. VII. The marine Bivalvia (Mollusca) of Greece*. N.C.M.R., ISBN 960-85952-0-7, Athens

68. Zenetos, A.; Vardala-Theodorou, E. & Alexandrakis, C. (2005). Update of the marine Bivalvia Mollusca checklist in Greek waters. *Journal of the Marine Biological Association of the United Kingdom*, Vol.85, No.4, (June 2005), pp. 993-998

08 GENOMICS AND GENOME SEQUENCING ADVANTAGES FOR FINFISH AQUACULTURE

1. INTRODUCTION: AN OVERVIEW

1.1 Genomics?

The basic dogma of molecular biology asserts that DNA is transcribed into RNA, which is then translated into proteins, the molecules responsible for all biological processes (Crick, 1970). A genome is the complete set of an organism's DNA, containing all its hereditary information. Genomics is the study of whole sets of genes within an organism. Genomics is a field of biotechnology that involves using genetic and molecular biology tools to map genes or sequence the DNA of specific organisms or entire genomes. Genetics focuses on individual genes or a small number of genes, while genomics analyzes all genes and non-coding elements in the DNA of an organism's genome. Although the term genomics was first used in 1987, the field is relatively new and is growing rapidly as new technologies for exploring genomes emerge. Genomics has wide-ranging applications in fields such as medicine, industry, and ecology, impacting global issues including cancer diagnosis, prevention, treatment, alternative energy sources, agriculture, conservation, and sustainable development. Genomics, based on DNA which is universal throughout living organisms, has the potential to connect different fields of study and produce synergistic outcomes through the development of multi-disciplinary methods to biological inquiries.

1.2 Genomics in aquaculture: Phenotypic vs. genotypic selection

Individuals with desirable observable traits are selectively bred to create new strains of plants or animals with qualities like faster development, larger size, enhanced robustness, or superior aesthetic appeal in conventional selective breeding. Farmers have significantly altered natural populations over thousands of years by selectively breeding animals to create various breeds, such as the numerous breeds of domestic dogs and different strains of cattle for dairy or beef purposes. Nevertheless, any breeding strategies focused on physical characteristics have inherent disadvantages. They are depleting resources and time due to their reliance on trial and error, the need for a large number of individuals with significant phenotypic variation, and the necessity of multiple generations of selective breeding. This process is particularly challenging for species with long generation times and high parental investment requirements. Additionally, animals are frequently required to be slaughtered in order to identify or quantify morphological traits like flesh color and tissue quality. Animals that have been challenged with disease (to assess immune function or disease resistance) are therefore not suitable for breeding. Most traits relevant to aquaculture are complex, meaning they are influenced by multiple genes and intricate interactions between various pathways. Additionally, some traits are influenced by environmental factors, making it challenging to select for multiple complex traits solely based on observable characteristics. The time and costs involved in classic phenotypic methods for selective breeding are substantial.

Phenotype-based selection directly chooses a heritable characteristic, while genotype-based selection, or marker-assisted selection (MAS), indirectly picks features based on DNA variability. Genomes of breeding populations are analyzed for numerous markers, or DNA tags, linked to genes that collectively contribute to a favorable phenotype for a complex characteristic. Genetic screening can be done utilizing high-throughput genomics systems that utilize technologies like polymerase chain reaction (PCR) or DNA sequence-based screening.

MAS has been widely applied in enhancing the genetic traits of various cultivated plant varieties like wheat, corn, and soy, as well as in improving breeding stocks in agricultural species such as swine and cattle. With the growing availability of genomics resources for aquatic species, there is a rise in research aimed at enhancing broodstocks for several important aquaculture species globally. Genomics, utilizing genomic technologies, allows for the comparison of organisms at the DNA level, rather than relying on appearance-based methods. This detailed viewpoint examines variations in genes (alleles) and diversity in non-coding sites in the genome, allowing for the dissection of complex features influenced by multiple genes or pathways into their individual components. Differences can be evaluated in great depth at several levels such as individual, family, population, species, and organism. Understanding genetic variables influencing complex features can aid in developing advanced breeding techniques with significant consequences for the aquaculture sector.

2. ATLANTIC SALMON: A MODEL AQUACULTURE SPECIES FOR MAS

The ultimate goal of any genomics program for a species is a fully assembled and annotated entire genome sequence. The benefits include the generation of a large amount of data, including the complete set of genes along with details like gene location and copy number. Additionally, the sequenced genome can serve as a reference for the same species and related species, facilitating re-sequencing and providing valuable information. Despite recent advancements in sequencing technology, acquiring a complete genome sequence is still a challenging, expensive, and time-consuming task. Fish species have diverse and complicated genomes due to their evolutionary antiquity and the large number of extant species, which exceeds 20,000. This complexity poses challenges for sequencing. Only five fish genomes have been reported so far: medaka, tiger pufferfish, green spotted pufferfish, zebrafish, and stickleback. However, others are currently being worked on. The sequences for their genomes and those of other creatures are publicly accessible on the Ensembl database at www.ensembl.org. The fish species were selected as model organisms for genetics study, not for their practicality in aquaculture. The medaka and zebrafish genomes were sequenced for developmental biology research, the stickleback genome for adaptive evolution studies, and the two pufferfish genomes are the smallest among vertebrates. Figure 1 depicts the evolutionary relationships among these species, including important aquaculture species, and highlights that the genome sequences currently accessible do not cover the entire range of teleosts.

Without access to a complete genome sequence, various genomics tools and resources can be created to offer detailed information about a genome. These tools can be utilized individually or in combination to gain valuable insights and can be applied for purposes like Marker-Assisted Selection (MAS) for the target species and related species. In the upcoming sections, we will discuss the techniques, development, and usefulness of genomics in aquaculture using the example of Atlantic salmon (Salmo salar). Atlantic salmon is an ideal fish species for genomics research due to its significant role in aquaculture, with around 1.5 million tonnes harvested globally in 2009 according to the Food and Agriculture Organization (FAO) Fishery Statistic, 2009. It holds considerable environmental, economic, and social value similar to other salmonids. Extensive genomics resources for Atlantic salmon have been created using conventional methods and approaches that can be applied to other genomes. A compelling case has been shown for acquiring the complete genome sequence of Atlantic salmon, a project that is now underway (Davidson et al., 2010). Atlantic salmon can be used as a reference genome for salmonid species since no other salmonid species have been sequenced yet. This allows for detailed comparisons with other salmonids like Pacific salmon, rainbow trout, and Arctic charr.

Atlantic salmon exemplifies the issues encountered in fish genetics. The ancestor of salmonids experienced a complete duplication of its genome between 20 and 120 million years ago (Allendorf & Thorgaard, 1984; Ohno, 1970). Atlantic salmon have four copies of each gene in their genome, which are evolving into genes with novel roles or non-coding DNA. The genome duplication led to an increase in both the size and repetitiveness of the genome. These traits, along with the absence of a closely linked reference sequence, make the sequencing and assembly of the Atlantic salmon genome highly difficult.

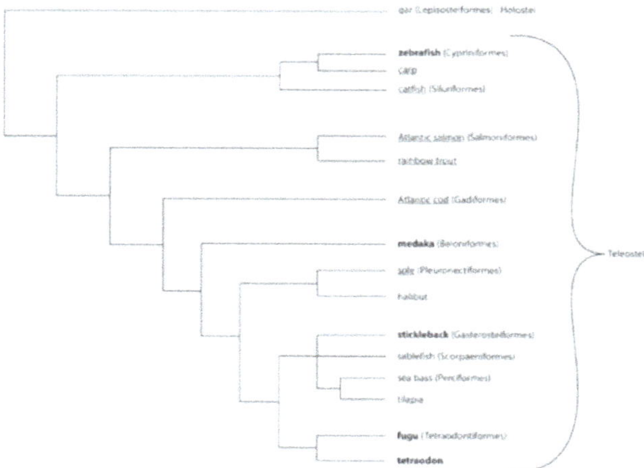

Fig. 1. Schematic representation of the phylogenetic relationships among fish species. Species in bold have publically available full genome sequences, while those that are underlined are currently being sequenced. Note that the full spectrum of teleosts is not represented by the genome sequences that are currently available. Species are listed by their common names with orders in parentheses. Gar is used as an outgroup.

3. GENOMICS RESOURCES

3.1 ESTs

An expressed sequence tag (EST) is a sequenced section of a transcribed, or expressed gene. ESTs are developed by isolating the mRNA, i.e., the transcribed DNA, from a tissue, which is then reverse-transcribed into cDNA (complementary DNA). Thus, an EST library is the full repertoire–essentially an inventory–of expressed genes within a genome in a given tissue at a given time. More than 499,000 ESTs have been identified for Atlantic salmon, the sequences of which are available publicly in an online database (Rise et al., 2004; Koop et al., 2008) as well as within the NCBI database.

Perhaps the most common use of ESTs is to study gene expression (described below). Other uses of ESTs include the identification of molecular markers, or DNA tags, which are described in detail in the subsequent section. EST-based markers have unique applications because they occur in expressed genes, as opposed to non-coding DNA. Thus, they can provide insight into allelic differences between individuals or populations, thereby facilitating an understanding of the relationship between phenotypes and particular genes.

3.2 Genetic markers

Genetic markers are brief DNA sequences linked to a particular region, or locus, on a chromosome. For a genetic marker to be valuable, it needs to exhibit significant variability (polymorphism) among people or populations. Markers are essential instruments in molecular biology with several practical applications. They are utilized for examining ancestry, defining discrete populations, forensic work, paternity testing, and tissue typing for transplants due to the inheritance of genotypes in markers, resulting in a unique combination of marker genotypes known as a "DNA fingerprint."

3.2.1 Microsatellite markers

Satellite markers are a group of DNA markers that are characterized by containing tandemly repeated sequences of nucleotides (e.g., ...ACACACAC....). Depending on their size, which is a function of the number of repeated bases in a repeat unit and the number of tandemly arrayed units, satellite markers are classed as microsatellites (short repeat unit) or minisatellites (long repeat unit). Microsatellite markers are heavily utilized in genomics. They act as DNA anchors or tags within a genome because: 1) like all molecular markers, their genotypes are heritable, 2) the repeat units make them more vulnerable to mutation than other types of DNA sequence (e.g., coding sequence), which means they are variable at all levels, from individual to species, and generally have several alleles, 3) their locations (loci) within the chromosome can be determined relative to one another, and 4) their small size (vs. minisatellites, which can be very long) facilitates ease of detection using PCR. The development of a suite of microsatellite markers is often one of the first steps in a genomics program for a species of interest.

The first microsatellites characterized for Atlantic salmon came from genomic libraries (McConnell et al. 1995; O'Reilly et al., 1998; Sanchez et al., 1996; Slettan et al., 1995). These were classified as "anonymous" or "Type II" markers as they were not associated with a particular gene. The mining of large EST databases allowed the identification of Type I microsatellites, which could be related directly to a gene (Anti Vasemagi et al., 2005; Ng et al., 2005,). More recently, the sequencing of the ends of Bacterial Artificial Chromosome (BAC) clones has provided a rich source of microsatellites (Danzmann et al., 2008; Phillips et al., 2009). Microsatellites have been used extensively to examine the population structure of Atlantic salmon (e.g., King et al., 2001) as well as for assigning parentage and relatedness of individuals in a mixed family stock (Norris et al., 2000), which precludes the necessity of rearing families in individual tanks or using physical tags. Many of the microsatellite markers that have been developed for one species of salmonid can be used in others (e.g., McGowan et al., 2004; Presa & Guyomard, 1996; Scribner et al., 1996), which enables genomic comparisons among salmonids such as Atlantic salmon, brown trout (*Salmo trutta*), rainbow trout and Arctic charr (Danzmann et al., 2005, 2008; Woram et al., 2004). Most aquaculture species have a set of microsatellites now, and the list is too extensive to reference here.

3.2.2 Single Nucleotide Polymorphisms (SNPs)

A single nucleotide polymorphism (SNP) (pronounced *snip*) is a sequence change in a single nucleotide – A, T, C or G - that occurs with sufficient frequency in a population to serve as a genetic marker. SNPs can be within genes or within non-coding regions of the genome. Because they are usually bi-allelic (i.e., exhibit only two alleles, such as an A or a T at a given locus), they are easily identified and studied. The development of novel sequencing technologies has made it possible to obtain thousands SNP markers for genotyping (see Molecular Ecology Resources (2011) 11: Supplement 1, for several recent articles on this subject). Moreover, new genotyping technologies enable the simultaneous analysis of a great number of these markers, which allows the construction of high density genetic maps (Goddard & Hayes, 2009). These techniques are now being widely used in species of agricultural importance and have led to the development of numerous dense SNP arrays, which are tools that are used to screen entire genomes for the genotypes of known SNPs (Matukumalli et al., 2009; Muir et al., 2008; Ramos et al., 2009).

There have been several attempts to identify SNPs in individual genes in Atlantic salmon (Ryynanen & Primmer, 2006) and other salmonids (Smith et al., 2005), but these were small- scale efforts. It was not until large EST databases became available that an intensive search was undertaken for SNPs that could be tied to actual genes (Hayes et al., 2007). Re- sequencing BAC-end sequences (described below) was another approach taken to identify SNPs for Atlantic salmon (Lorenz et al., 2010), but this process was highly labor intensive relative to the information gained. A next generation sequencing strategy (Van Tassell et al., 2008) has also been used to identify SNPs in Atlantic salmon; however, the repetitive nature of the genome made it difficult to distinguish between putative SNPs arising from duplicated regions and those from a unique locus. Recently a custom SNP-array (iSelect by Illumina) containing approximately 6,500 SNP markers was developed for Atlantic salmon. This has been used to search for generic genetic differences between farmed and wild Atlantic salmon (Karlsson et al., 2011) and to assess the level of heterozygosity in Tasmanian Atlantic salmon (Dominik et al., 2010). In addition, the SNP-array was used to construct a relatively dense SNP-based linkage map (described below) (Lien et al., 2011). For other aquaculture species the availability of SNPs is still somewhat limited, with only a large number of SNPs being reported for catfish (*Ictalurus* sp.), (Wang et al., 2010) rainbow trout (Sanchez et al., 2009) and Atlantic cod (*Gadus morhua*) (Hubert et al., 2010; Moen et al., 2008b).

3.3 Linkage maps

Genetic markers are commonly used to create a genetic map, sometimes called a linkage map, of an organism's genome. Markers that are physically close to each other on a chromosome, or 'linked', are likely to be inherited together. By analyzing the frequency of inherited alleles at various genetic loci, one can determine the recombination distances between markers, which indicate the relative placements of markers in a genome. Adequate markers on a linkage map should result in the number of linkage groups matching the number of chromosomes in a genome. Linkage maps are highly valuable tools in genomics and serve as the basis for QTL mapping.

Initial linkage maps for Atlantic salmon were created with a limited number of microsatellite markers (50–60) (Gilbey et al., 2004; Moen et al., 2004). Through BAC-end sequencing, the number of microsatellite markers was increased to around 2,000. This allowed for the creation of a comprehensive genetic map encompassing the 29 pairs of chromosomes in the European strain of Atlantic salmon (Danzmann et al., 2008). More maps were created utilizing Single Nucleotide Polymorphisms (SNPs) obtained from Expressed Sequence Tags (ESTs) and Bacterial Artificial Chromosome (BAC) end sequences. These maps were then integrated into a comprehensive SNP map containing around 5,600 SNPs. Linkage maps have been created for rainbow trout, brown trout, coho salmon, and Arctic charr in previous studies. Linkage maps have been developed for various aquaculture species such as catfish, Japanese flounder, European sea bass, tilapia, common carp, olive flounder, and Atlantic cod.

3.4 Bacterial Artificial Chromosome (BAC) libraries

3.4.1 BACs and BAC libraries

A bacterial artificial chromosome (BAC) clone is comprised of a special plasmid and a large segment of genomic DNA from the species of interest. Specifically, the entire genome of a selected individual is cut into pieces of approximately 150,000–300,000 base pairs, each of which is inserted into the plasmid, thereby generating thousands of closed, circular "hybrid DNA constructs". Thus, a BAC library refers to all of the BACs, which in combination comprise the entire genome, for a representative individual of a species. Each BAC is then inserted ("transformed") into live *E. coli* cells; thus, the foreign (non-bacterial) DNA can be replicated as if it were a bacterial chromosome using the cellular machinery found within the bacterial cell. Ideally there should be redundancy of the segments of the genome represented in a BAC library such that overlaps of individual clones occur. The ability to identify which BACs share a segment of the genome enables a network called a physical map to be constructed. Therefore, the physical map represents the genome, and has traditionally been used to assist with sequence assembly following whole genome sequencing. BAC libraries also facilitate the identification of target sequences or regions, as the library can be "screened" for genetic markers or other sequence-specific probes. An Atlantic salmon BAC library comprising approximately 300,000 clones and providing an 18- fold coverage of the genome is available (Thorsen et al., 2005), and it was used to generate a physical map (Ng et al., 2005). BAC libraries have been constructed for several aquaculture species including rainbow trout (Palti et al., 2009), catfish (Xu et al., 2007), common carp (Li et al., 2011b) and barramundi (*Lates calcarifer*) (Xia et al., 2010).

3.4.2 BAC-end sequencing

Once a BAC library has been generated, typically the next step is to select a subset of BACs for end sequencing. Because the sequence of the plasmid vector within the BAC is known, the bacterial sequences at the junctions of the genomic DNA and vector DNA can act as sequence primers. Thereby, the first 500–1,000 nucleotides (based on Sanger sequencing) of the genomic DNA inserts–i.e., the BAC-ends–can be determined. End-sequencing a large subset of BACs can provide extensive insight into the full genome sequence. For example, the 207,869 BAC-end sequences for the Atlantic salmon BAC library cover approximately 3.5% of the whole genome sequence. This glance into a genome is very powerful, as it can provide information about the complexity of the genome (i.e., repeat content), and the BAC- end sequences can be a source of molecular markers (as described above for Atlantic salmon (Danzmann et al., 2008; Lorenz et al., 2010)). Finally, the BAC-end sequences can be used for comparative synteny analyses by aligning them against other, fully sequenced genomes, which can provide insight into the gene content of the BACs, thus providing a partial, putative annotation of segments of the genome (Sarropoulou et al., 2007). This is particularly powerful if the two genomes being compared are closely related, as one can act as a reference genome (e.g., common carp and zebrafish, which are both Cypriniformes) (Xu et al., 2011). Even when there is no obvious reference genome there is value in this exercise as it often suggests syntenic regions among phylogenetically distantly related species, and can lead to candidate genes being suggested for specific traits (see below for an example with ISA resistance in the MAS section). The Atlantic salmon BAC end sequences have been made available in a publicly accessible searchable database (www.asalbase.org) along with comparative genomic information for four of the five published fish genomes. Similar work is ongoing with aquaculture species such as catfish (Liu et al., 2009), rainbow trout (Genet et al., 2011) and sea bream (Kuhl et al., 2011).

3.4.3 Integration of physical and linkage maps and chromosomes

Once both a physical map (i.e., a BAC library is assembled into overlapping, contiguous BACs), and a linkage map (generated using the relationships between variable markers) exist for a species, the next step is to integrate the two maps. This procedure is relatively straight forward if the linkage map contains many markers from BAC-ends, as illustrated by the integration of the Atlantic salmon physical and linkage maps (Danzmann et al., 2008). However, this is seldom the case and considerable effort is normally required to carry out this process (e.g., see preliminary data for rainbow trout) (Palti et al., 2011). The integration of the linkage and physical maps then allows BACs containing particular markers to be selected for fluorescent *in situ* hybridization on chromosomes so that the each linkage group can be assigned to a particular chromosome. This information is now available for several salmonid species including Atlantic salmon (Phillips et al., 2009) and rainbow trout (Phillips et al., 2006).

3.5 Quantitative trait loci (QTL)

Generally, the characteristics that are desirable for selection are complex, or 'quantitative', which means that they are not simply governed by a single gene, but rather are controlled by a suite of genes and regulatory factors that interact to produce a phenotypic characteristic. The resulting phenotype usually falls somewhere on a spectrum. This is not always the case; traits such as genetic disorders can be determined by a single gene (e.g., cystic fibrosis and Duchenne muscular dystrophy), but most phenotypic traits associated with health, growth, meat quality and the capacity for utilizing new and ecologically sustainable food items are complex traits that are strongly influenced by the environment, and thus show a range of phenotypes.

A deep biological understanding of phenotypic patterns, ranging from intracellular molecular signatures to whole individual physiological, morphological and behavioral characteristics of genes of commercial importance is synonymous with an understanding of the functioning of the underlying gene regulatory networks and how these are molded by the environment. This calls for the identification of genes and characterization of their functional repertoires in various systemic settings. Specifically, the first step in understanding the mechanisms that govern complex traits is to determine the heritability of the trait – i.e., the extent to which it is determined genetically and by the environment. Heritability estimates vary tremendously depending on the population, the experimental design and the researchers. Garcia de Leaniz et al. provide the best available review of heritability estimates for Atlantic salmon (Garcia de Leaniz et al., 2007). What is clear is that the genes do play a role in the variation detected in the majority of economically important production traits.

The two most common approaches to identify genes associated with complex traits are to examine candidate genes, based on previous knowledge of a biochemical process, or by mapping quantitative trait loci (QTL). We will not discuss the candidate gene approach here, but rather concentrate on QTL analysis. QTL are regions of the genome that contain, or are linked to, genes that contribute to a particular trait (Lynch & Walsh, 1998). Note that the term "QTL" does not refer to a particular gene, but rather its putative location in the genome. This association, or linkage, is determined using the genetic markers within a linkage map, and how alleles at these markers co-segregate with the trait in question. Thus, finding functional markers to assist in MAS is largely dependent on the variability and polymorphic level of the species. In most agricultural examples, there are highly inbreed strains that make MAS usable. When a species is highly variable, such as the case with fish and most marine species, the QTL markers have to be very close to the causal variant to be usable in MAS breeding studies. The more variable the species, the closer the marker has to be to be used in selective breeding. Given sufficiently high linkage, the markers that define QTL can feed directly into marker assisted selection (MAS). The number, positions and the magnitude of the QTL affecting a trait are determined by statistical associations between marker alleles or genotypes and particular trait phenotypes. QTL analysis, therefore, relies heavily on robust, dense linkage maps as well as large families that are accurately pedigreed and carefully described with respect to phenotype. Indeed, a limiting factor for effective QTL analysis is the difficulty in characterizing "phenomes", the full set of phenotypes of an individual (Houle et al., 2010).

Growth is one of the most economically important traits for aquaculture species. It is no surprise therefore, that several QTL have been identified for body weight and condition factor in Atlantic salmon (Baranski et al., 2010; Boulding et al., 2008; Moghadam et al., 2007; Reid et al., 2005) rainbow trout (Haidle et al., 2008; Martyniuk et al., 2003; O'Malley et al., 2003; Wringe et al., 2010), coho salmon (McClelland & Naish, 2010; O'Malley et al., 2010) and Arctic charr (Moghadam et al., 2007; Kuttner et al., 2011). QTL for several other traits in salmonids have been mapped, including upper temperature tolerance (Jackson et al., 1998; Perry et al., 2001; Somorjai et al., 2003), spawning time (O'Malley et al., 2003), developmental rate (Nichols et al., 2008), and resistance to pathogens (Gheyas et al., 2010; Gilbey et al., 2006; Houston et al. 2008, 2009, 2010; Jones et al., 2002; Moen et al. 2007, 2009a). Given that many of the microsatellite markers derived from one salmonid species amplify the DNA from other salmonid species, it has been relatively straightforward to carry out comparative analyses of the rainbow trout, Arctic charr and Atlantic salmon genomes (Danzmann et al., 2005).

Most of the QTL studies described above were based on a limited number of microsatellite markers; thus, many regions of the genome were sparsely represented. Therefore, it was not possible to obtain precise and complete information about the number and locations of most of these QTL, and to identify specific genes and the underlying genetic variation, or to obtain markers with sufficiently robust associations with the QTL to warrant their integration into MAS programs. Cost has been the main factor limiting the number of microsatellites that could reasonably be used. However, with the advent of SNP arrays, this problem has largely been overcome for agriculture animals such as cattle, pigs and chicken (Groenen et al., 2009; Matukumalli et al., 2009; Ramos et al., 2009) and great progress is being made for some aquaculture species such as Atlantic salmon (Kent M et al., in preparation).

3.6 Expression profiling (microarrays and qPCR)

The concept of gene expression was introduced above in the discussion on expressed sequence tags, or ESTs and EST libraries, which are assemblages of all small sequences of all the expressed genes in a tissue at a particular time point. The term "expression profiling" refers to the examination of the genes expressed in an individual, or group of individuals, under specific conditions. This is based on the fact that gene expression varies with genotype (i.e., the suite of genes and their alleles that are present within an individual's genome) and environment, which cues the expression of particular genes as well as the extent of expression. Expression profiling studies generally compare groups (treatment groups) of individuals that differ in some way, such as genetic background, or environmental surroundings. Thereby, one can examine which genes are relatively over- and under-expressed across treatment groups using statistical analyses, thus gaining insight into the genes that govern particular traits that differ between the subjects.

There are two approaches to studying gene expression that are now standard: real-time PCR (also called quantitative real time PCR or qPCR, or sometimes abbreviated qrtPCR), and microarray analysis. The former is used to examine the expression of a single gene, or a few genes at a time. The technique follows the general principles of regular PCR, in which DNA (in this case, having been reverse-transcribed from RNA, which is present in amounts that are directly proportional to the extent of gene expression) is amplified from a template strand, with the key difference that the amount of product is measured in real time. This is accomplished in one of two ways: 1) the DNA components in the reaction mixture are tagged with dyes that fluoresce when integrated into the DNA molecule, and the level of fluorescence (i.e., corresponding to the amount of product) is measured after each PCR cycle, or 2) sequence-specific probes are similarly tagged, and fluoresce when hybridized to the complementary DNA target, thus providing a direct measurement of the amount of target present after each PCR cycle. The benefits of qPCR are that the process is gene-specific and the results are relatively accurate and reproducible. The cons are that only a small number of genes within a genome can be examined at once, and that sequence-specific PCR primers must be designed, which is dependent on knowing the sequence of the gene *a priori*. This is further complicated when examining gene families or duplicated genes, as it is often not feasible to design primers specific for each gene, a problem of particular relevance when there is no genome sequence for the study organism.

Whereas qPCR targets single genes at a time, DNA microarrays are specifically designed to examine the gene expression at thousands of genes at once. A microarray (also referred to as a "chip") is comprised of thousands of single-stranded DNA spots, or "features" attached in known locations to a surface, such as a glass slide, with each spot corresponding to a gene. To analyze gene expression using a microarray, cDNA (again, reverse-transcribed from RNA) that has been fluorescently tagged is washed over the slide, and complementary strands will hybridize to corresponding spots on the microarray. Thus, when the level of fluorescence is measured, the intensity of each spot will correspond to extent of the expression of that particular gene. Statistical tests are then used to determine the relative expression of each gene on the microarray for the individual/tissue/environmental condition being tested. This method is extremely powerful given the number of genes that can be assessed at once. For example, several microarrays have been designed using the Atlantic salmon EST library, including an initial 3.7K (i.e., 3700 spots) array (Rise et al., 2004), followed by 16K (von Schalburg et al., 2005) and a 32K cDNA (Koop et al., 2008) arrays, as well as two recent 44K oligonucleotide arrays. These resources, which are collectively referred to as the GRASP (Genomic Research on All Salmon Project) arrays, have been used extensively for gene expression studies throughout the salmonid genomics world. For example, recent studies have used the GRASP arrays to look for genetic markers of immune responses in Atlantic salmon (LeBlanc et al., 2010), as well as of thermal stress in rainbow trout (Lewis et al., 2010) and Arctic charr (Quinn et al., 2011a, 2011b), and to predict spawning failure in wild sockeye populations (Miller et al., 2011).

Despite their utility, however, microarrays (particularly those based on cDNA clones) have inherent drawbacks, such that they are resource intensive and highly susceptible to technical variations as well as statistical pitfalls (Martin et al., 2007). The latter may result in Type I errors (false positive results), a particular concern when using a large microarray, as the false-positive rate increases as the number of spots increases. Thus, often microarray data are confirmed, or validated, by selecting a few genes to test using qPCR. Another challenge is the potential for cross-hybridization between similar transcripts, which is a problem of particular concern for duplicated genomes, or for large gene families, such as that of the hemoglobin genes, which contains multiple members of high sequence similarity (Quinn et al., 2010). This can mute out evidence of differential expression between individual transcripts, and instead, produce the general trends for all highly similar genes on the microarray. In addition, this phenomenon may cause information to be lost if the elevated expression of a small number of family members is spread among all similar genes on the microarray such that no spots meet the statistical filters assigned. The use of transcript- specific oligonucleotide arrays may overcome some of these challenges. For these reasons, profiling using microarrays is sometimes described as an exploratory tool, for which the data must be interpreted with a caution, or which must be followed up or correlated with other types of analysis (Lewis et al., 2010; Quinn et al., 2011a).

3.7 Bioinformatics and genomics databases

Genomics technologies have advanced, leading to the production of large quantities of biological data. The demand for computer systems and procedures to handle large amounts of data has driven the creation of bioinformatics, which focuses on data analysis. Bioinformatics is the utilization of computer science and information technology in the field of biology to enhance the comprehension of biological processes. Bioinformatics and genomics are closely linked. The challenges in acquiring and analyzing genomics results often arise in the bioinformatics processes due to the large amount of data generated by high-throughput genomics technologies beyond the capacity of existing computing tools. Professionals with expertise in both biology and computer science are in high demand. The primary objectives of bioinformatics in genomics are to create computational tools for managing data (such as databases and interactive online tools), manipulating data (like map assembly, sequence/genome assembly, and protein structure alignment), and mining data (including pattern recognition, genome annotation, sequence comparisons, predictions of gene expression, and genome-wide association studies). Since the essential objectives of genomics study are similar among all genomes, conventional computational techniques can typically be used to several genomes with small adjustments to suit the specific characteristics of the species being studied.

Genomes initiatives typically adhere to the principle of "open data," making most genomes data available in publicly accessible databases online. Most peer-reviewed academic publications require that sequencing and gene expression data be accessible online before considering any results for publication. Open access data not only prevents repetitive effort and facilitates study groups to benefit from the work of others but also enables for cross-species comparisons, especially within the Salmonidae family. Online, public databases like NCBI, Ensembl, and GEO offer a chance to predict genes in closely related species with limited genomic resources, such as Arctic charr. Moreover, a species-specific database will be created for most species involved in extensive genomics programs. The Atlantic salmon database, found at www.asalbase.org, includes linkage and physical maps, BAC-end sequences, and marker information. The tools provided allow for comparative synteny comparisons between Atlantic salmon and other sequenced species. Asalbase's next task is to include the sequence scaffolds into the entire genome sequence. Over the past decade, many research labs have transitioned from using traditional lab equipment to utilizing large computer networks and powerful servers. This shift has been made possible by the abundance of genomics data accessible online, enabling experiments like comparative synteny analyses, evolutionary studies, and gene expression meta-analyses to be carried out solely using computers, a process known as "in silico". This enhances research efficiency, while also decreasing costs and the need for live specimens or research animals.

4. MAS IN AQUACULTURE

4.1 Examples of QTL-based markers for aquaculture

Over the last decade breeders have begun to include molecular genetics in their strategies for genetically improving salmonids for aquaculture production efficiency. Markers that exhibit strong associations with QTL that play a substantial role in a desirable trait can be integrated into MAS programs. That is, rather than using trial and error to cross individuals that exhibit phenotypic traits of interest, breeders can purposely select individuals that exhibit the marker associated with the desirable QTL. This fine-scale, targeted approach to artificial selection has the potential to be substantially more precise, efficient and effective than traditional, phenotype based approaches.

One example of an economically important trait for which there is a QTL is Infectious Salmon Anemia (ISA) resistance in Atlantic salmon. Previous studies determined that there is a genetic component to ISA resistance in Atlantic salmon (Gjøen & Bentsen 1997, Ødegård et al., 2007) and a QTL for this trait was mapped (Moen et al., 2004). A more recent study confirmed this QTL and showed that it explains 6% of the phenotypic variation and 32-47% of the additive genetic variation (Moen et al., 2007). A comparative genomics analysis of the most likely QTL region suggested several candidate genes for ISA resistance (Li et al., 2011a). These genes remain to be validated based on the association of polymorphisms within them and the trait. However, it is anticipated that a combination of fine mapping and genomic sequence information will identify the causative allele that confers resistance and that this will be incorporated into a MAS breeding program.

Another excellent example of the potential of MAS involves resistance to a pathogen that causes Infectious Pancreatic Necrosis (IPN) in Atlantic salmon. A major QTL for IPN resistance in Atlantic salmon has been identified and confirmed (Gheyas et al., 2010; Houston et al., 2008; Moen et al., 2009a). A four-marker haplotype has been identified that strongly segregates with IPN resistance in Norwegian Atlantic salmon broodstock, providing a solid framework for linkage-based MAS in this population (Moen et al., 2009a).

These results support the prediction that selective breeding for resistance ISA and IPN (caused by viruses) as well as furunculosis (caused by a bacterium) should be possible given the heritabilities for these traits (Kjoglum et al., 2008). However, the speed at which selective breeding for these traits can be performed efficiently and effectively will depend on the ability to identify the underlying genetic basis for the resistance, and this requires a reference sequence for Atlantic salmon. Other traits of interest for QTL identification for MAS in aquaculture include tissue quality, the ability to thrive on commercial diets (i.e., increased vegetable oil content, pathway analysis for vaccine development, and ability/suitability for high-density rearing conditions. Overall, integrating MAS into broodstock development programs can provide an effective and efficient, science-based means of minimizing animal stress while maximizing growth, health and sustainability

4.2 Challenges facing MAS for aquaculture

Currently, given the lack of fully sequenced genomes for aquaculture species, genomics- based selection, or MAS is fully dependent on a having a suite of well-developed genomics tools for the genome of the species of interest. In addition to the cost associated with the development of these tools, the complex nature of fish genomes combined with the large number of diverse fish species of interest for aquaculture complicates this goal. However, the availability of fully sequenced reference genomes for a few key species groups would go a long way to fill in the major under-represented gaps in the fish phylogeny. As illustrated in Figure 1, there is no reference genome from a closely related species for either the Salmoniformes or the Gadiformes. However, work on the genomes of Atlantic salmon (Davidson et al., 2010) and Atlantic cod (Johansen et al., 2009) is in progress, and the results are eagerly anticipated by the aquaculture community that breeds these and related species.

5. ROLE OF WHOLE GENOME SEQUENCING (WGS) IN GENOMICS

5.1 Benefits of whole genome sequences

The availability of a complete genome sequence for Atlantic salmon will have a major impact on all sectors of the international salmonid community. For the aquaculture industry it will provide a complete suite of genetic markers for the identification of the genes and alleles responsible for production traits. Companies will be able to develop tailored broodstock using nucleotide or allele assisted selection rather than the more general marker assisted selection. In conjunction with traditional breeding practices, this approach promises rapid gains that will make companies who embrace this technology more competitive.

5.2 Techniques and strategies for whole genome sequencing

Traditional method for sequencing entire genomes involves dividing the genome into smaller segments and sequencing each segment using Sanger sequencing as discussed in Hutchison's 2007 assessment. The traditional method is considered the best for sequence and assembly quality, but its high cost, manpower requirements, and slow pace have led to a need for alternative DNA sequencing methods. Recently, multiple innovative high-throughput sequencing systems, often known as 'next generation' sequencing methods, have become available in the market (Metzker, 2010). Most of these technologies aim to sequence a complete human genome for less than $1,000. However, their ability to sequence whole genomes de novo, meaning for the first time without a reference sequence for comparison, is still uncertain. The complexity of fish genomes, characterized by their diversity, high repetition, and overall complexity, poses a challenge for present technologies because to their restricted sequence read lengths and issues with certain types of repetitive sequences. Next-generation sequencing platforms can generate large amounts of sequence data quickly and at a lower cost compared to traditional Sanger sequencing. However, the challenge lies in creating computational tools to process the data and assemble short sequence reads into continuous DNA sequences when no reference genome is present (Quinn et al., 2008).

Next generation sequencing technologies have numerous implications for genomics, in addition to their potential for whole genome sequencing. One prevalent application of next-generation sequencing in human genomics is re-sequencing, which involves sequencing the genomes of individuals with a complete reference sequence. This method could be used in aquaculture to identify allelic variations among individuals of the same species that exhibit variances in specific traits such as disease resistance or growth. Next generation sequencing is utilized for transcriptome sequencing, which involves sequencing all transcribed RNA in a species, also known as transcriptomics. This is a type of expression analysis where the number of gene transcripts can be compared to each other, giving an indication of gene expression levels at a specific moment. This strategy is becoming increasingly widespread as the cost of next-generation sequencing tools falls. Next generation sequencing is a useful method for identifying molecular markers such as microsatellites, SNPs, and RAD markers, which are markers found within RAD tags or the sequences surrounding specific DNA sites targeted by restriction enzymes. RADseq is the process of using next-generation sequencing technology to sequence RAD tags for identifying RAD markers. Its popularity is rapidly growing due to its ability to detect individual-specific markers and create an indexed marker library for an organism. As next-generation technologies improve and become more accessible, we should expect significant improvements in genomics methodologies, focusing on high-throughput technologies and bioinformatics analyses. The progress will offer extensive possibilities for incorporating genomics into aquaculture breeding strategies.

6. AT A GLANCE

Genomics has emerged as a potent discipline with wide-ranging consequences across health, pharmacology, industry, and ecology. Genomics applied to aquaculture offers significant potential to enhance output, efficiency, animal welfare, sustainability, reduce inbreeding, and increase sector profitability. Current genomics tools have significantly enhanced traits like growth and disease resistance in fish, particularly in aquaculture species. However, their effectiveness is hindered by the absence of complete genome sequences for some crucial species, which could serve as reference sequences for other aquaculture species. As the genetic sequences of Atlantic salmon, rainbow trout, and Atlantic cod become more accessible in the coming years, advancements in technology and computational tools will likely enhance the potential for genomics to drive significant progress in aquaculture.

REFERENCES

1. Allendorf, F.W. & Thorgaard, G.H. (1984). Tetraploidy and the Evolution of Salmonid Fishes. in Evolutionary Genetics of Fishes, Plenum Press, pp. 55, ISBN 978- 0306415203, New York, USA.

2. Anti Vasemagi, Jan Nilsson, Craig R. Primmer (2005). Seventy-five EST-linked Atlantic Salmon (*Salmo salar* L.) Microsatellite Markers and their Cross-amplification in Five Salmonid Species. Molecular Ecology Notes, Vol. 5, No. 2, pp. 282-288, ISSN 1471- 8286.

3. Baranski, M., Moen, T. & Vage, D.I. (2010). Mapping of Quantitative Trait Loci for Flesh Colour and Growth Traits in Atlantic Salmon (*Salmo salar*). Genetics, Selection, Evolution, Vol. 42, No. 1, pp. 17. ISSN 1297-9686.

4. Boulding, E.G., Culling, M., Glebe, B., Berg, P.R., Lien, S. & Moen, T. (2008). Conservation Genomics of Atlantic Salmon: SNPs Associated with QTLs for Adaptive Traits in Parr from Four Trans-Atlantic Backcrosses. Heredity, Vol. 101, No. 4, pp. 381-391, ISSN 0018-067X.

5. Chan, E.Y. (2009). Next-Generation Sequencing Methods: Impact of Sequencing Accuracy on SNP Discovery. In Single Nucleotide Polymorphisms, Humana Press, pp. 95-111, ISBN 978-1-60327-411-1.
6. Chistiakov, D.A., Hellemans, B., Haley, C.S., Law, A.S., Tsigenopoulos, C.S., Kotoulas, G., Bertotto, D., Libertini, A. & Volckaert, F.A. (2005). A Microsatellite Linkage Map of the European Sea Bass *Dicentrarchus labrax L.* Genetics, Vol. 170, No. 4, pp. 1821- 1826, ISSN 0016-6731.
7. Coimbra, M.R.M., Kobayashi, K., Koretsugu, S., Hasegawa, O., Ohara, E., Ozaki, A., Sakamoto, T., Naruse, K. & Okamoto, N. (2003). A genetic Linkage Map of the Japanese Flounder, *Paralichthys olivaceus.* Aquaculture, Vol. 220, No. 1-4, pp. 203, ISSN 0044-8486.
8. Crick, F. (1970). Central Dogma of Molecular Biology. Nature, Vol. 227, No. 5258, pp. 561- 563, ISSN 0028-0836.
9. Danzmann, R.G., Cairney, M., Davidson, W.S., Ferguson, M.M., Gharbi, K., Guyomard, R., Holm, L.E., Leder, E., Okamoto, N., Ozaki, A., Rexroad, C.E.,3rd, Sakamoto, T., Taggart, J.B. & Woram, R.A.(2005). A Comparative Analysis of the Rainbow Trout Genome with 2 Other Species of Fish (Arctic charr and Atlantic salmon) Within the Tetraploid Derivative Salmonidae Family (subfamily: Salmoninae). Genome, Vol. 48, No. 6, pp. 1037-1051, ISSN 0831-2796.
10. Danzmann, R.G., Davidson, E.A., Ferguson, M.M., Gharbi, K., Koop, B.F., Hoyheim, B., Lien, S., Lubieniecki, K.P., Moghadam, H.K., Park, J., Phillips, R.B. & Davidson, W.S. (2008). Distribution of Ancestral Proto-Actinopterygian Chromosome Arms Within the Genomes of 4R-derivative Salmonid Fishes (Rainbow trout and Atlantic salmon). BMC Genomics, Vol. 9, No.1, ISSN 1471-2164.
11. Davidson, W.S., Koop, B.F., Jones, S.J., Iturra, P., Vidal, R., Maass, A., Jonassen, I., Lien, S. & Omholt, S.W. (2010). Sequencing the Genome of the Atlantic Salmon (*Salmo salar*). Genome Biology, Vol. 11, No. 9, pp. 403, ISSN 1465-6906.
12. Dominik, S., Henshall, J.M., Kube, P.D., King, H., Lien, S., Kent, M.P. & Elliott, N.G.(2010). Evaluation of an Atlantic Salmon SNP Chip as a Genomic Tool for the Application in a Tasmanian Atlantic Salmon (*Salmo salar*) Breeding Population. Aquaculture, Vol. 308, Suppl. 1, pp. S56, ISSN 0044-8486.
13. Garcia de Leaniz, C., Fleming, I.A., Einum, S., Verspoor, E., Jordan, W.C., Consuegra, S., Aubin-Horth, N., Lajus, D., Letcher, B.H., Youngson, A.F., Webb, J.H., VÃ¸llestad, L.A., Villanueva, B., Ferguson, A. & Quinn, T.P. (2007). A Critical Review of Adaptive Genetic Variation in Atlantic Salmon: Implications for Conservation. Biological Reviews, Vol. 82, No. 2, pp. 173-211, ISSN 1469-185X.

14. Genet, C., Dehais, P., Palti, Y., Gao, G., Gavory, F., Wincker, P., Quillet, E. & Boussaha, M. (2011). Analysis of BAC-end Sequences in Rainbow Trout: Content Characterization and Assessment of Synteny Between Trout and Other Fish Genomes. BMC Genomics, Vol. 12, No. 1, ISSN 1471-2164.

15. "genomics" Mirriam-Webster.com, Mirriam-Webster, 2011. Web. 27 June 2011.

16. Gharbi, K., Gautier, A., Danzmann, R.G., Gharbi, S., Sakamoto, T., Hoyheim, B., Taggart, J.B., Cairney, M., Powell, R., Krieg, F., Okamoto, N., Ferguson, M.M., Holm, L.E. & Guyomard, R. (2006). A Linkage Map for Brown Trout (*Salmo trutta*): Chromosome Homeologies and Comparative Genome Organization With Other Salmonid Fish. Genetics, Vol. 172, No. 4, pp. 2405-2419, ISSN 0016-6731.

17. Gheyas, A.A., Haley, C.S., Guy, D.R., Hamilton, A., Tinch, A.E., Mota-Velasco, J.C. & Woolliams, J.A. (2010). Effect of a Major QTL Affecting IPN Resistance on Production Traits in Atlantic Salmon. Animal Genetics, Vol. 41, No. 6, pp. 666-668, ISSN 1365-2052.

18. Gilbey, J., Verspoor, E., McLay, A. & Houlihan, D. (2004). A Microsatellite Linkage Map for Atlantic Salmon (*Salmo salar*). Animal Genetics, Vol. 35, No. 2, pp. 98-105, ISSN 1365-2052.

19. Gilbey, J., Verspoor, E., Mo, T.A., Sterud, E., Olstad, K., Hytterod, S., Jones, C. & Noble, L. (2006). Identification of Genetic Markers Associated with *Gyrodactylus salaris* Resistance in Atlantic Salmon *Salmo salar*. Diseases of Aquatic Organisms, Vol. 71, No. 2, pp. 119-129, ISSN 0177-5103.

20. Gjøen, H.M. & Bentsen, H.B. (1997). Past, Present, and Future of Genetic Improvement in Salmon Aquaculture. ICES Journal of Marine Science: Journal du Conseil, Vol. 54, No. 6, pp. 1009-1014, ISSN 1095-9289.

21. Goddard, M.E. & Hayes, B.J. (2009). Mapping Genes for Complex Traits in Domestic Animals and Their use in Breeding Programmes. Nature Reviews. Genetics, Vol. 10, No. 6, pp. 381-391, ISSN 1471-0056.

22. Groenen, M.A., Wahlberg, P., Foglio, M., Cheng, H.H., Megens, H.J., Crooijmans, R.P., Besnier, F., Lathrop, M., Muir, W.M., Wong, G.K., Gut, I. & Andersson, L. (2009). A High-density SNP-based Linkage Map of the Chicken Genome Reveals Sequence Features Correlated with Recombination Rate. Genome Research, Vol. 19, No. 3, pp. 510-519, ISSN 1088-9051.

23. Gupta, P.K., Mir, R.R., Mohan, A. & Kumar, J. (2008). Wheat Genomics: Present Status and Future Prospects. International Journal of Plant Genomics, Vol. 2008, pp.36, ISSN 1687-5389.

24. Guyomard, R., Mauger, S., Tabet-Canale, K., Martineau, S., Genet, C., Krieg, F. & Quillet, E. (2006). A Type I and Type II Microsatellite Linkage Map of Rainbow Trout (*Oncorhynchus mykiss*) with Presumptive Coverage of All Chromosome Arms. BMC Genomics, Vol. 7, No. 1, ISSN 1471-2164.

25. Haidle, L., Janssen, J., Gharbi, K., Moghadam, H., Ferguson, M. & Danzmann, R. (2008). Determination of Quantitative Trait Loci (QTL) for Early Maturation in Rainbow Trout (*Oncorhynchus mykiss*). Marine Biotechnology, Vol. 10, No. 5, pp. 579-592, ISSN 1436-2228.

26. Hayes, B.J., Nilsen, K., Berg, P.R., Grindflek, E. & Lien, S. (2007). SNP Detection Exploiting Multiple Sources of Redundancy in Large EST Collections Improves Validation Rates. Bioinformatics, Vol. 23, No. 13, pp. 1692-1693, ISSN 1367-4803.

27. Houle, D., Govindaraju, D.R. & Omholt, S. (2010). Phenomics: The Next Challenge. Nature Reviews. Genetics, Vol. 11, No. 12, pp. 855-866, ISSN 1471-0056.

28. Houston, R.D., Bishop, S.C., Hamilton, A., Guy, D.R., Tinch, A.E., Taggart, J.B., Derayat, A., McAndrew, B.J. & Haley, C.S. (2009). Detection of QTL Affecting Harvest Traits in a Commercial Atlantic Salmon Population. Animal Genetics, Vol. 40, No. 5, pp. 753- 755, ISSN 1365-2052.

29. Houston, R.D., Gheyas, A., Hamilton, A., Guy, D.R., Tinch, A.E., Taggart, J.B., McAndrew, B.J., Haley, C.S. & Bishop, S.C. (2008). Detection and Confirmation of a Major QTL Affecting Resistance to Infectious Pancreatic Necrosis (IPN) in Atlantic Salmon (*Salmo salar*). Developments in Biologicals, Vol. 132, pp. 199-204, ISSN 1424-6074.

30. Houston, R.D., Haley, C.S., Hamilton, A., Guy, D.R., Mota-Velasco, J.C., Gheyas, A.A., Tinch, A.E., Taggart, J.B., Bron, J.E., Starkey, W.G., McAndrew, B.J., Verner-Jeffreys, D.W., Paley, R.K., Rimmer, G.S., Tew, I.J. & Bishop, S.C. (2010). The susceptibility of Atlantic Salmon Fry to FreshwaterI pancreatic Necrosis is Largely Explained by a Major QTL. Heredity, Vol. 105, No. 3, pp. 318-327, ISSN 0018-067X.

31. Houston, R.D., Haley, C.S., Hamilton, A., Guy, D.R., Tinch, A.E., Taggart, J.B., McAndrew, B.J. & Bishop, S.C. (2008). Major Quantitative Trait Loci Affect Resistance to Infectious Pancreatic Necrosis in Atlantic Salmon (*Salmo salar*). Genetics, Vol. 178, No. 2, pp. 1109-1115, ISSN 0016-6731.

32. Hubert, S., Higgins, B., Borza, T. & Bowman, S. (2010). Development of a SNP Resource and a Genetic Linkage Map for Atlantic Cod (*Gadus morhua*). BMC Genomics, Vol. 11, No. 1, ISSN 1471-2164.

33. Hutchison, C.A.,3rd (2007). DNA Sequencing: Bench to Bedside and Beyond. Nucleic Acids Research, Vol. 35, No. 18, pp. 6227-6237, ISSN 1362-4962.

34. Johansen, S.D., Coucheron, D.H., Andreassen, M., Karlsen, B.O., Furmanek, T., Jorgensen, T.E., Emblem, A., Breines, R., Nordeide, J.T., Moum, T., Nederbragt, A.J., Stenseth, N.C. & Jakobsen, K.S. (2009). Large-scale Sequence Analyses of Atlantic Cod. New Biotechnology, Vol. 25, No. 5, pp. 263-271, ISSN 1871-6784.

35. Jones, C.S., Lockyer, A.E., Verspoor, E., Secombes, C.J. & Noble, L.R. (2002). Towards Selective Breeding of Atlantic Salmon for Sea Louse Resistance: Approaches to Identify Trait Markers. Pest Management Science, Vol. 58, No. 6, pp. 559-568, ISSN 1526-4998.

36. Kang, J.H., Kim, W.J. & Lee, W.J. (2008). Genetic Linkage Map of Olive Flounder, *Paralichthys olivaceus*. International Journal of Biological Sciences, Vol. 4, No. 3, pp. 143-149, ISSN 1449-2288.

37. Karlsson, S., Moen, T., Lien, S., Glover, K.A. & Hindar, K. (2011). Generic Genetic Differences Between Farmed and Wild Atlantic Salmon Identified from a 7K SNP- chip. Molecular Ecology Resources, Vol. 11, Suppl. 1, pp. 247-253, ISSN 1755-098X.

38. Kim, D.H., Kim, K.H., Van, K., Kim, M.Y. & Lee, S.H. (2010). Fine Mapping of a Resistance Gene to Bacterial Leaf Pustule in Soybean. Theoretical and Applied Genetics.Theoretische und Angewandte Genetik, Vol. 120, No. 7, pp. 1443-1450, ISSN 0040-5752.

39. King, T.L., Kalinowski, S.T., Schill, W.B., Spidle, A.P. & Lubinski, B.A. (2001). Population Structure of Atlantic Salmon (*Salmo salar* L.): A Range-wide Perspective from Microsatellite DNA Variation. Molecular Ecology, Vol. 10, No. 4, pp. 807-821, ISSN 0962-1083.

40. Kjoglum, S., Larsen, S., Bakke, H.G. & Grimholt, U. (2008). The Effect of Specific MHC Class I and Class II Combinations on Resistance to Furunculosis in Atlantic Salmon (*Salmo salar*). Scandinavian Journal of Immunology, Vol. 67, No. 2, pp. 160-168, ISSN 0300-9475.

41. Koop, B.F., von Schalburg, K.R., Leong, J., Walker, N., Lieph, R., Cooper, G.A., Robb, A., Beetz-Sargent, M., Holt, R.A., Moore, R., Brahmbhatt, S., Rosner, J., Rexroad, C.E.,3rd, McGowan, C.R. & Davidson, W.S. (2008). A Salmonid EST Genomic Study: Genes, Duplications, Phylogeny and Microarrays. BMC Genomics, Vol. 9, No. 1, ISSN 1471-2164.

42. Kucuktas, H., Wang, S., Li, P., He, C., Xu, P., Sha, Z., Liu, H., Jiang, Y., Baoprasertkul, P., Somridhivej, B., Wang, Y., Abernathy, J., Guo, X., Liu, L., Muir, W. & Liu, Z. (2009). Construction of Genetic Linkage Maps and Comparative Genome

Analysis of Catfish Using Gene-associated Markers. Genetics, Vol. 181, No. 4, pp. 1649-1660, ISSN 0016-6731.

43. Kuhl, H., Sarropoulou, E., Tine, M., Kotoulas, G., Magoulas, A. & Reinhardt, R. (2011). A Comparative BAC Map for the Gilthead Sea Bream (*Sparus aurata* L.). Journal of Biomedicine & Biotechnology, Vol. 2011, 7 pp, ISSN 1110-7243

44. Kuttner, E., Moghadam, H.K., Skulason, S., Danzmann, R.G. & Ferguson, M.M. (2011). Genetic Architecture of Body Weight, Condition Factor and Age of Sexual Maturation in Icelandic Arctic Charr (*Salvelinus alpinus*). Molecular Genetics and Genomics, Vol. 286, No. 1, pp. 67-79, ISSN 1617-4623.

45. LeBlanc, F., Laflamme, M. & Gagne, N. (2010). Genetic Markers of the Immune Response of Atlantic Salmon (*Salmo salar*) to Infectious Salmon Anemia Virus (ISAV). Fish & Shellfish Immunology, Vol. 29, No. 2, pp. 217-232, ISSN 1050-4648.

46. Lee, B.Y., Lee, W.J., Streelman, J.T., Carleton, K.L., Howe, A.E., Hulata, G., Slettan, A., Stern, J.E., Terai, Y. & Kocher, T.D. (2005). A Second-generation Genetic Linkage Map of Tilapia (*Oreochromis spp.*). Genetics, Vol. 170, No. 1, pp. 237-244, ISSN 0016-6731.

47. Lewis, J.M., Hori, T.S., Rise, M.L., Walsh, P.J. & Currie, S. (2010). Transcriptome Responses to Heat Stress in the Nucleated Red Blood Cells of the Rainbow Trout (*Oncorhynchus mykiss).* Physiological Genomics, Vol. 42, No. 3, pp. 361-373, ISSN 1094-8341.

48. Li, J., Boroevich, K.A., Koop, B.F. & Davidson, W.S. (2011a). Comparative Genomics Identifies Candidate Genes for Infectious Salmon Anemia (ISA) Resistance in Atlantic Salmon (*Salmo salar*). Marine Biotechnology, Vol. 13, No. 2, pp. 232-241, ISSN 1436-2228.

49. Li, Y., Xu, P., Zhao, Z., Wang, J., Zhang, Y. & Sun, X.W. (2011b). Construction and Characterization of the BAC Library for Common Carp *Cyprinus Carpio* L. And Establishment of Microsynteny with Zebrafish *Danio Rerio.* Marine Biotechnology, Vol. 13, No. 4, pp. 706-712, ISSN 1436-2228.

50. Lien, S., Gidskehaug, L., Moen, T., Hayes, B.J., Berg, P.R., Davidson, W.S., Omholt, S.W., & Kent, M.P. (2011). A dense SNP-based linkage map for Atlantic salmon (*Salmo salar*) reveals extended chromosome homeologies and striking difference in sex- specific recombination rates. BMC Genomics (in press).

51. Liu, G., Jennen, D.G., Tholen, E., Juengst, H., Kleinwachter, T., Holker, M., Tesfaye, D., Un, G., Schreinemachers, H.J., Murani, E., Ponsuksili, S., Kim, J.J., Schellander, K. & Wimmers, K.(2007). A Genome Scan Reveals QTL for Growth, Fatness, Leanness and Meat Quality in a Duroc-Pietrain Resource Population. Animal Genetics, Vol. 38, No. 3, pp. 241-252, ISSN 1365-2052.

52. Liu, H., Jiang, Y., Wang, S., Ninwichian, P., Somridhivej, B., Xu, P., Abernathy, J., Kucuktas, H. & Liu, Z. (2009). Comparative Analysis of Catfish BAC end Sequences with the Zebrafish Genome. BMC Genomics, Vol. 10, No. 1, ISSN 1471-2164.

53. Lorenz, S., Brenna-Hansen, S., Moen, T., Roseth, A., Davidson, W.S., Omholt, S.W. & Lien, S. (2010). BAC-based Upgrading and Physical Integration of a Genetic SNP Map in Atlantic Salmon. Animal Genetics, Vol. 41, No. 1, pp. 48-54, ISSN 1365-2052.

54. Lynch, M. & Walsh, B. (1998), Genetics and Analysis of Quantitative Traits, Sinauer Associates Inc., ISBN 0-87893-481-2, Sunderland, USA. Martin, L.J., Woo, J.G., Avery, C.L., Chen, H.S., North, K.E., Au, K., Broet, P., Dalmasso, C., Guedj, M., Holmans, P., Huang, B., Kuo, P.H., Lam, A.C., Li, H., Manning, A., Nikolov, I., Sinha, R., Shi, J., Song, K., Tabangin, M., Tang, R. & Yamada, R. (2007). Multiple Testing in the Genomics Era: Findings From Genetic Analysis Workshop 15, Group 15. Genetic Epidemiology, Vol. 31, Suppl. 1, pp. S124-131, ISSN 0741- 0395.

55. Martyniuk, C.J., Perry, G.M.L., Mogahadam, H.K., Ferguson, M.M. & Danzmann, R.G. (2003). The Genetic Architecture of Correlations Among Growth-related Traits and Male Age at Maturation in Rainbow Trout. Journal of Fish Biology, Vol. 63, No. 3, pp. 746-764, ISSN 1095-8649.

56. Matukumalli, L.K., Lawley, C.T., Schnabel, R.D., Taylor, J.F., Allan, M.F., Heaton, M.P., O'Connell, J., Moore, S.S., Smith, T.P., Sonstegard, T.S. & Van Tassell, C.P. (2009). Development and Characterization of a High Density SNP Genotyping Assay for Cattle. PloS One, Vol. 4, No. 4, pp. e5350, ISSN 1932-6203.

57. McClelland, E.K. & Naish, K.A. (2010). Quantitative Trait Locus Analysis of Hatch Timing, Weight, Length and Growth Rate in Coho Salmon, *Oncorhynchus kisutch*. Heredity, Vol. 105, No. 6, pp. 562-573, ISSN 0018-067X.

58. McClelland, E.K. & Naish, K.A. (2008). A Genetic Linkage Map for Coho Salmon (*Oncorhynchus kisutch*). Animal Genetics, Vol. 39, No. 2, pp. 169-179, ISSN 1365-2052.

59. McConnell, S.K., O'Reilly, P., Hamilton, L., Wright, J.M. & Bentzen, P. (1995). Polymorphic Microsatellite Loci from Atlantic Salmon (*Salmo salar*): Genetic Differentiation of North American and European Populations. Canadian Journal of Fisheries and Aquatic Sciences, Vol. 52, No. 9, pp. 1863-1872, ISSN 1205-7533.

60. McGowan, C.R., Davidson, E.A., Woram, R.A., Danzmann, R.G., Ferguson, M.M. & Davidson, W.S. (2004). Ten Polymorphic Microsatellite Markers from

Arctic Charr (*Salvelinus alpinus*): Linkage Analysis and Amplification in Other Salmonids. Animal Genetics, Vol. 35, No. 6, pp. 479-481, ISSN 1365-2052.

61. Metzker, M.L. (2010). Sequencing Technologies - The Next Generation. Nature Reviews. Genetics, Vol. 11, No. 1, pp. 31-46, ISSN 1471-0056.

62. Miller, K.M., Li, S., Kaukinen, K.H., Ginther, N., Hammill, E., Curtis, J.M., Patterson, D.A., Sierocinski, T., Donnison, L., Pavlidis, P., Hinch, S.G., Hruska, K.A., Cooke, S.J., English, K.K. & Farrell, A.P. (2011) "Genomic Signatures Predict Migration and Spawning Failure in Wild Canadian Salmon. Science, Vol. 331, No. 6014, pp. 214- 217, ISSN 0036-8075.

63. Moen, T., Baranski, M., Sonesson, A.K. & Kjoglum, S. (2009a), "Confirmation and Fine- mapping of a Major QTL for Resistance to Infectious Pancreatic Necrosis in Atlantic Salmon (*Salmo salar*): Population-level Associations Between Markers and Trait", BMC Genomics, Vol.10, No.1, ISSN 1471-2164.

64. Moen, T., Delghandi, M., Wesmajervi, M., Westgaard, J. & Fjalestad, K. (2009b), "A SNP/Microsatellite Genetic Linkage Map of the Atlantic Cod (*Gadus morhua*)", Animal Genetics, Vol. 40, No.6, pp. 993-996, ISSN 1365-2052.

65. Moen, T., Fjalestad, K.T., Munck, H. & Gomez-Raya, L. (2004). A Multistage Testing Strategy for Detection of Quantitative Trait Loci Affecting Disease Resistance in Atlantic Salmon. Genetics, Vol. 167, No. 2, pp. 851-858, ISSN 0016-6731.

66. Moen, T., Hayes, B., Baranski, M., Berg, P.R., Kjoglum, S., Koop, B.F., Davidson, W.S., Omholt, S.W. & Lien, S. (2008a). A Linkage Map of the Atlantic Salmon (*Salmo salar*) Based on EST-derived SNP Markers. BMC Genomics, Vol. 9, No.1, ISSN 1471-2164.

67. Moen, T., Hayes, B., Nilsen, F., Delghandi, M., Fjalestad, K.T., Fevolden, S.E., Berg, P.R. & Lien, S. (2008b). Identification and Characterisation of Novel SNP Markers in Atlantic Cod: Evidence for Directional Selection. BMC Genetics, Vol. 9, No.1, ISSN 1471-2156.

68. Moen, T., Hoyheim, B., Munck, H. & Gomez-Raya, L. (2004). A Linkage Map of Atlantic Salmon (*Salmo salar*) Reveals an Uncommonly Large Difference in Recombination Rate Between the Sexes. Animal Genetics, Vol. 35, No. 2, pp. 81-92, ISSN 1365-2052.

69. Moen, T., Sonesson, A.K., Hayes, B., Lien, S., Munck, H. & Meuwissen, T.H. (2007). Mapping of a Quantitative Trait Locus for Resistance Against Infectious Salmon Anaemia in Atlantic Salmon (*Salmo salar*): Comparing Survival Analysis with Analysis on Affected/Resistant Data. BMC Genetics, Vol. 8, No.1, ISSN 1471-2156.

70. Moghadam, H.K., Poissant, J., Fotherby, H., Haidle, L., Ferguson, M.M. & Danzmann, R.G. (2007). Quantitative Trait Loci for Body Weight, Condition Factor and Age at Sexual Maturation in Arctic Charr (*Salvelinus alpinus*): Comparative Analysis with Rainbow Trout (*Oncorhynchus mykiss*) and Atlantic Salmon (*Salmo salar*). Molecular Genetics and Genomics, Vol. 277, No. 6, pp. 647-661, ISSN 1617-4623.

71. Muir, W.M., Wong, G.K., Zhang, Y., Wang, J., Groenen, M.A., Crooijmans, R.P., Megens, H.J., Zhang, H., Okimoto, R., Vereijken, A., Jungerius, A., Albers, G.A., Lawley, C.T., Delany, M.E., MacEachern, S. & Cheng, H.H. (2008). Genome-wide Assessment of Worldwide Chicken SNP Genetic Diversity Indicates Significant Absence of Rare Alleles in Commercial Breeds. Proceedings of the National Academy of Sciences of the United States of America, Vol. 105, No. 45, pp. 17312- 17317, ISSN 0027-8424.

72. Nelson, J.S. (2006). Fishes of the World, John Wiley and Sons Inc., ISBN 978-0-471-25031-9, New Jersey, USA.

73. Ng, S.H., Artieri, C.G., Bosdet, I.E., Chiu, R., Danzmann, R.G., Davidson, W.S., Ferguson,

74. Nichols, K.M., Young, W.P., Danzmann, R.G., Robison, B.D., Rexroad, C., Noakes, M., Phillips, R.B., Bentzen, P., Spies, I., Knudsen, K., Allendorf, F.W., Cunningham, B.M., Brunelli, J., Zhang, H., Ristow, S., Drew, R., Brown, K.H., Wheeler, P.A. & Thorgaard, G.H. (2003). A Consolidated Linkage Map for Rainbow Trout (*Oncorhynchus mykiss*). Animal Genetics, Vol. 34, No. 2, pp. 102-115, ISSN 0268- 9146.

75. Nichols, K., Felip-Edo, A., Wheeler, P. & Thorgaard, G. (2008). The Genetic Basis of Smoltification-related Traits in *Oncorhynchus mykiss*. Genetics, Vol. 179, No. 3, pp. 1559-1575, ISSN 0016-6731.

76. Norris, A.T., Bradley, D.G. & Cunningham, E.P. (2000). Parentage and Relatedness Determination in Farmed Atlantic Salmon (*Salmo salar*) Using Microsatellite Markers. Aquaculture, Vol. 182, No. 1-2, pp. 73, ISSN 0044-8486.

77. Ødegård, J., Olesen, I., Gjerde, B. & Klemetsdal, G. (2007). Positive Genetic Correlation Between Resistance to Bacterial (Furunculosis) and Viral (Infectious Salmon Anaemia) Diseases in Farmed Atlantic Salmon (*Salmo salar*). Aquaculture, Vol. 271, No. 1-4, pp. 173, ISSN 0044-8486.

78. Ohno, S. (1970). Evolution by Gene Duplication, Springer-Verlag, ISBN 0-04-575015-7, New York, USA. O'Malley, K.G., McClelland, E.K. & Naish, K.A. (2010). Clock Genes Localize to Quantitative Trait Loci for Stage-specific Growth in Juvenile Coho Salmon, *Oncorhynchus kisutch*. The Journal of Heredity, Vol. 101, No. 5, pp. 628-632, ISSN 0022-1503.

79. O'Malley, K.G., Sakamoto, T., Danzmann, R.G. & Ferguson, M.M. (2003). Quantitative Trait Loci for Spawning Date and Body Weight in Rainbow Trout: Testing for Conserved Effects Across Ancestrally Duplicated Chromosomes. Journal of Heredity, Vol. 94, No. 4, pp. 273-284, ISSN 1465-7333.

80. O'Reilly, P.T., Herbinger, C. & Wright, J.M. (1998). Analysis of Parentage Determination in Atlantic Salmon (*Salmo salar*) Using Microsatellites. Animal Genetics, Vol. 29, No. 5, pp. 363-370, ISSN 1365-2052.

81. Palti, Y., Genet, C., Luo, M.C., Charlet, A., Gao, G., Hu, Y., Castano-Sanchez, C., Tabet- Canale, K., Krieg, F., Yao, J., Vallejo, R.L. & Rexroad, C.E.,3rd (2011). A First Generation Integrated Map of the Rainbow Trout Genome. BMC Genomics, Vol. 12, No. 1, ISSN 1471-2164.

82. Palti, Y., Luo, M.C., Hu, Y., Genet, C., You, F.M., Vallejo, R.L., Thorgaard, G.H., Wheeler, P.A. & Rexroad, C.E.,3rd (2009). A First Generation BAC-based Physical Map of the Rainbow Trout Genome. BMC Genomics, Vol. 10, No. 1, ISSN 1471-2164.

83. Perry, G.M., Danzmann, R.G., Ferguson, M.M. & Gibson, J.P. (2001). Quantitative Trait Loci for Upper Thermal Tolerance in Outbred Strains of Rainbow Trout (*Oncorhynchus mykiss*). Heredity, Vol. 86, No. 3, pp. 333-341, ISSN 0018-067X.

84. Phillips, R.B., Keatley, K.A., Morasch, M.R., Ventura, A.B., Lubieniecki, K.P., Koop, B.F., Danzmann, R.G. & Davidson, W.S. (2009). Assignment of Atlantic salmon (Salmo salar) Linkage Groups to Specific Chromosomes: Conservation of Large Syntenic Blocks Corresponding to Whole Chromosome Arms in Rainbow Trout (*Oncorhynchus mykiss*). BMC Genetics, Vol. 10, No. 1, ISSN 1471-2156.

85. Phillips, R.B., Nichols, K.M., DeKoning, J.J., Morasch, M.R., Keatley, K.A., Rexroad, C.,3rd, Gahr, S.A., Danzmann, R.G., Drew, R.E. & Thorgaard, G.H. (2006). Assignment of Rainbow Trout Linkage Groups to Specific Chromosomes. Genetics, Vol. 174, No. 3, pp. 1661-1670, ISSN 0016-6731.

86. Presa, P. & Guyomard, R. (1996). Conservation of Microsatellites in Three Species of Salmonids. Journal of Fish Biology, Vol. 49, No. 6, pp. 1326-1329, ISSN 1095-8649. Quinn, N.L., Boroevich, K.A., Lubieniecki, K.P., Chow, W., Davidson, E.A., Phillips, R.B., Koop, B.F. & Davidson, W.S. (2010). Genomic Organization and Evolution of the Atlantic Salmon Hemoglobin Repertoire. BMC Genomics, Vol. 11, No.1, ISSN 1471- 2164.

87. Quinn, N.L., Levenkova, N., Chow, W., Bouffard, P., Boroevich, K.A., Knight, J.R., Jarvie, T.P., Lubieniecki, K.P., Desany, B.A., Koop, B.F., Harkins, T.T. & Davidson, W.S. (2008). Assessing the Feasibility of GS FLX Pyrosequencing for Sequencing the Atlantic Salmon Genome. BMC Genomics, Vol. 9, No. 1, ISSN 1471-2164.

88. Quinn, N.L., McGowan, C.R., Cooper, G.A., Koop, B.F. & Davidson, W.S. (2011a). Identification of Genes Associated with Heat Tolerance in Arctic Charr Exposed to Acute Thermal Stress. Physiological Genomics, Vol. 43, No. 11, pp 685-96, ISSN 1094-8341.

89. Quinn, N.L., McGowan, C.R., Cooper, G.A., Koop, B.F. & Davidson, W.S. (2011b). Ribosomal Genes and Heat Shock Proteins as Putative Markers for Chronic, Sub-lethal Heat Stress in Arctic Charr: Applications for Aquaculture and Wild Fish. Physiological Genomics, Jul 12 2011 [epub ahead of print].

90. Ramos, A.M., Crooijmans, R.P., Affara, N.A., Amaral, A.J., Archibald, A.L., Beever, J.E., Bendixen, C., Churcher, C., Clark, R., Dehais, P., Hansen, M.S., Hedegaard, J., Hu, Z.L., Kerstens, H.H., Law, A.S., Megens, H.J., Milan, D., Nonneman, D.J., Rohrer, G.A., Rothschild, M.F., Smith, T.P., Schnabel, R.D., Van Tassell, C.P., Taylor, J.F., Wiedmann, R.T., Schook, L.B. & Groenen, M.A. (2009). Design of a High Density SNP Genotyping Assay in the Pig Using SNPs Identified and Characterized by Next Generation Sequencing Technology. PloS One, Vol. 4, No. 8, pp. e6524, ISSN 1932-6203.

91. Reid, D.P., Szanto, A., Glebe, B., Danzmann, R.G. & Ferguson, M.M. (2005). QTL for Body Weight and Condition Factor in Atlantic Salmon (Salmo salar): Comparative Analysis with Rainbow Trout (*Oncorhynchus mykiss*) and Arctic Charr (*Salvelinus alpinus*). Heredity, Vol. 94, No. 2, pp. 166-172, ISSN 0018-067X.

92. Rexroad, C.E.,3rd, Rodriguez, M.F., Coulibaly, I., Gharbi, K., Danzmann, R.G., Dekoning, J., Phillips, R. & Palti, Y. (2005). Comparative Mapping of Expressed Sequence Tags Containing Microsatellites in Rainbow Trout (*Oncorhynchus mykiss*). BMC Genomics, Vol. 6, No.1, ISSN 1471-2164.

93. Rise, M.L., von Schalburg, K.R., Brown, G.D., Mawer, M.A., Devlin, R.H., Kuipers, N., Busby, M., Beetz-Sargent, M., Alberto, R., Gibbs, A.R., Hunt, P., Shukin, R., Zeznik, J.A., Nelson, C., Jones, S.R., Smailus, D.E., Jones, S.J., Schein, J.E., Marra, M.A., Butterfield, Y.S., Stott, J.M., Ng, S.H., Davidson, W.S. & Koop, B.F. (2004). Development and Application of a Salmonid EST Database and cDNA Microarray: Data Mining and Interspecific Hybridization Characteristics. Genome Research, Vol. 14, No. 3, pp. 478-490, ISSN 1088-9051.

94. Ryynanen, H.J. & Primmer, C.R. (2006). Single Nucleotide Polymorphism (SNP) Discovery in Duplicated Genomes: Intron-primed Exon-crossing (IPEC) As a Strategy for Avoiding Amplification of Duplicated Loci in Atlantic Salmon (*Salmo salar*) and Other Salmonid Fishes. BMC Genomics, Vol. 7, No.1, ISSN 1471-2164.

95. Sakamoto, T., Danzmann, R.G., Gharbi, K., Howard, P., Ozaki, A., Khoo, S.K., Woram, R.A., Okamoto, N., Ferguson, M.M., Holm, L.E., Guyomard, R. & Hoyheim, B. (2000). A Microsatellite Linkage Map of Rainbow Trout (*Oncorhynchus mykiss*) Characterized by Large Sex-specific Differences in Recombination Rates. Genetics, Vol. 155, No. 3, pp. 1331-1345, ISSN 0016-6731.

96. Sanchez, C.C., Smith, T.P., Wiedmann, R.T., Vallejo, R.L., Salem, M., Yao, J. & Rexroad, C.E.,3rd (2009). Single Nucleotide Polymorphism Discovery in Rainbow Trout by Deep Sequencing of a Reduced Representation Library. BMC Genomics, Vol. 10, No.1, ISSN 1471-2164.

97. Sanchez, J.A., Clabby, C., Ramos, D., Blanco, G., Flavin, F., Vazquez, E. & Powell, R. (1996). Protein and Microsatellite Single Locus Variability in *Salmo Salar* L. (Atlantic salmon). Heredity, Vol. 77, No. 4, pp. 423-432, ISSN 0018-067X.

98. Sarropoulou, E., Franch, R., Louro, B., Power, D.M., Bargelloni, L., Magoulas, A., Senger, F., Tsalavouta, M., Patarnello, T., Galibert, F., Kotoulas, G. & Geisler, R. (2007). A Gene-based Radiation Hybrid Map of the Gilthead Sea Bream *Sparus aurata* Refines and Exploits Conserved Synteny with *Tetraodon nigroviridis*. BMC Genomics, Vol. 8, No.1, ISSN 1471-2164.

99. Scribner, K.T., Gust, J.R. & Fields, R.L. (1996). Isolation and Characterization of Novel Salmon Microsatellite Loci: Cross-species Amplification and Population Genetic Applications. Canadian Journal of Fisheries and Aquatic Sciences, Vol. 53, No. 4, pp. 833-841, ISSN 1205-7533.

100. Slettan, A., Olsaker, I. & Lie, O. (1995). A Polymorphic Dinucleotide Repeat Microsatellite in Atlantic Salmon, *Salmo salar* (SSOSL436). Animal Genetics, Vol. 26, No. 5, pp. 368, ISSN 1365-2052.

101. Smith, C.T., Elfstrom, C.M., Seeb, L.W. & Seeb, J.E. (2005). Use of Sequence Data From Rainbow Trout and Atlantic Salmon for SNP Detection in Pacific Salmon. Molecular Ecology, Vol. 14, No. 13, pp. 4193-4203, ISSN 0962-1083.

102. Somorjai, I.M., Danzmann, R.G. & Ferguson, M.M. (2003). Distribution of Temperature Tolerance Quantitative Trait Loci in Arctic Charr (*Salvelinus alpinus*) and Inferred Homologies in Rainbow Trout (*Oncorhynchus mykiss*). Genetics, Vol. 165, No. 3, pp. 1443-1456, ISSN 0016-6731.

103. Sun, X. & Liang, L. (2004). A Genetic Linkage Map of Common Carp (*Cyprinus carpio L.*) and Mapping of a Locus Associated with Cold Tolerance. Aquaculture, Vol. 238, No. 1- 4, pp. 165, ISSN 0044-8486.

104. Thorsen, J., Zhu, B., Frengen, E., Osoegawa, K., de Jong, P.J., Koop, B.F., Davidson, W.S. & Hoyheim, B. (2005). A Highly Redundant BAC Library of Atlantic Salmon (*Salmo salar*): An Important Tool for Salmon Projects. BMC Genomics, Vol. 6, No. 1, ISSN 1471-2164.

105. Jackson, T.R, Ferguson, M.M, Danzmann, R.G, Fishback, A.G, Ihssen, P.E, O'Connell, M. & Crease, T.J.(1998). Identification of Two QTL Influencing Upper Temperature Tolerance in Three Rainbow Trout (*Oncorhynchus mykiss*) Half-sib Families. Heredity, Vol. 80, No. 2, pp. 143-151, ISSN 0018-067X.

106. Tuberosa, R., Salvi, S., Sanguineti, M.C., Landi, P., Maccaferri, M. & Conti, S. (2002). Mapping QTLs Regulating Morpho-physiological Traits and Yield: Case Studies, Shortcomings and Perspectives in Drought-stressed Maize. Annals of Botany, Vol. 89 Spec No, pp. 941-963, ISSN 0305-7364.

107. Van Tassell, C.P., Smith, T.P., Matukumalli, L.K., Taylor, J.F., Schnabel, R.D., Lawley, C.T., Haudenschild, C.D., Moore, S.S., Warren, W.C. & Sonstegard, T.S.(2008). SNP Discovery and Allele Frequency Estimation by Deep Sequencing of Reduced Representation Libraries. Nature Methods, Vol. 5, No. 3, pp. 247-252, ISSN 1548- 7091.

108. Veerkamp, R.F. & Beerda, B. (2007). Genetics and Genomics to Improve Fertility in High Producing Dairy Cows. Theriogenology, Vol. 68, Suppl. 1, pp. S266-273, ISSN 0093- 691X.

109. von Schalburg, K.R., Rise, M.L., Cooper, G.A., Brown, G.D., Gibbs, A.R., Nelson, C.C., Davidson, W.S. & Koop, B.F. (2005). Fish and Chips: Various Methodologies Demonstrate Utility of a 16,006-gene Salmonid Microarray. BMC Genomics, Vol. 6, No. 1, ISSN 1471-2164.

110. Waldbieser, G.C., Bosworth, B.G., Nonneman, D.J. & Wolters, W.R. (2001). A Microsatellite- Based Genetic Linkage Map for Channel Catfish, *Ictalurus punctatus*. Genetics, Vol. 158, No. 2, pp. 727-734, ISSN 0016-6731.

111. Wang, S., Peatman, E., Abernathy, J., Waldbieser, G., Lindquist, E., Richardson, P., Lucas, S., Wang, M., Li, P., Thimmapuram, J., Liu, L., Vullaganti, D., Kucuktas, H., Murdock, C., Small, B.C., Wilson, M., Liu, H., Jiang, Y., Lee, Y., Chen, F., Lu, J., Wang, W., Xu, P., Somridhivej, B., Baoprasertkul, P., Quilang, J., Sha, Z., Bao, B., Wang, Y., Wang, Q., Takano, T., Nandi, S., Liu, S., Wong, L., Kaltenboeck, L., Quiniou, S., Bengten, E., Miller, N., Trant, J., Rokhsar, D., Liu, Z. & Catfish Genome Consortium (2010). Assembly of 500,000 Inter-specific Catfish Expressed Sequence Tags and Large Scale Gene-associated Marker Development for Whole Genome Association Studies. Genome Biology, Vol. 11, No. 1, pp. R8, ISSN 1465-6906.

112. Woram, R.A., McGowan, C., Stout, J.A., Gharbi, K., Ferguson, M.M., Hoyheim, B., Davidson, E.A., Davidson, W.S., Rexroad, C. & Danzmann, R.G. (2004). A Genetic Linkage Map for Arctic Char (Salvelinus alpinus): Evidence for Higher Recombination Rates and Segregation Distortion in Hybrid Versus Pure Strain Mapping Parents. Genome, Vol. 47, No. 2, pp. 304-315, ISSN 0831-2796.

113. Wringe, B., Devlin, R., Ferguson, M., Moghadam, H., Sakhrani, D. & Danzmann, R. (2010). Growth-related Quantitative Trait Loci in Domestic and Wild Rainbow Trout (*Oncorhynchus mykiss*). BMC Genetics, Vol. 11, No. 1, ISSN 1471-2156.

114. Xia, J.H., Feng, F., Lin, G., Wang, C.M. & Yue, G.H. (2010). A First Generation BAC-based Physical Map of the Asian Seabass (*Lates calcarifer*). PloS One, Vol. 5, No. 8, pp. e11974, ISSN 1932-6203.

115. Xu, P., Li, J., Li, Y., Cui, R., Wang, J., Wang, J., Zhang, Y., Zhao, Z. & Sun, X. (2011). Genomic Insight Into the Common Carp (Cyprinus carpio) Genome by Sequencing Analysis of BAC-end Sequences. BMC Genomics, Vol. 12, No. 1, ISSN 1471-2164.

116. Xu, P., Wang, S., Liu, L., Thorsen, J., Kucuktas, H. & Liu, Z. (2007). A BAC-based Physical Map of the Channel Catfish Genome. Genomics, Vol. 90, No. 3, pp. 380, ISSN 0888- 7543.

09 A NEW METHOD FOR MANAGING LIPID OXIDATION AND MELANOSIS IN AQUACULTURED FISH AND CRUSTACEANS: APPLICATION OF EDIBLE MUSHROOM (FLAMMULINA VELUTIPES) EXTRACT IN VIVO

1. INTRODUCTION: AN OVERVIEW

The global demand for fish and fisheries products has been rising in correlation with the growth of the world population. More aquaculture technologies have been introduced in the fisheries industry to suit the increased demand. According to the Food and Agriculture Organization (FAO, 2006), the global aquaculture production of fish, crustaceans, molluscs, and other aquatic animals has been steadily increasing, vrising from 3.9%, 27.1%, and 32.4% of total production by weight in 1970, 2000, and 2004, respectively. Aquaculture is growing faster than any other sectors that provide animal food. The FAO (2006) stated that the aquaculture industry has risen by an average of 8.8% annually since 1970, while capture fisheries and terrestrial farmed meat production methods have only grown by 1.2% and 2.8% respectively during the same period globally. Production from aquaculture has significantly exceeded population increase, with per capita supply from aquaculture rising from 0.7 kg in 1970 to 7.1 kg in 2004, showing an average annual growth rate of 7.1%.

The recognized constraints of aquaculture and capture fisheries production, along with the growing disparity between fish supply and demand for human consumption, emphasize the crucial role of postharvest technologies in the business. Postharvest losses are an unacceptable waste of our limited natural resources.

Postharvest losses of fish manifest in many ways during handling, processing, and preservation. Substantial economic losses result from spoiling of fish and crustaceans, leading to a fall in market value or necessitating reprocessing, which in turn raises the cost of the final product. Inadequate handling and processing techniques can decrease nutritional levels. Converting significant amounts of fish captures into fish meal for animal feeds may be viewed as a detriment to human food security in specific circumstances. FAO (2006) estimates that fish losses due to rotting amount to 10 to 12 million tons per year. Yearly, accounting for around 10% of the entire production from both capture fisheries and aquaculture. Postharvest losses in fisheries might be one of the greatest among all commodities in the food production system. Effective preservation techniques can greatly minimize losses, especially during the many stages of handling, processing, distribution, and marketing of fishing products.

2. FISHERIES POSTHARVEST TECHNOLOGY

Fisheries resources provide sustenance and livelihood for many people. Yet, the ability to gather additional resources is currently restricted due to overexploitation, while the global demand for food protein is on the rise. Implementing efficient postharvest technology can narrow the disparity between output and demand and lead to the optimal exploitation of aquatic resources. This pertains to the many procedures and methods employed in the postharvest management, processing, and marketing of aquatic products.

2.1 Methods for handling and preserving fresh seafood products

Seafood products are among the world`s most perishable commodities, and their spoilage begins soon after the death of the fish. Improper handling could hasten spoilage, resulting in gradual development of undesirable qualities in seafood products. Thus, effective handling and preservation techniques need to be used to address the problems of postharvest losses.

Many developments have taken place in postharvest technology. Freezing at sea was a major development in the advance nations in the 1950`s along with the bulk chilling in chilled and refrigerated seawater systems on board fishing vessels and no new major technologies have been recently introduced but only a consolidation and anticipated development of current ones (Hermes, 1998). The industry tends to be more perceptive of the necessity of satisfying the requirements and demands of consumers, particularly with regards to the safety and quality of seafood products.

2.1.1 Chilling

Chilling is mostly utilized in the business to maintain the freshness of seafood following harvest. Chilling is the process of lowering the temperature of marine items to a point below -2 to -4°C for superchilling or above 0 to 5°C, which is the freezing point of water (Hermes, 1998). Chilling methods include wet icing, cooled saltwater, ice slurry, refrigerated air, dry ice, and gel ice mats. The most popular and effective method of cooling fish catch is through icing, which is influenced by direct contact between melted ice and the fish. Adequate ice and correct positioning of ice and the product are necessary to facilitate quick cooling. Chilled seawater or slush ice is utilized, which consists of seawater and ice. The quantity of ice required is determined by the starting temperature of the water and fish, the dimensions of the container, the effectiveness of its insulation, and the duration of the journey. Refrigerated air is utilized in large commercial boats by circulating refrigerated air through a finned evaporator and a fan located at the cold room's end. The cooling of fish using dry ice is influenced by the ice's evaporation process. However, direct contact between the dry ice and the product should be avoided to prevent cold burns (Shoemaker, 1991 as referenced by Hermes, 1998). This method is typically utilized for air freight. These chilling methods do not prevent rotting but significantly reduce its rate. They are employed solely to postpone spoilage of the fish harvest, not for extended storage of the product.

2.1.2 Freezing

Freezing seafood is necessary when alternative methods like refrigeration are not suited for extended storage. Under optimal conditions, it can maintain the product's frozen state for several months without significantly altering its quality. Cryopreservation is a technique used to halt the degradative processes of microbes and enzymes, either partially or completely. Microorganisms cease reproduction at temperatures about -10°C and lower, but enzyme activity is often regulated when the temperature drops below freezing, to about -1°C (Cornell, 1995 as referenced by Hermes, 1998). Water in the flesh freezes between -1°C and -3°C, with most of it turning into ice during the process. At a temperature of -5°C, almost 75% of the water in fish muscles is solidified. Three primary methods for freezing seafood products in the fishing business are air-blast freezing, contact or plate freezing, and spray or immersion freezing. Air-blast freezing is the process of exposing the product to a continuous flow of cold air. Consistent freezing is achieved when the temperature and velocity of the air above the product remain constant. This freezing technology is very adaptable and therefore beneficial for creating individually quick frozen (IQF) items such as crustaceans, fish fillets, and value-added products. Contact or plate freezing involves the product making direct contact with freezing metal plates, through which a cold fluid is circulated. It is commonly used to freeze products like entire fish, fillets, shrimps, and other seafood items into blocks. The substance directly contacts the fluid refrigerant, like liquid nitrogen, in the spray- or immersion-freezing procedure. This technology is typically utilized for manufacturing high-value and specialty IQF items. The decision on freezing systems will be based on cost, function, and feasibility due to the potentially high expenses associated with this preservation method.

1.1.3 Preservatives

Various problems are encountered with seafood products even when they are kept in cold storage. Frozen tuna becomes dark brown during cold storage because of oxidation of hemoglobin in the blood and myoglobin (Mb) in the meat. In crustaceans, browning or blackening of frozen shrimps and prawns is also observed. These problems prompted the food industry to look for ways to control quality deterioration in seafood products, and the use of preservatives has been one of the immediate responses.

The addition of antioxidants to products is one of the most widely studied methods for controlling discoloration and lipid oxidation in fish meat and other meat products. Antioxidants have many modes of action which include sequestering catalytic metal ions, decreasing oxygen concentration, quenching singlet oxygen and superoxide anion, decomposing primary oxidation products to non-volatile compounds, preventing first-chain initiation by scavenging initially generated radicals and chain breaking. The chain-breaking mechanism has been studied for several antioxidants (Roginsky & Lissi, 2005). In this mechanism, the antioxidant donates a hydrogen atom to a lipid peroxyl radical and forms an antioxidant radical, the antioxidant radical subsequently either combines with other lipid peroxyl radical or another antioxidant radical to terminate the reaction.

Several categories of antioxidants can be used for food applications. However, selecting antioxidants for foods is a major concern in the industry because of strict regulations. Generally, the antioxidants involved in food additives must be effective at low doses, must not affect sensory flavor, and must not toxic. Therefore, natural antioxidants are generally preferred for food applications.

3. LIPID OXIDATION IN SEAFOOD PRODUCTS

Lipids are one of the important structural and functional components of foods. They provide energy to humans and essential nutrients such as eicosapentaenoic acid; docosahexaenoic acid; and fat-soluble vitamins including, vitamins A, D, E, and K. Lipids are generally defined as "fatty acids and their derivatives, and substances related biosynthetically or functionally to these components" (Christie, 1987). They have been known to significantly affect food quality even though they constitute a minor component of food. Lipids not only impart flavor, odor, texture, and color to foods but also contribute to the feeling of satiety and help in making food products palatable.

However, constant exposure of lipids, particularly unsaturated fatty acids, to air could adversely affect food quality. The susceptibility of lipids to oxidation is one of the main causes of quality deterioration in various types of fresh food products as well as in processed foods. Lipid oxidation is a perennial problem for both the food industry and the consumers. Lipid oxidation has been believed to be one of the factors limiting the shelf life of foods, particularly that of many complex products (Jacobsen, 1999). It is a complex process whereby lipids, particularly polyunsaturated fatty acids, are degraded via free radical formation, causing the deterioration of flavor, texture, color, aroma, taste, consistency, nutritional benefits, and to some extent, the safety of foodstuffs for human consumption. Notably, consumers` preferences for foods are being influenced by such factors. Thus, lipid oxidation is a decisive factor in the useful processing and storage of food products.

Various effects of lipid oxidation on food properties are briefly summarized in Figure 1. Lipid oxidation itself is primarily the formation of reactive compounds like hydroperoxides (HPO) and peroxy radicals. The primary products very often undergo further reactions to form more stable compounds such as hydroxy acids or epoxides. It has been reported that compounds like hydroxy acids can contribute to bitter taste (Grosch et al., 1992). In complex food systems, the interaction of lipid HPO and secondary oxidation products with proteins and other components significantly impact oxidative and flavor stability and texture during processing, cooking, and storage (Erickson, 1992). Oxidized lipids can react with amines, amino acids, and proteins to form brown macromolecular products (Frankel, 1998). Color formation is known to be primarily influenced by the degree of fatty acid unsaturation, water activity, oxygen pressure, and the presence of phenolic compounds (Pan, 2004).

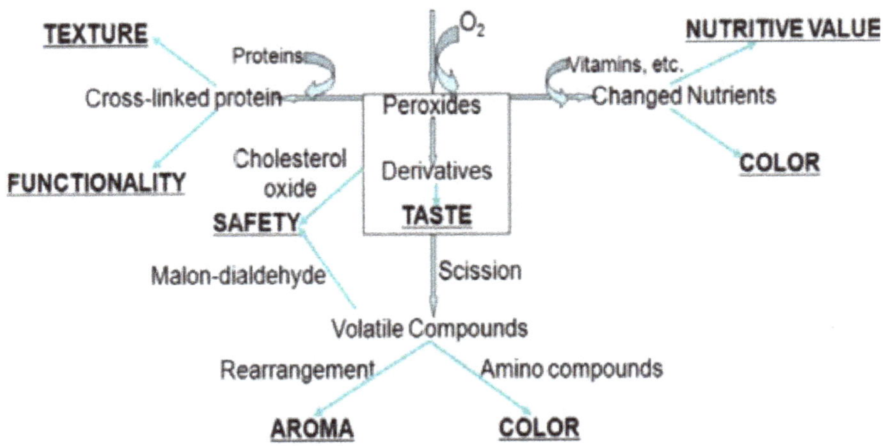

Fig. 1. Reactions of polyunsaturated fatty acids leading to quality and nutritional changes in foods (Erickson, 1992).

Some of the known factors that promote or inhibit lipid oxidation in foods are shown in Table 1. Metal, metallo-proteins, and enzymes are important factors affecting lipid oxidation in raw materials. Water activity, lipid interactions, proteins, and sugars are important elements affecting the food quality of processed foods.

Table 1. Lipid oxidation factors in food (Frankel, 1998).

Type of lipid fatty acid	Non-polar, polar, fatty acids, sterols, terpenes, chain length, unsaturation, cis-trans isomers, free or bound
Catalysts	Enzymes, heme compounds, trace metals
State and access of oxygen	Triplet, singlet or radical, package
Light	Frequency, intensity, sensitizers
Temperature	Denaturating, non-denaturating
pH	Dissociation, denaturing
Inhibitors	Antioxidants, chelators, enzymes, enzyme inhibition or activation

3.1 Control of lipid oxidation by natural antioxidants

Fish tissue contains many endogenous antioxidants that can contribute to stabilizing the natural color of fish meat by acting against either lipid oxidation or Mb oxidation in the meat. These antioxidants may act directly or indirectly to inhibit the initiation and propagation steps of lipid oxidation. Antioxidants that interfere with the propagation step by converting free radicals to stable compounds are generally referred to as primary antioxidants, and those interfering with the initiation step are considered as secondary antioxidants or preventive inhibitors. Secondary antioxidants include both oxygen scavengers and chelators.

The endogenous antioxidants found in aquatic food products include the tocopherols (Syvaoja & Salminen, 1985), ubiquinol (Petillo et al., 1998), carotenoids (Miki, 1991), ascorbate, organic acids, glutathione peroxidase (Watanabe et al., 1996), peroxide dismutase (Aksnes & Njaa, 1981), catalase, peroxidases (Kanner & Kinsella, 1983), ferroxidases (Kanner et al., 1988), nucleotides, peptides, amino acids (Boldyrev et al., 1987), and phospholipids (Ohshima et al., 1993). These natural antioxidants are usually extracted from fruits, vegetables, and other edible materials in nature.

Mushrooms have been known as a potential source of antioxidants. Numerous studies have shown that certain mushroom extracts (ME) have antioxidative activity *in vitro* and *in vivo* (Mau et al., 2002; Jang et al., 2004; Cheung & Cheung, 2005; Elmastas et al., 2007). Active compounds, including ergothioneine (ERT), certain phenolic compounds, and saccharides, have been found in extracts from different mushroom species (Wasser, 2002; Quang et al., 2006; Dubost et al., 2007).

3.2 Prevention of fish meat discoloration and lipid oxidation by dietary supplementation with ME

Mushrooms are widely cultivated because of their known uses and health benefits. However, accumulation of industrial waste has been an issue with the increasing production of mushrooms worldwide. Although spent medium is still underutilized, it is normally used as compost or raw material for extracting soluble sugars (Makishima et al., 2006). Among the mushroom species that are cultivated, *Flammulina velutipes* has been known as a medicinal mushroom and has also been cultured as an edible mushroom on a large scale in Japan and other Asian countries (Wasser, 2002).

Jang et al. (2004) reported the antioxidative properties of the *F. velutipes* extracts against the oxidation of cod liver oil in oil-in-water emulsions. In addition, water extracts from this mushroom have been reported to inhibit the oxidation of oxymyoglobin (MbO2) isolated from cattle meats (Ashida & Sato, 2005). The activity of phenolic compounds depends on their structure and is relative to the number and location of the hydroxyl (OH) groups involved (Dziedric & Hudson, 1984). Recently, ERT is another potent antioxidant which usually exists in mushrooms. It has been known as a powerful scavenger of hydroxyl radicals (.OH) and an inhibitor of .OH generation from hydrogen peroxide; .OH generation is catalyzed by iron or copper ions (Akanmu et al., 1991). Arduini et al. (1990) showed that ferrylmyoglobin (ferrylMb), which was formed when deoxymyoglobin (deoxyMb) and metmyoglobin (metMb) were exposed to hydrogen peroxide, was reduced to metMb in the presence of ERT.

Bao et al. (2008) found that *F. velutipes* contains 300 ug/g ERT and can prevent lipid oxidation as well as stabilize the color of beef and fish meats during low-temperature storage. In addition, Bao et al. (2009a) evaluated the antioxidative activity of hydrophilic extract prepared from solid waste medium obtained during *F. velutipes* cultivation with regard to value-added utilization of the waste for recovery of bioactive compounds. The extract contained 341 ug/mL ERT and was used for stabilizing the lipids and the color of the dark meat of two year-old yellowtail, *Seriola quinqueradiata* through feeding prior to harvest (Bao et al., 2009a).

The DPPH radical scavenging activity of the extracts correlated with their ERT content is shown in Figure 2. In this study, the effective content of ERT was 0.3 ug in the extract in which the DPPH radicals were scavenged by 50% of the original.

Changes in the HPO content of the yellowtail dark muscle during chilled storage are shown in Figure 3. The HPO content in the dark muscle of the control yellowtail increased with prolongation of storage time, whereas HPO accumulation in the ME-supplemented fish was significantly suppressed. This effect is believed to be because of the radical scavenging behavior of ERT (Franzoni et al., 2006). ERT is also known to not only protect organs against lipid peroxidation but also conserve endogenous glutathione and α-tocopherol (Deiana et al., 2004). The same study also showed that feeding yellowtails with the ME remarkably delayed metMb formation in the dark muscle during chilled storage, as shown in Figure 4.

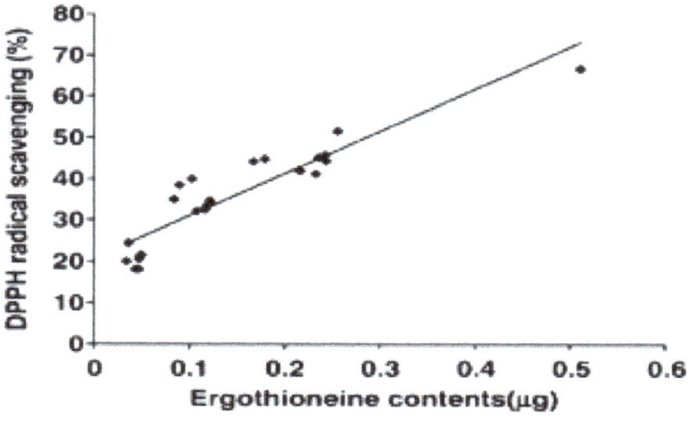

Fig. 2. DPPH radical scavenging activity of the mushroom (*Flammulina velutipes*) extract. Results are presented in terms of mean ± standard deviation (*n* = 3) (Bao et al., 2009a).

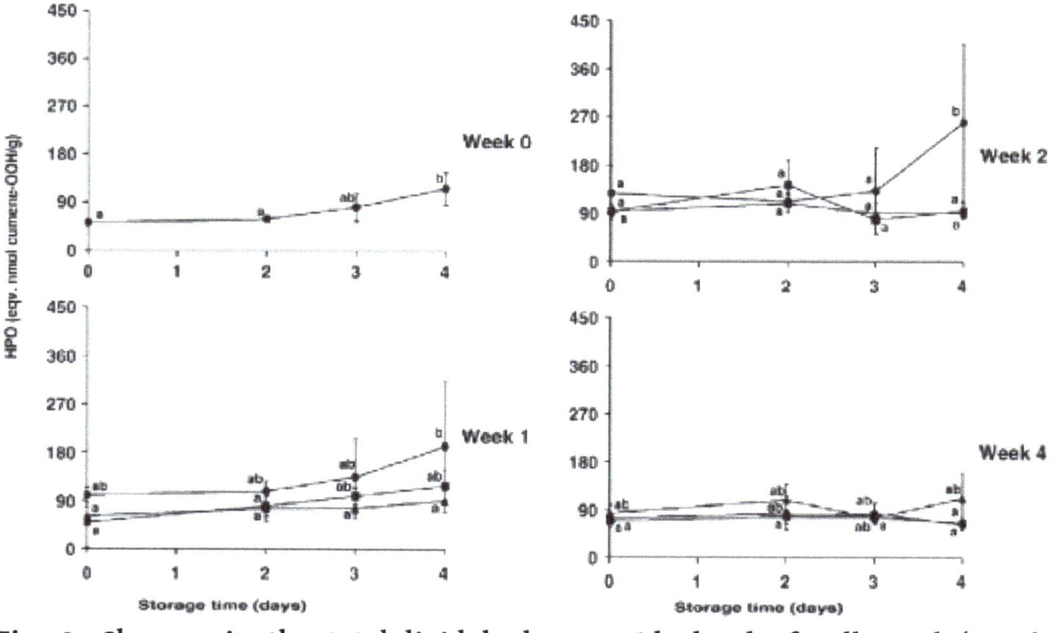

Fig. 3. Changes in the total lipid hydroperoxide level of yellowtail (*Seriola quinqueradiata*) dark muscle during chilled storage. (●), fed a diet without mushroom extract; (■), fed with diet containing 1% concentrated mushroom extract; (▲), fed with diet containing 10% mushroom extract. Data are presented in terms of mean ± standard deviation (n = 5). The values with different superscript letters represent significant difference ($P < 0.05$) (Bao et al., 2009a).

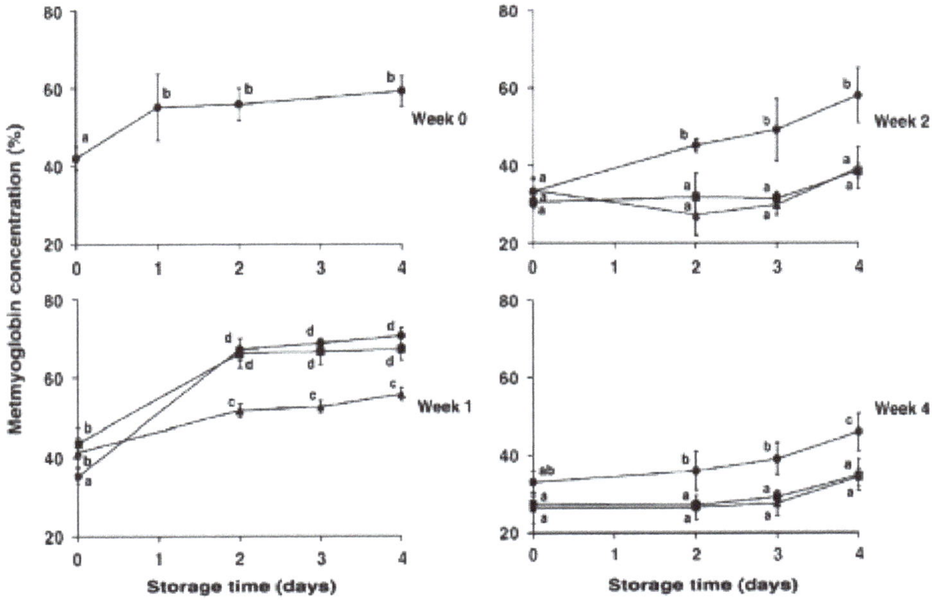

Fig. 4. Changes in the metmyoglobin content of yellowtail (*Seriola quinqueradiata*) dark muscle during chilled storage. (●), fed a diet without mushroom extract; (■), fed with diet containing 1% concentrated mushroom extract; (▲), fed with diet containing 10% mushroom extract. Data are presented in terms of mean ± standard deviation (n = 5). The values with different superscript letters represent significant difference (P <0.05) (Bao et al., 2009a).

Because ERT conserves endogenous antioxidants such as glutathione and α-tocopherol, it enhances the color stability of fish meat during postharvest handling and chilled storage. In addition, discoloration of meats is known to result from interaction between Mb and lipid oxidation (Renerre, 1990). Mb exists in 3 forms in fresh meat: deoxyMb, oxyMb, and metMb. metMB is an undesirable form not only because of its brown color but also because of its catalytic effect during the oxidation of unsaturated lipids (Love & Pearson, 1971). Grunwald & Richards (2006) reported that the reaction between metMb and lipid HPO generates ferrylMb, which can abstract a hydrogen atom from lipid (LH) to form an alkyl radical (L.). In the presence of oxygen, the alkyl radical then generates a peroxyl radical (LOO.), which can in turn abstract a hydrogen atom from another unoxidized lipid, resulting in the production of lipid HPO. The lipid HPOs generated are continuously involved in the redox reactions of Mb to generate radicals. Once underway, the alkoxyl radical (LO.) formed rearranges to generate an epoxyradical (epoxyL.), which reacts with molecular oxygen to form LOO.. The epoxyL. can also react with ferrylMB to form metMB, which continuously plays a role as an intermediate for oxidizing lipids in meat. In this case, the antioxidants that are present in the ME and accumulated in the meat could have acted as chain-breaking antioxidants. The radicals generated in meat from the interaction between Mb and lipid oxidation could be scavenged by the antioxidants, and the ferrylMb produced from this process could be reduced by donating electrons from the thiol groups of the antioxidants (Romero et al., 1992).

The proposed mechanism for ERT-delayed oxidation of lipid and Mb in minced big eye tuna (*Thunnus obesus*) meat to which *F. velutipes* extract has been added is shown in Figure 5.

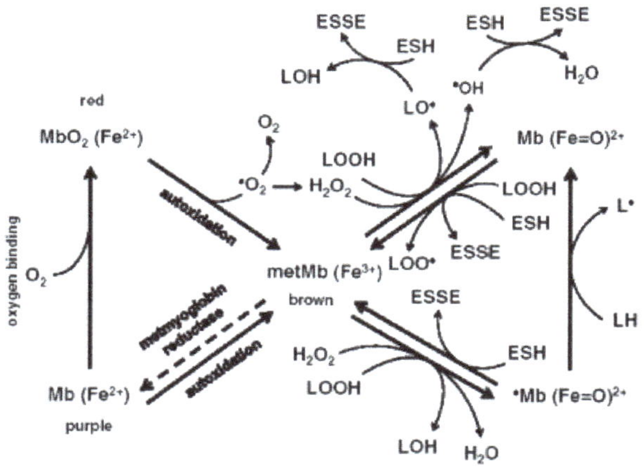

ESH, ergothioneine; ESSH, dimer of ergothioneine; LO, lipid; LO•, alcoxyl radical, LOOH, lipid hydroperoxide, LOO, Peroxy radical; Mb, myoglobin; Mb(Fe=O)2+, ferryl myoglobin; •Mb(Fe=O)2+, ferryl myoglobin radical; MbO2, oxymyoglobin; Mb•, myoglobin radical; metMb, metmyoglobin.

Fig. 5. Proposed mechanism for ERT-delayed oxidation of lipid and myoglobin in minced big eye tuna *(Thunnus obesus)* meat to which *F. velutipes* extract has been added (Bao et al., 2009b).

Thus, the delay in metMb formation in the dark muscle of yellowtail fed with ME might result from both suppression of lipid oxidation and conservation of endogenous antioxidants. The *in vivo* study of Bao et al. (2009a) was the first model to show that dietary supplementation with ME can be used on an industrial scale for preventing discoloration and lipid oxidation in the dark muscle of yellowtail.

4. MELANOSIS IN CRUSTACEANS

The development of melanosis or blackspot during the postharvest period of crustaceans is a well-known postmortem phenomenon attributed to the polymerization of phenol into an insoluble black pigment, melanin. Phenol polymerization is mainly initiated by the action of an enzymatic complex called polyphenoloxidase (PPO; also known as tyrosinase). Severe blackspot formations can cause tremendous economic losses because of the high value commanded by these aquatic products in the marketplace (Kim et al., 2002). Its prevention has been a challenge to the industry, especially for food scientists.

Intensive studies on PPO in crustaceans could help understand the mechanisms underlying the development of melanosis in crustaceans; the properties of PPO; their substrates and inhibitors; and the physical, biological and chemical factors that affect each of these parameters. Complete understanding of this phenomenon and the mechanisms underlying it may help provide a scientific approach to prevent melanosis and slow its rate, thus extending the shelf life and acceptability of the product.

4.1 PPO in melanosis development

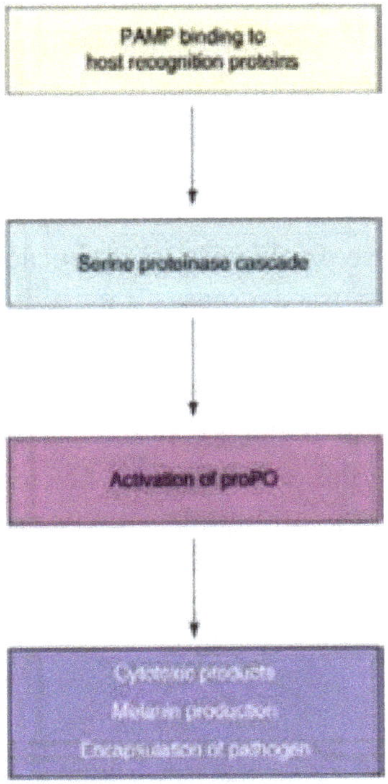

Fig. 6. Induction of the prophenoloxidase (proPO) cascade in the invertebrate immunity. Pathogen-associated molecular patterns (PAMPs; e.g., peptidoglycans, lipopolysaccharides, and β-1,3-glucans) are bound by host recognition proteins. This initiates a serine proteinase cascade that leads to the conversion of zymogenic proPO into catalytically active phenoloxidase and ultimately results in the generation of cytotoxic products and encapsulation of the pathogen (Cerenius et al., 2008).

PPO plays an important physiological role in crustaceans, particularly in the process of sclerotization. The melanization cascade has been reported to be closely associated with the occurrence of factors that stimulate cellular defence by aiding phagocytosis and encapsulation reactions (Cerenius et al., 2008). Thus, melanization is an important immune response in crustaceans, similar to that observed in plants. In plants, the compounds produced as a result of the polymerization of quinones have been reported to exhibit both antibacterial and antifungal activities, whereas in crustaceans, PPO is thought to be involved in wound healing and sclerotization of the cuticle. In live crustaceans, the activation of prophenoloxidase (proPO) to PPO requires proteases and microbial activators such as polysaccharide-binding proteins (García-Carreño et al., 2008). This activation system plays an important role in the primary immune response, cuticle sclerotization, and injury healing in crustaceans (Buda & Shafer 2005; Martínez-Álvarez et al., 2005; José-Pablo et al., 2009). Simple diagram on the induction of the proPO cascade in the invertebrate immunity has been shown in Figure 6.

Postharvest PPO-catalyzed blackening of the shell in crustaceans adversely affects both quality and the consumers' acceptance of these products. Crustaceans such as lobsters, shrimps, and crabs are extremely vulnerable to enzymatic blackening or melanosis. Although the occurrence of melanosis in these aquatic products does not necessarily mean that they are unfit for human consumption, consumers tend to be selective regarding these products, mainly because browning in the carapace connotes spoilage; thus, melanosis decreases the product's market value.

Enzymatic browning or melanosis takes place in the presence of oxygen when tyrosinase and their polyphenolic substrates are mixed. Tyrosinase catalyzes 2 fundamental reactions: 1) hydroxylation at the *o*-position adjacent to an existing hydroxyl group of phenolic compounds (monophenolase activity) and 2) oxidation of diphenol to *o*-benzoquinones (diphenolase activity).

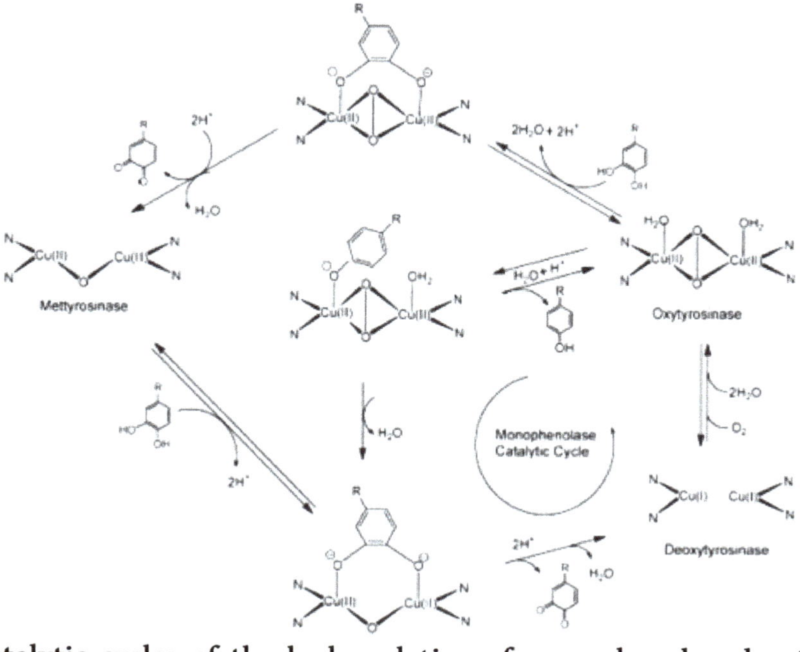

Fig. 7. Catalytic cycles of the hydroxylation of monophenol and oxidation of *o*-dipehol to *o*- quinone by tyrosinase (Kim and Uyama, 2005).

The subunits of tyrosinase have been reported to differ with respect to chemical, physical and kinetic properties, which were believed to be responsible for the relative affinities of the enzymes for both mono- and diphenolic substrates (Kim et al., 2002). The aforementioned considerations have led to many studies on the molecular mechanism underlying the monophenolase and diphenolase activity of tyrosinase. Many studies have been performed on the monophenolase activity of tyrosinase, based on 3 forms of the enzyme. The monophenolase cycle is shown in Figure 7, wherein monophenol reacts with the oxy form and binds at the axial position of one of the coppers ions of this oxy form. It has been reported that rearrangement through the trigonal bipyramidal intermediate leads to binding of peroxide and thereby generates a coordinated *o*-diphenol that is oxidized to *o*-quinone, resulting in a deoxy form ready for dioxygen binding. In the diphenolase cycle, both the oxy and met forms react with *o*-diphenol, oxidizing it to *o*-quinone. However, monophenol can compete with *o*- diphenol for binding to the met form site, thereby inhibiting its reduction (Kim & Uyama, 2005). Diphenolases have received much attention because of their high catalytic rate and their association with quinone formation that in turn leads to melanin production.

4.2 Control of melanosis

The fundamental step in melanosis is the transformation of an *o*-diphenol such as 3,4- dihydroxyphenylalenine (L-DOPA) to the corresponding *o*-quinone, which undergoes further oxidation to form brown or black pigments (Martinez & Whitaker, 1995). *o*-Quinones are known to be powerful electrophiles that can be attacked by water, other polyphenols, amino acids, peptides, and proteins, leading to Michael-type products (Mayer et al., 1990).

Many studies have been performed on controlling or inhibiting the activity of PPO in foods, and various mechanisms and techniques have been proposed and developed over the years to mitigate concerns regarding undesirable products of enzyme activity. This involves elimination of one or more of the important components involved in the enzymatic reaction, such as oxygen, copper, substrate, and the enzyme itself. There are also known compounds that react with the product of enzyme activity, thus inhibiting the formation of colored compounds. Many other techniques are also being applied to prevent melanosis in foods, such as processing methods, different types and kinds of inhibitors, and a molecular approach to control PPO, as presented in the following section.

4.2.1 Processing

Traditionally, heating is one of the most popular method to destroy microorganism and to inactivate enzymes. The catalytic activity of PPO is inhibited at temperatures ranging from 70ºC to 90ºC (Vamos-Vigyoso, 1981). Lee et al. (1988) reported that blanching green beans at temperatures above 82 ºC inactivated its enzymes, including PPO.

Similar to heat treatment, low-temperature treat can also help control enzyme-catalyzed reactions. At low temperatures, the kinetic energy of the reactant molecules decreases, subsequently leading to a decrease in the mobility and reaction necessary for the formation of enzyme-substrate complexes and their products (Kim et al., 2002). Thus, low-temperature preservation and storage are used during distribution and retailing to control the development of melanosis in food products. Temperatures as low as -18ºC can be used to inactivate enzymes. However, this changes the physical attributes of the products.

Dehydration can also be used to control enzyme catalyzed reactions that cause browning in foods. Generally, water greatly influences enzyme activity because it acts as a solvent or a reactant (Ashie & Simpson, 1996). Water activity can be controlled by physical drying (i.e., dehydration or freeze drying) or by chemical methods (i.e., addition of water-binding agents such as sugars and salts).

Irradiation is another method used for controlling enzymatic browning in foods. It uses ionizing radiation to inactivate microorganisms as well as enzymes. Irradiation is being increasingly recognized as a method for reducing postharvest losses, maintaining food quality, as well as ensuring the hygienic quality of food products.

The rearrangement or destruction of noncovalent bonds such as hydrogen bonds, hydrophobic interactions, and ionic bonds of the tertiary protein structure can lead to enzyme denaturation. This can be achieved by high-pressure treatment that reduces molecular spacing and increases interchain reactions. However, this varies depending the enzymes; the nature of the substrate; and the temperature and duration of high-pressure processing (Cheftel, 1992).

Other processing methods that have been developed and used to control the enzymatic browning of food products included the treatment of food with supercritical carbon dioxide, ultrafiltration, or ultrasonication (Kim et al., 2002). Treatment with supercritical carbon dioxide inactivates the enzyme through carbonic acid production, which causes a significant pH reduction. In contrast, ultrafiltration removes PPO during the process, whereas ultrasonication causes severe shear stresses, thereby promoting enzyme denaturation.

4.2.2 Inhibitors

Melanosis can be inhibited by targeting the enzyme, the substrate, and/or the reaction products involved in the process. Inhibition that targets the enzyme includes metal chelators and compounds that consist of carboxylic acids of the benzoic and cinnamic series. These compounds can be competitive owing to their structural similarity with phenolic substrates. Inhibition that targets the substrate can be performed by removing either oxygen or the phenolic substrate from the reaction. Inhibition targeting the products involves compounds that can react with *o*-quinones to produce a colorless addition product.

These PPO inhibitors can be classified into 6 categories that include: 1) reducing agents such as sulfites, ascorbic acid and its analogs, cysteine, and glutathione; 2) chelating agents such as phosphates, EDTA, and organic acids; 3) acidulants such as citric acid and phosphoric acid; 4) enzyme inhibitors such as aromatic carboxylic acids, aliphatic alcohols, anions, peptides, and substituted resorcinols; 5) treatments with enzymes such as oxygenases, *o*- methyltransferase and proteases; and 6) complexing agents such as cyclodexrins (McEvily et al., 1992). To date, there have been many studies on the different compounds used to inhibit melanosis in foods.

4.2.3 Molecular approach

The nature of PPO and its role in the immune system of plants and crustaceans have been extensively reviewed. One possible approach to reduce the activity of PPO and the resultant enzymatic browning reactions is to characterize and inactivate the PPO coding genes. This can be done by generating antisense RNAs specific for PPO.

Antisense genes have been successfully used for altering plant processes and improving crops. It involves blocking gene expression of the plant enzymes involved in a certain process. The process is based on blocking the flow of information from DNA via RNA to protein by introducing an RNA strand complementary to the sequence of the target mRNA. The antisense approach involves the insertion of a gene or a significant part of it into the cell in a reverse orientation. This approach has been used to increase the shelf life of fruits (Fray & Grierson, 1993), and commercial applications of this technology now include alteration of flower color, viral resistance induction, and fruit ripening. It has been reported that the lack of bruising sensitivity in transgenic potatoes without any side effects opens up the possibility of preventing melanosis in wide varieties of food crops and even to crustaceans, without the use of any physical and chemical treatments. Thus, by using this antisense technology, it is possible to develop fruits, vegetables and crustaceans that are resistant to enzymatic browning.

4.3 Application of ERT-rich edible mushroom (*Flammulina velutipes*) extract for controlling melanosis

As mentioned earlier, many technologies and techniques have been developed for the inhibition of melanosis in food products, and some more new approaches are under study. However, these alternatives must be evaluated in terms of their impact on the overall food quality, effectiveness, cost, and regulatory status. Because of health concerns, only a number of inhibitors have been accepted for food application, based on the basis of government health regulations.

A relatively new approach involves the design and development of another application of ERT-rich edible mushroom (*F. velutipes*) extract for inhibiting melanosis in commercially important crustaceans.

4.3.1 Dietary supplementation

The extract from the edible mushroom *F. velutipes* has been found to significantly inhibit mushroom PPO activity; prevent browning in apples; and delay melanosis in shrimps which usually develops during storage (Jang et al., 2002; 2003). These observations suggest that the mushroom extract contains certain compounds contributing to such actions. In section 3, it has been mentioned that ME containing ERT at the level of 3.03 ± 0.07 mg/mL has been reported to show remarkable DPPH radical scavenging activity and suppress lipid oxidation in bigeye tuna meat (Bao et al., 2008). ME have been reported to stabilize the fresh color of tuna meat during ice storage (Bao et al., 2009a). These results strongly suggest that ERT could be one of the major compounds in the ME that effectively inhibited mushroom PPO activity, apple browning, and melanosis in shrimps in the previous studies (Jang et al., 2002; 2003).

The *in vivo* application of L-ERT as an antioxidant in rats (Deiana et al., 2004) and the effects of feeding fish with ERT-rich ME have been studied; however, before this study, no study had been performed to determine the efficacy of extracts that contains this compound, when fed to crustaceans to control melanosis. Therefore, a study was performed on the use of a hydrophilic extract prepared from the fruiting body of *F. velutipes* for controlling melanosis and lipid oxidation in cultured kuruma shrimps (*Marsupenaeus japonicus*) *in vivo*.

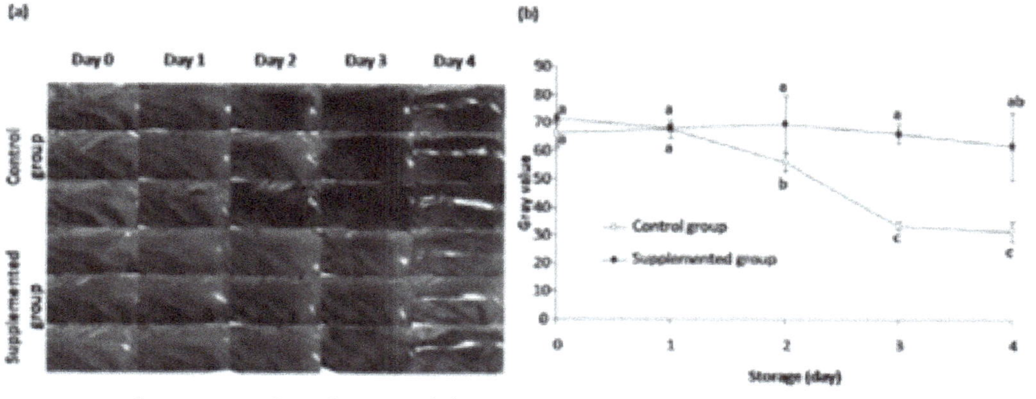

Fig. 8. Development of melanosis (a) and changes in the mean gray values of the carapace area (b) of kuruma shrimps (*Marsupenaeus japonicus*) during ice storage. Results are presented interms of mean±standard deviation ($n = 3$). The values with different superscript letters represent significant difference between groups for different storage periods ($p < 0.05$) (Encarnacion et al., 2010).

In a study performed by Encarnacion et al. (2010), the hot water ME containing 2.05 ± 0.24 mg/mL ERT remarkably inhibited mushroom PPO. The PPO activity is usually found to be higher in the carapace area during postmortem of crustaceans (José-Pablo et al., 2009) and this study showed that the development of melanosis in the carapace area of the shrimps during ice storage was relatively controlled in the ME-fed group than in the control (Fig. 8). The accumulation of ERT from the ME added in the diet during feeding could have contributed to this effect.

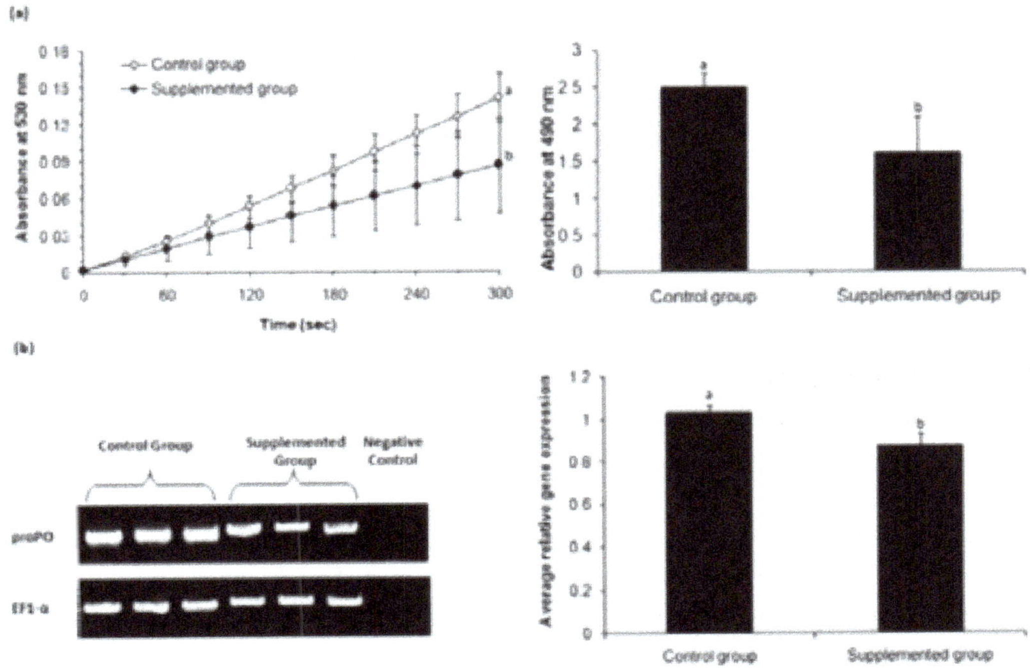

Fig. 9. (a) Enzymatic oxidation of catechol and L-DOPA (b) in the hemolymph of kuruma shrimps (*Marsupenaeus japonicus*) after 7 days of feeding. Results are presented in terms of mean ± standard deviation ($n = 4$). The values with different superscript letters represent significant difference ($p < 0.05$) at the end of the reaction period for each assay. (b) Gene expression analysis of prophenoloxidase (proPO) transcripts in the hemocytes of hemolymphs of kuruma shrimp (*M. japonicus*) after 7 days of feeding (representative gel and quantitative analyses of band intensities obtained using the ImageJ software. Elongation factor1-α gene (EF1-α) was used as the standard for computing relative gene expression level for each sample band. Results are presented in terms of mean ± standard deviation ($n = 3$). The values with different superscript letters represent significant difference ($p < 0.05$) (Encarnacion et al., 2010).

In the same study, the PPO activities of hemolymph from shrimps fed a diet containing the ME was significantly lower than that of the control group, and the expression of the proPO genes in the hemocytes of shrimps fed with the ME was relatively lower than that in the hemocytes of the control samples, showing that the ME inhibited PPO activation in the hemolymphs of the supplement-fed shrimps (Fig. 9). Because PPO activity mainly depends on the activation of the proPO system, decreasing the expression of proPO genes in hemocytes consequently reduces the activity of PPO.

The results of a study performed by Amparyup et al. (2009) indicated the significant role of the proPO activating system as one of the major immune responses in shrimps and also reported that inhibiting proPO activation in shrimps could cause death. However, the study did not show any significant differences in mortality between the supplement-fed shrimps the control samples. A study by Leclerc et al. (2006) reported that microbial infections in *Drosophila* did not require the activation of the proPO system for survival thus, raising questions regarding the precise function of phenoloxidase activation and its level of participation in the immune defense system of invertebrates because they use 2 broad but interacting categories of defense responses against pathogens, namely, the cellular and hormonal responses.

4.3.2 Immersion technique

Because feeding shrimps with MEs is time consuming and requires tedious technical work, the possibility of immersing live full-grown crustaceans in an ME solution for inhibiting postharvest melanosis was considered. Martínez-Álvarez et al. (2005) reported the efficacy of immersing shrimp in 0.05% hexylresorcinol (HR) for preventing postharvest melanosis. *In vivo* treatment with antimelanosic agents may be more effective than postmortem treatment because of greater absorption of antimelanosic agents through the gills, increased accumulation in the hemolymph, and enhanced distribution throughout the body. Therefore, the effects of *in vivo* treatment with *F. velutipes* ME, sodium ascorbate (ascorbic acid, AA), sodium sulfate (SS), and HR on postharvest melanosis in crustaceans were compared.

The study conducted by Encarnacion et al. (2011a) showed that immersing live *M. japonicus* shrimps in ME effectively inhibited postharvest melanosis in the shrimps (Fig. 10). Moreover, the development of melanosis in the carapace decreased with increase in the concentration of ME in the immersion solution. These effects may be due to greater absorption and accumulation of ERT in the shrimps with increase in the concentration of ME in the immersion solution. At 0.5 % ME concentration in pure seawater, melanosis was effectively inhibited in shrimp samples after 2 days of ice storage; the effect was the same as that obtained with 500 ppm HR and better than obtained with 500 ppm AA.

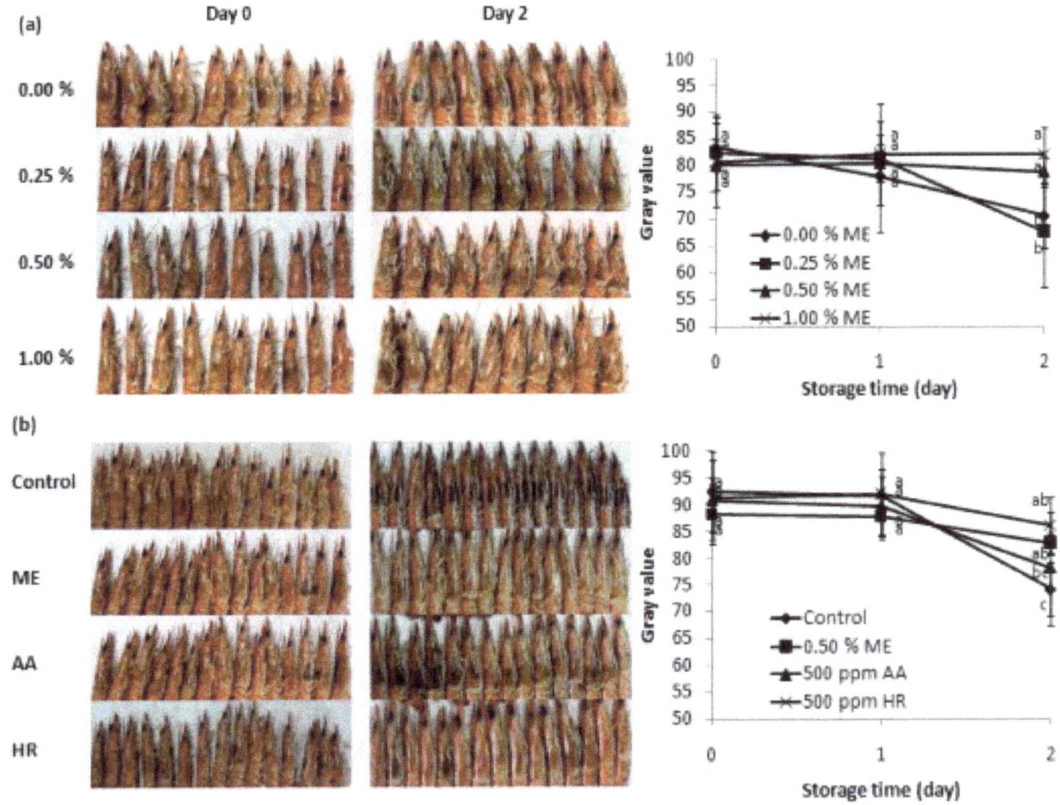

Fig. 10. Development of melanosis and changes in the mean gray value of the carapace of *Marsupenaeus japonicus* shrimps before and after ice storage: (a) dose-dependent effect of ME, (b) comparison with other antimelanosic compounds. Results are presented in terms of mean ± standard deviation ($n = 3$). The superscript letters above each data point represent statistically significant differences ($p < 0.05$) (Encarnacion et al., 2011a).

The study performed by Encarnacion et al. (2011b) also used this technique for the red queen crab, *Chionoecetes japonicus,* and the result clearly showed that treating live crabs with purified seawater containing 1.0% ME effectively prevented melanosis during 4 days of ice storage; and the effect obtained the same as that with 500 ppm SS or with 500 ppm HR solutions (Fig. 11).

ERT accumulation in shrimp tissue may inhibit PPO activity in the carapace during the postharvest period. Because thiols, such as ERT, are powerful nucleophiles that can chelate Zn2+ and Cu2+ (Hanlon, 1971; Park et al., 2006), the mechanism underlying the inhibitory effect of ERT could be due to its Cu2+ chelating activity. The mechanism of ERT could also be attributed to its Cu2+ chelating activity; thus, melanosis in the shrimps immersed in the ME solution was inhibited.

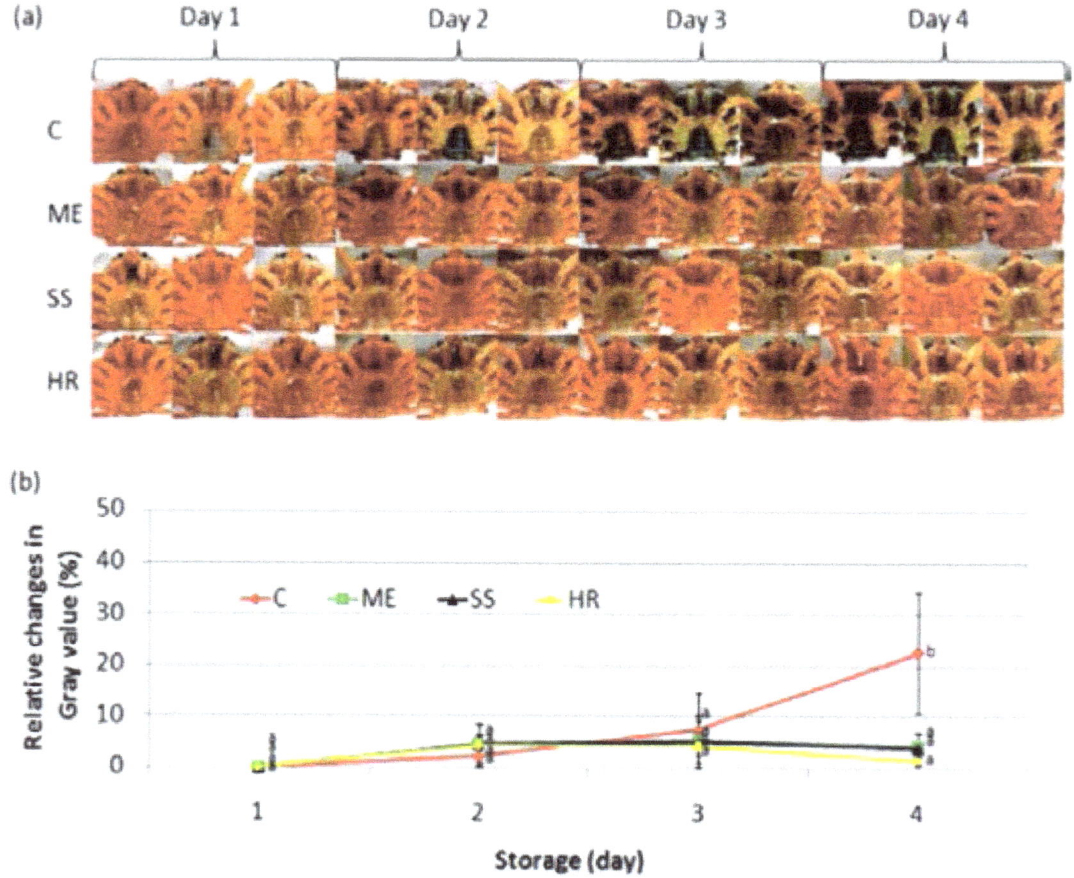

Fig. 11. (a) Digital photographs of the development of melanosis in *Chionoecetes japonicus* crabs pretreated with purified seawater (C), 1.0% mushroom extract (ME), 500 ppm sodium sulphite (SS), and 500 ppm 4-hexyl-1,3-benzenediol (HR) solutions during ice storage. (b) Relative changes in the mean gray value of the carapace of *C. japonicus* during ice storage, analyzed using the ImageJ software. Results are presented in terms of mean (SD) (n = 3). The superscript letters above each data point represent statistically significant differences ($p < 0.05$) (Encarnacion et al., 2011b).

In addition to inhibiting PPO activity, ME also suppressed proPO gene expression in the hemolymph of *M. japonicus* (Fig. 12) and *C. japonicus* (Fig. 13). The same result was obtained in the feeding trial. Because immersing shrimp in ME significantly reduced PPO activity in the hemolymph, it is possible that the ME containing ERT inhibited PPO activation. The observation that proPO gene expression in the hemocytes of ME-treated shrimps was lower than that in control shrimps supports this hypothesis. Thus, decreasing the expression of proPO reduces the amount that can be proteolytically activated into PPO (Adachi et al., 2003).

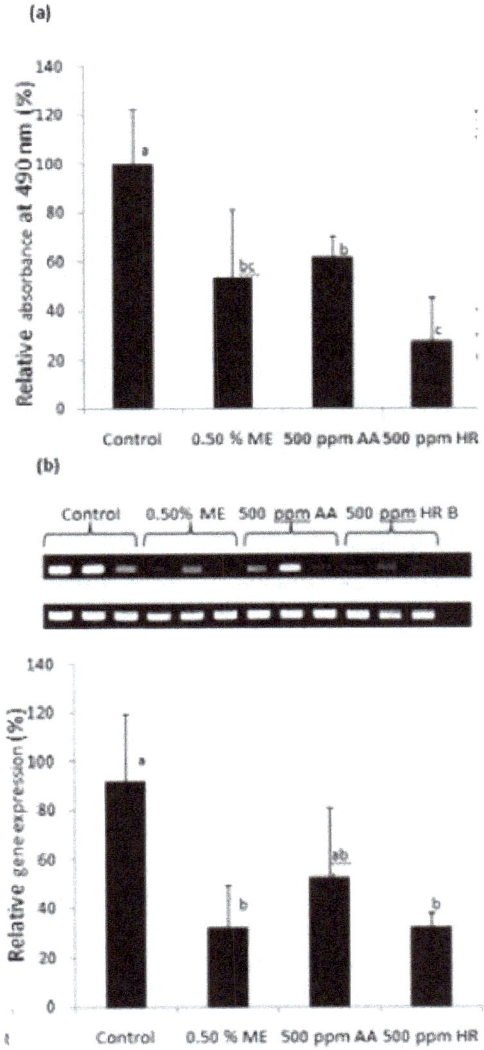

Fig. 12. (a) Enzymatic oxidation of L-DOPA in the hemolymph of kuruma shrimps (*M. japonicus*) immersed in control or treatment solutions. In each graph, the average absorbances of samples are shown relative to the average absorbance of the control (defined as 100%). Results are presented in terms of mean ± SD (n = 5). (b) Gene expression analysis of prophenoloxidase (proPO) transcripts in the hemocytes of *M. japonicus* shrimp immersed in control or treatment solutions. The top and bottom panels show representative gels and quantitative analyses of the band intensities obtained using the ImageJ software, respectively. The expression level of the elongation factor 1α (EF1-α) gene was used as the reference for calculating the relative expression level for each band. B denotes a blank sample (negative control). Results are presented in terms of mean ± SD (n = 3). The superscript letters above each bar represent statistically significant differences ($p < 0.05$) (Encarnacion et al., 2011a).

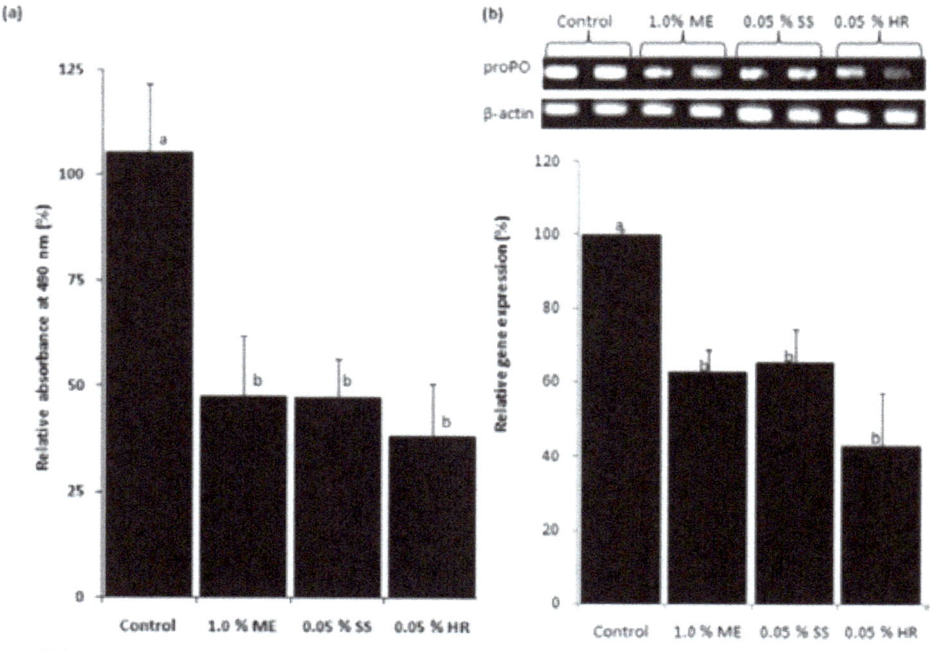

Fig. 13. (a) Enzymatic oxidation of L-DOPA in the hemolymph of red queen crabs (*Chionoecetes japonicus*) immersed in control or treatment solutions. In each graph, the average absorbances of samples are shown relative to the average absorbance of the control (defined as 100%). Results are presented in terms of mean ± SD (*n* = 5). (b) Gene expression analysis of prophenoloxidase (proPO) in the hemocytes of *C. japonicus* immersed in purified seawater (Control), 1.0% mushroom extract (ME), 500 ppm sodium sulphite (SS), or 500 ppm 4-hexyl-1,3-benzenediol (HR) solutions, performed using reverse transcription–polymerase chain reaction. (b) Quantitative analysis of the band intensities obtained using the ImageJ software. The expression level of the β-actin gene was used as the reference for carculating the relative expression level for each band. Results are presented in terms of mean (SD). The superscript letters above each bar represent statistically significant differences ($p < 0.05$) (Encarnacion et al., 2011b).

The efficacy of ME in inhibiting postharvest melanosis is comparable to the efficacy of existing antimelanosic compounds, such as HR, AA and SS in this regard. HR forms an inactive complex with PPO that is incapable of catalyzing melanosis development (Guererro-Beltran et al., 2005). Therefore, PPO activity significantly decreases in the presence of HR (Encarnacion et al., 2010; 2011). HR possesses antimicrobial activity in addition to its antimelanosic activity. The antimicrobial activity of HR (Martínez-Álvarez et al., 2005) may be related to the decrease in proPO gene expression because invading microorganisms trigger a proteolytic cascade culminating in the release of active PPO (Leclerc et al., 2006). Because AA did not significantly affect proPO gene expression, its mechanism of inhibiting postharvest melanosis is most likely to be due to its reducing power. The slight melanosis that occurred in AA-immersed shrimps after 2 days of ice storage may be due to the oxidation of AA, which prevented it from reducing *o*-quinones, thereby allowing melanin development (Guererro-Beltran et al., 2005). Sulfites are known to inhibit both enzymatic and nonenzymatic browning reactions. Bisulphite is reported to inhibit melanosis by 2 mechanisms: 1) by reacting with intermediate quinones in the melanosis reaction that forms sulfoquinones; and 2) by irreversibly reacting with PPO, causing complete inactivation (Ferrer et al., 1989). Therefore, PPO activity also significantly decreases in the presence of SS. SS and HR posses antimicrobial activities in addition to their antimelanosic activities, SS and HR are also antimicrobial agents. The antimicrobial activity of SS (McFeeters et al., 2004; Martínez-Álvarez et al., 2005) may also be related to the decrease in proPO gene expression.

4.3.3 Mechanism of how the extract inhibits melanosis

The mushroom hot water ME remarkably inhibited mushroom PPO. Commercial L-ERT had the same effect on mushroom PPO. Mushroom PPO activity remarkably declined with increase in the concentrations of L-ERT (Fig. 14). PPO inhibition by the ME was also depended on L-ERT concentration. These results suggest that extracts containing ERT as one of their active compounds potently inhibit mushroom PPO activity.

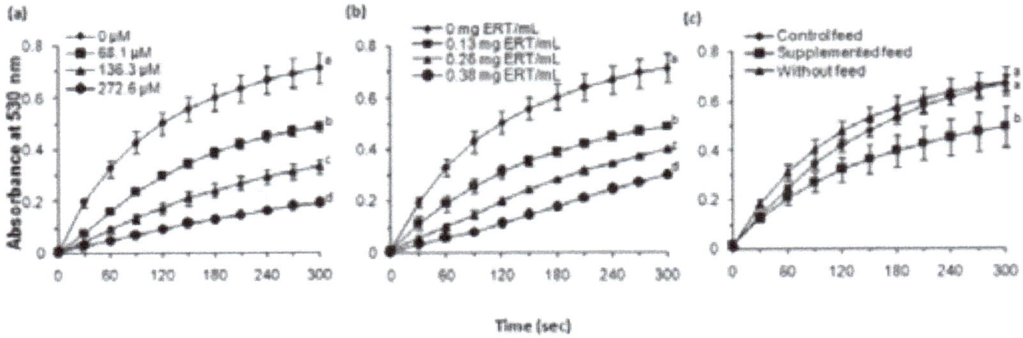

Fig. 14. Inhibitory effects of commercial L-ergothioneine (a), mushroom (*F. velutipes*) extract (b), and residues of diets used in the feeding trial (c) on the activity of mushroom polyphenoloxidase. Results are presented in terms of mean ± standard deviation (*n*=3). The values with different superscript letters represent significant difference (*p*<0.05) at the end of the reaction period (Encarnacion et al., 2010).

In vitro experiments performed by Encarnacion et al. (2010) also showed that the transcript expression of the proPO genes in the HLS was lower than that in the L-ERT- and *p*- amidinophenyl methanesulfonyl fluoride hydrochloride (*p*-APMSF)-treated HLSs (Fig. 15). The PPO activity in the L-ERT-treated HLS was also remarkably low. ERT could have been involved in inhibiting transcriptional factors in the cascade system, leading to a decrease in proPO gene expression. Maeda et al. (2007) reported that in human melanosis, hydroperoxy traxastane-type triterpene decreased the protein levels of PPO and its related proteins in B16 melanoma cells because of inhibition of the transcription factor melanocyte-type isoform of the microphthalmia-associated transcription factor, leading to a decrease in the PPO gene and its related genes. In the case of crustaceans, it is generally believed that the enzyme responsible for the activation of proPO to PPO is a serine protease, or commonly called as proPO-activating enzyme (PPAE). This enzyme is believed to be involved in the final step in

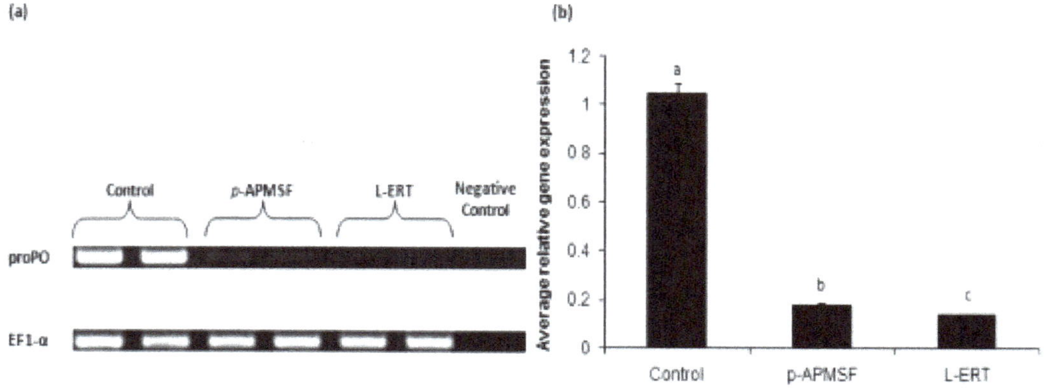

Fig. 15. Inhibition of the activation of the proPO system in hemocyte lysate supernatant by commercial L-ergothioneine (L-ERT) and *p*-amidinophenyl methanesulfonyl fluoride hydrochloride (*p*-APMSF). Polyphenoloxidase activity (a) and peptidase activity (b). 7- amino-4-methylcoumarin (AMC) was used as the standard solution in the peptidase activity assay. The final concentration of each inhibitor in the reaction system was 1 mM. Results are presented in terms of mean ± standard deviation (n = 3). The values with different superscript letters represent significant difference between groups for different storage periods ($p < 0.05$) (Encarnacion et al., 2010).

the proPO cascade leading to PPO activation and is itself tightly regulated (Buda & Shafer, 2005). The results of the peptidase activity assay indicated that L-ERT inhibited serine protease activity but that this effect was not as strong as that of *p*-APMSF (Fig. 16b). However, the presence of L-ERT could possibly affect the overall activation of the proPO system because it inhibited PPO activity (Fig.16a). At least 4 mechanisms underlying PPAE regulation have been recognized: gene induction, activation by another protease, requirement for noncatalytic serine protease homologues as cofactors, and inactivation by serine protease inhibitors (Wang & Jiang, 2004). Thus, other proPO activating factors such as serine protease homologues and other proteases could have been inhibited by L-ERT. The inhibition of PPO activity in the hemolymphs of the supplement-fed shrimp supports this phenomenon. Moreover, accumulation of ERT in the shrimp muscles by feeding or in the hemolymph by the immersion technique could directly inhibit PPO activity in the carapace during the postharvest period. The thiol (SH) group, present in compounds such as ERT, is a powerful nucleophile, that tends to chelate Zn2+ and Cu2+ (Park et al., 2006). The latent PPO activity of hemocyanin, a copper-binding protein, in whiteleg shrimp *Penaeus vannamei* has been shown to be involved in postmortem melanosis. The mechanism underlying the inhibitory effect of ERT could also be attributed to its Cu2+ chelating activity, leading to the inhibition of melanosis in ME-fed and ME-immersed shrimps and crabs.

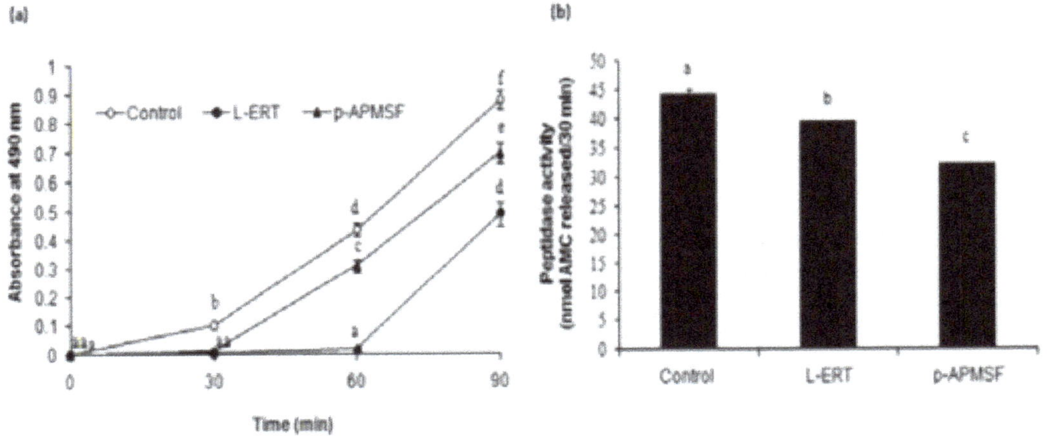

Fig. 16. Gene expression analysis of proPO transcripts in hemocyte lysate supernatanttreatedwithcommercialL-ergothioneine(L-ERT)and*p*-amidinophenyl methanesulfonyl fluoride hydrochloride (*p*-APMSF). Representative gel (a) and quantitative analyses of band intensities obtained using the ImageJ software (b). Elongation factor1-α gene (EF1-α) was used as the standard for computing the relative gene expression level for each sample band. The final concentration of the inhibitor in the reaction system was 1mM (Encarnacion et al., 2010).

5. AT A GLANCE

Mushroom trimmings are an economical and expandable source of metabolizable energy (ME) and contain substantial levels of ergothioneine (ERT), which has demonstrated antioxidant and melanosis-inhibiting capabilities. Yellowtail fish (Seriola quinqueradiata) fed a meal with marine oil extract (ME) exhibited decreased myoglobin and lipid oxidation in dark muscles while stored at low temperatures. The feeding trial results are corroborated by data from comparable studies conducted on other aquacultured fish species like jack mackerel and rainbow trout (data not displayed). Submerging live M. japonicus shrimps and C. japonicus crabs in an F. velutipes ME solution with a high concentration of ERT successfully prevented postharvest melanosis in both shrimps and crabs. This outcome was similar to the effects of melanosis-inhibiting substances like HR, AA, and SS. Experiments conducted in a controlled environment using externally supplied ERT demonstrated its ability to suppress PPO activity and proPO gene expression in shrimp HLS. ME with ERT is a viable natural alternative to pharmaceutical melanosis inhibitors. Agents employed to inhibit postharvest melanosis in shrimps and other crustaceans. Moreover, utilizing biowaste as the primary source for extracting antioxidants or melanosis-inhibiting substances like ERT for culinary and industrial purposes could aid in mitigating the increasing issue of agricultural waste.

REFERENCES

1. Adachi, K.; Hirata, T.; Nishioka, T. & Sakaguchi, M. (2003). Hemocyte components in crustaceans convert hemocyanin into phenoloxidase-like enzyme. *Comparative Biochemistry and Physiology Part B,* Vol. 134, No. 1, (January 2003), pp. 135-41, ISSN 1096-4959
2. Akanmu, D.; Cecchini, R.; Aruoma, O. I. & Halliwell, B. (1991). The antioxidant action of ergothioneine. *Archives of Biochemistry and Biophysics,* Vol. 288, No. 1, (July 1991), pp. 10-16, ISSN 0003-9861
3. Aksnes, A. & Njaa, L. R. (1981). Catalase, glutathione peroxidase and superoxidase dismutase in different fish species. *Comparative Biochemistry and Physiology Part B,* Vol. 69, No.4, pp. 893-896, ISSN 1096-4959
4. Amparyup, P.; Charoensapsri, W. & Tassanakajon, A. (2009). Two prophenoloxidases are important for the survival of *Vibrio harveyi* challenged shrimp *Penaeus monodon*. *Developmental and Comparative Immunology*, Vol. 33, No. 2, (February 2009), pp. 247-256, ISSN 0145-305X
5. Arduini, A.; Eddy, L. & Hocstein, P. (1990). The reduction of ferryl myoglobin by ergothioneine: A novel function for ergothioneine. *Archives of Biochemistry and Biophysics,* Vol. 281, No. 1, (August 1990), pp. 41-43, ISSN 0003-9861
6. Ashida, S. & Sato, M. (2005). Screening of edible plants for reducing activity by monitoring their effects on the oxidation of oxymyoglobin. *Food Science and Technology Research,* Vol. 11, No. 3, (October 2005), pp. 349-354, ISSN 1344-6606
7. Ashie, I.N.A. & Simpson, B.K. (1996). α;2 Macroglobulin inhibition of endogenous proteases in fish muscle. *Journal of Food Science,* Vol. 61, No. 2, (March, 1996) pp. 357-361, ISSN 1750-3841
8. Bao, H.N.D.; Ushio, H. & Ohshima, T. (2008). Antioxidative activity and antidiscoloration efficacy of ergothionine in mushroom (*Flammulina velutipes*) extract added to beef and fish meats. *Journal of Agricultural and Food Chemistry,* Vol. 56, No. 21, (October 2008), pp. 10032-10040, ISSN 0021-8561
9. Bao, H.N.D.; Ushio, H. & Ohshima, T. (2009a). Antioxidative activities of mushroom (*Flammulina velutipes*) extract added to bigeye tuna meat: dose-

dependent efficacy and comparison with other biological antioxidants. *Journal of Food Science*, Vol. 74, No. 2, (March 2009) pp. 162-169, ISSN 1750-3841

10. Bao, H.N.D.; Shinomiya, Y.; Ikeda, H. & Ohshima, T. (2009b). Preventing discoloration and lipid oxidation in dark muscle of yellowtail by feeding an extract prepared from mushroom (*Flammulina velutipes*) cultured medium. *Aquaculture,* Vol. 295, No. 4, (October 2009), pp. 243-249, ISSN 0044-8486
11. Boldyrev, A. A.; Dupin, A. M.; Bumin, A. Y.; Balbizhaev, M. A. & Severin, S. E. (1987). The antioxidative properties of carnosine, a natural histidine containing dipeptide. *Biochemistry International,* Vol. 15, No. 6, (June 1987), pp. 1105-1113, ISSN 0158-5231
12. Buda, E.S. & Shafer, T.H. (2005). Expression of a serine proteinase homolog prophenoloxidase-activating factor from the blue crab, *Callinectes sapidus*. *Comparative Biochemistry and Physiology Part B*, Vol. 140, No. 4, (April 2005), pp. 521- 531 ISSN 1096-4959
13. Cerenius, L.; Lee, B.L. & Söderhäll, K. (2008). The proPO-pros and cons for its role in invertebrate immunity. *Trends in Immunology,* Vol. 29, No.6, (June 2008), pp. 263- 271, ISSN 1471-4906
14. Christie, W. W. (1987). Lipids: their structures and occurrence, In: *High Performance Liquid Chromatography and Lipids. A Practical Guide*, W. W. Christie (Ed.), 1-20, Paragon Press, ISBN 008-03421-2-4, Oxford, USA
15. Cheftel, J. C. (1992). High-pressure, microbial inactivation and food preservation. *Food Science and Technology International,* Vol. 1, No. 2-3, (August 1992), pp. 75-90, ISSN 1532-1738
16. Cheung, L. M. & Cheung, P. C. K. (2005). Mushroom extracts with antioxidant activity against lipid peroxidation. *Food Chemistry,* Vol. 89, No. 3, (February 2005), pp. 403-409, ISSN 0308-8146
17. Deiana, M.; Rosa, A.; Casu, V.; Piga, R.; Dessí, M. & Aruoma, O.I. (2004). L-ergothioneine modulates oxidative damage in the kidney and liver of rats in vivo: studies upon the profile of polyunsaturated fatty acids. *Clinical Nutrition,* Vol. 23, No. 2, (April 2004), pp. 183-193, ISSN 0261-5614
18. Dubost, N.J.; Beelman, R.; Peterson, D. & Royse, D. (2006). Identification and quantification of ergothionine in cultivated mushrooms using liquid chromatography-mass spectroscopy. *International Journal of Medicinal Mushroom*, Vol. 8, No.3, (July 2006), pp. 215-222, ISSN 1521-9437
19. Dubost, N. J.; Ou, B. & Beelman, R. B. (2007). Quantification of polyphenols and ergothioneine in cultivated mushrooms and correlation to total antioxidant capacity. *Food Chemistry*, Vol. 105, No. 2, (February 2007), pp. 727- 735, ISSN 0308- 8146

20. Dziedric, S. Z. & Hudson, B. J. F. (1984). Phenolic acids and related compounds as antioxidants for edible oils. *Food Chemistry,* Vol. 14, No. 1, (January 1984), pp. 45-51, ISSN 0308-8146

21. Elmastas, M.; Isildak, O.; Turkekul, I. & Temur, N. (2007). Determination of antioxidant activity and antioxidant compounds in wild edible mushrooms. *Journal of Food Composition Analysis,* Vol. 20, No. 3-4, (May 2007), pp. 337-345, ISSN 0889-1575

22. Encarnacion, A. B.; Fagutao, F.; Hirono, I.; Ushio, H. & Ohshima, T. (2010) Effects of ergothioneine from mushrooms (*Flammulina velutipes*) on melanosis and lipid oxidation of kuruma shrimp (*Marsupenaeus japonicus*). *Journal of Agricultural and Food Chemistry*, Vol. 58, No. 4, (January 2010), pp. 2577-2585, ISSN 0021-8561

23. Encarnacion, A. B.; Fagutao, F.; Hirayama, J.; Tereyama, M.; Hirono, I. & Ohshima T. (2011a). Ergothioneine from edible mushroom (*Flammulina velutipes*) as a potent inhibitor of melanosis in kuruma shrimp (*Marsupenaeus japonicus*). *Journal of Food Science*, Vol. 1, No. 76, (January/February 2011), pp. 52-58, ISSN 1750-3841

24. Encarnacion, A. B.; Fagutao, F.; Shozen, K.; Hirono, I., & Ohshima, T. (2011b) Biochemical intervention of ergothioneine-rich edible mushroom (*Flammulina velutipes*) extract inhibits melanosis in the crab (*Chionoecetes japonicus*). *Food Chemistry,* Vol. 127, No. 4, (August 2011), pp. 1594-1599, ISSN 0308-8146

25. Erickson, M.C. (1992). Changes in lipid oxidation during cooking of refrigerated minced channel catfish muscle. In: *Lipid oxidation in food,* A.J. St. Angelo (Ed.), 344-351, American Chemical Society, ISBN 084-12246-1-7, USA

26. Ferrer, O.; Otwell, W. & Marshall, M. (1989). Effect of bisulfate on lobster shell phenoloxidase. *Journal of Food Science.* Vol, 54, No. 2, (March 1989), pp. 478-480, ISSN 1750-3841

27. Food and Agriculture Organization (FAO) of the United Nations. (2006). Aquaculture, In: *The State of World Fisheries and Aquaculture 2006,* 11.05.2011, Available from: http://www.fao.org/docrep/009/A0699e/A0699e00.htm

28. Frankel, E.N. (1998). Antioxidants, In: *Lipids Oxidation*, E. N. Frankel, (Ed), 209-258, The Oily Press, ISBN 095-14171-9-3, Dundee, Scothland

29. Franzoni, F.; Colognato, R.; Galetta, F.; Laurenza, I.; Barsotti, M.; Stefano, R. D.; Bocchetti, R.; Regoli, F.; Carpi, A.; Balbarini, A.; Migliore, L. & Santoro, G. (2006). An *in vitro* study on the free radical scavenging capacity of ergothioneine comparison with reduced glutathione, uric acid and trolox. *Biomedicine and Pharmacotherapy*, Vol. 60, No. 8, (September 2006), pp. 45-47, ISSN 0753-3322

30. Fray, R.G. & Grierson D. (1993). Molecular genetics of tomato fruit ripening. *Trends in Genetics,* Vol. 9, No.12, (December 1993), pp. 438-443, ISSN 168-9525

31. Fu, H; Shieh, D. & Ho, C. (2002). Antioxidant and free radical scavenging activities of edible mushrooms. *Journal Food Lipids* Vol.9, No. 1, (March 2002), pp. 35-46, ISSN 1065-7258

32. García-Carreño, F.L.; Cota, K. & Navarrete del Toro, M.A. (2008). Phenoloxidase activity of hemocyanin in whiteleg shrimp *Penaeus vannamei*: conversion, characterization of catalytic properties, and role in postmortem melanosis. *Journal of Agricultural and Food Chemistry,* Vol. 56, No. 15, (July 2008), pp. 6454-6459, ISSN 0021-8561

33. Grosch, W.; Konopka, U. C. & Guth, H. (1992). Characterization of off-flavors by aroma extract dilution analysis, In: *Lipid oxidation in food,* A.J. St. Angelo (Ed), 266-279, American Chemical Society, ISBN 084-12246-1-7, USA

34. Grunwald, E. W. & Richarsd, M. P. (2006). Mechanisms of heme protein-mediated lipid oxidation using hemoglobin and myoglobin variants in raw and heated washed muscle. *Journal of Agricultural and Food Chemistry,* Vol. 54, No. 21, (September 2006), pp. 8271- 8280, ISSN 0021-8561

35. Guererro-Beltran, J.; Swanson, B. & Barbosa-Cánovas, G. (2005). Inhibition of polyphenoloxidase in mango puree with 4-hexylresorcinol, cysteine and ascorbic acid. *LWT-Food Science and Technology,* Vol. 38, No.6, (September 2005) pp. 625-630, ISSN 0023-6438

36. Hanlon, D. (1971). Interaction of Ergothioneine with metal ions and metalloenzymes. *Journal of Medicinal Chemistry,* Vol. 14, No. 1, (January 1971). pp. 1084-1087, ISSN 0023-6438

37. Hermes, J.E. (1998). *Fish Processing in the Tropics*, Tawid Publications, ISBN 971-91395-7-9, Quezon City, Philippines.

38. Jacobsen, C. (1999). Sensory impact of lipid oxidation in complex food systems. *Lipids*, Vol. 101, No. 5, (May 1999), pp. 484-490, ISSN 0024-4201

39. Jang, M.S.; Sanada, A.; Ushio, H.; Tanaka, M. & Ohshima, T. (2002). Inhibitory effect of 'enokitake' mushroom extracts on polyphenol oxidase and prevention of apple browning. *LWT-Food Science and Technology,* Vol. 35, No. 8, (December 2002), pp. 697-702, ISSN 0023-6438

40. Jang, M.S.; Sanada, A.; Ushio, H.; Tanaka, M. & Ohshima, T. (2003). Inhibitory effect of enokitake extract on melanosis of shrimp. *Fisheries Science*, Vol. 69, No. 2, (April 2003), pp. 379-384, ISSN 0919-9268

41. Jang, M.S.; Eun, J.B.; Ushio H. & Ohshima T. (2004). Antioxidative properties of mushroom (*Flammulina velutipes*) on the oxidation of cod liver oil in emulsion.

Food Science Biotechnology, Vol. 13, No. 2, (June 2004), pp. 215-218, ISSN 1226-7708

42. José-Pablo, Z.; Martínez-Álvarez, O.; Montero, P. & Gómez-Guillén, M. (2009). Characterization and tissue distribution of polyphenoloxidase of deepwater pink shrimp (*Parapenaeus longirostris*). *Food Chemistry*, Vol. 112, No. 1, (January, 2009), pp. 104-111, ISSN 0308-8146
43. Kanner, J. & Kinsella, J. E. (1983). Initiation of Lipid oxidation by a peroxidase/ hydrogen peroxide/halide system. *Lipids*, Vol. 18, No. 3, (May 1983), pp. 204-210, ISSN 0024-4201
44. Kanner, J.; Sofer, F.; Harel, S. & Doll, L. (1988). Antioxidant activity of ceruloplasmin in muscle membrane and *in situ* lipid peroxidation. *Journal of Agricultural and Food Chemistry*, Vol. 36, No. 3, (May 1988), pp. 415-417, ISSN 0021-8561
45. Kim, J.; Marshall, M.R. & Wei, C. (2002). Polyphenoloxidase, In: *Seafood Enzymes: Utilization and Influence on Postharvest Seafood Quality*, N. Haard, & B. Simpson (Ed.), 271-315, Marcel Dekker Inc., ISBN 082-47032-6-X, New York, USA
46. Kim, Y. J. & Uyama H. (2005). Tyrosinase inhibitors from natural and synthetic sources: structure, inhibition mechanism and perspective for the future. *Cellular and Molecular Life Science*, Vol. 62, No. 15, (August 2005), pp. 1707-23, ISSN 1420-682X
47. Leclerc, V.; Pelte, N.; El Chamy, L.; Martinelli, C.; Ligoxygakis, P.; Hoffmann, J.A. & Reichhart, J.M. (2006). Prophenoloxidase activation is not required for survival to microbial infections in *Drosophila*. *EMBO reports*, Vol. 7, No. 1, (July 2006), pp. 231- 235, ISSN 1469-3178
48. Lee, C.Y.; Smith, N.L. & Hawbecker, D.E. (1988). Enzyme activity and quality of frozen green beans as affected by blanching and storage. *Journal of Food Quality*, Vol. 11, No. 4, (November, 1988), pp. 278-287, ISSN 0146-9428
49. Love, J. D. & Pearson, A. M. (1971). Lipid oxidation in meat and meat products-A review. *Journal of American Oil Chemist Society*, Vol. 48, No. 9, (September, 1971), pp. 547-549, ISSN 0003-021X
50. Maeda, K.; Naitou, T.; Umishio, K.; Fukuhara, T. & Motoyama, A. (2007). A novel melanin inhibitor: hydroperoxy traxastane-type triterpene from flowers of *Arnica montana*. *Biological and Pharmaceutical Bulletin*, Vol. 30, No. 5, (May 2007), pp. 873-879, ISSN 0918-6158
51. Makishima, S.; Nozaki, K.; Mizuno, M.; Netsu, E.; Shinji, K.; Shibayama, T.; Kanda, T. & Amano, Y. (2006). Recovery of soluble sugars from waste medium for Enokitake *(Flammulina velutipes)* mushroom cultivation with hydrothermal

reaction and enzyme digestion. *Journal of Applied Glycoscience,* Vol. 53, No. 4, (October 2006), pp. 261-266, ISSN 1344-7882

52. Martínez-Álvarez, O.; López-Caballero, M.E; Montero, P. & Gómez-Guillén, M. (2005). A 4- hexylresorcinol-based formulation to prevent melanosis and microbial growth in chilled tiger prawns (*Marsupenaeus japonicas*) from aquaculture. *Journal of Food Science,* Vol. 70, No. 9, (November 2005), pp. 415-422, ISSN 1750-3841
53. Martinez, M. & Whitaker, J. (1995). The biochemistry and control of enzymatic browning. *Trends in Food Science and Technology,* Vol. 6, No. 6, (June 1995), pp. 195-200, ISSN 0924-2244
54. Mau, J. L.; Lin, H. C. & Song, S. F. (2002). Antioxidant properties of several specialty mushrooms. *Food Research International,* Vol. 35, No. 6, (November 2002), pp. 519-526, ISSN 0963-9969
55. Mayer, M. P.; Beyer, P. & Kleinig, H. (1990). Quinone compounds are able to replace molecular oxygen as terminal electron acceptor in phytoene desaturation in chromoplasts of *Narcissus pseudonarcissus* L. *European Journal of Biochemistry,* Vol. 191, No. 2, (July 1990), pp. 359-363, ISSN 0014-2956
56. McEvily, A.J.; Iyengar, R. & Otwell, W.S. (1992). Inhibition of enzymatic browning in foods and beverages. *Critical Review on Food Science and Nutrition,* Vol. 32, No. 3, (October 1992), pp. 253-273, ISSN 1040-8398
57. McFeeters, R.; Barrangou, L.; Barish, A. & Morrison, S. (2004). Rapid softening of acidified peppers: effect of oxygen and sulphite. *Journal of Agricultural and Food Chemistry,* Vol. 52, No. 14, (June 2004), pp. 4554-4557, ISSN 0021-8561
58. Miki, W. (1991). Biological functions and activities of animal carotenoids. *Pure and Applied Chemistry,* Vol. 63, No. 1, (May 1990), pp. 141-146, ISSN 0033-4545
59. Ohshima, T.; Fujita, Y. & Koizumi, C. (1993). Oxidative stability of sardine and mackerel lipids with reference to synergism between lipids and α-tocopherol . *Journal of American Oil Chemist Society,* Vol. 70, No. 3, (March 1993), pp. 269-276, ISSN 000-021X
60. Pan, X.; Hideki, U. & Ohshima, T. (2004). Photo-oxidation of lipids impregnated on the surface of dried seaweed (*Porphyra yezoenosis* Ueda): Hydroperoxide distribution. *Journal of American Oil Chemist Society*, Vol. 81, No. 8, (August 2004), pp. 765-771, ISSN 003-021X
61. Park, Y.D; Lyou, Y.J.; Hahn, H.S.; Hahn M.J. & Yang, J.M. (2006) Complex inhibition of tyrosinase by thiol-composed Cu2+ chelators: a clue for designing whitening agents. *Journal of Biomolecular Structure and Dynamics*, Vol. 24, No. 2, (October 2006), pp. 131-137, ISSN 0739-110

62. Petillo, D.; Hultin, H. O.; Krzynowek, J. & Autio, W. R. (1998). Kinetics of antioxidant loss in mackerel light and dark muscle. *Journal of Agricultural Food Chemistry,* Vol. 46, No.10, (October 1998), pp. 4128-4137, ISSN 0021-8561

63. Quang, D. N.; Hashimoto, T. & Asakawa, Y. (2006). Inedible mushroom: A good source of biologically active substances. *The Chemistry Record,* Vol. 6, No. 2, (February 2006) pp. 79-99, ISSN 1527-8999

64. Renerre, M. (1990). Review: Factors involved in the discoloration of beef meat. *International Journal of Food Science and Technology,* Vol. 25, No. 6, (December 1990), pp. 613-630, ISSN 1365-2621

65. Roginsky, V. & Lissi, E. A. (2005). Review of methods to determine chain-breaking antioxidant in food. *Food Chemistry*, Vol. 92, No. 2, (September 2005), pp. 235-254, ISSN 0308-8146

66. Romero, F. T.; Ordonez, I.; Arduini, A. & Cadenas, E. (1992). The reactivity of thiols and disulfides with different redox states of myoglobin. Redox and addition reactions and formation of thiyl radial intermediates. *Journal of Biology and Chemistry,* Vol. 267, No. 3, (January 1992), pp. 1680-1688, ISSN 0021-9258

67. Syvaoja, E. L. & Salminen, K. (1985). Tocopherols and tocotrienols in finfish foods: Fish and fish products. *Journal of American Oil Chemist Society,* Vol. 62, No. 8, (August 1995), pp. 1245-1248, ISSN0021-9258

68. Vamos-Vigyaso, L. (1981). Polyphenol oxidase and peroxidasein fruits and vegetables. *Critical Review on Food Science and Nutrition,* Vol. 14, No. 1 (January 1981), pp. 44-50, ISSN 1040-8398

69. Wang, Y. & Jiang, H. (2004). Purification and characterization of *Manduca sexta* serpin-6: a serine proteinase inhibitor that selectively inhibits polyphenoloxidase-activating proteinase-3. *Insect Biochemistry and Molecular Biology*, Vol. 34, No. 4, (April 2004), pp. 387-395, ISSN 0965-1748

70. Wasser, S. P. (2002). Medicinal mushrooms as a source of anti tumor and immunomodulating polysaccharides. *Applied Microbiology and Biotechnology,* Vol. 60, No. 3, (March 2002), pp. 258-274, ISSN 0175-7598

71. Watanabe, F.; Goto, M.; Abe, K. & Nakano, Y. (1996). Glutathione peroxidase activity during storage of fish muscle. *Journal of Food Science,* Vol. 61, No. 4, (July 1996), pp. 734-735, ISSN 1750-3841

10 IMPROVING LARVAL CULTURE AND REARING TECHNIQUES ON COMMON SNOOK (CENTROPOMUS UNDECIMALIS)

1. INTRODUCTION

The common snook or *Centropomus undecimalis* (Bloch) is a diadromus, stenothermic, euryhaline, estuarine-dependent species found in the tropical and sub-tropical western Atlantic Ocean from about 34° N to about 25° S latitude (Howells *et al.*, 1990). The snook physiology is characterized by a distinct lateral line, high divided dorsal fin, sloping forehead, a large mouth, a protruding lower jaw and a yellow pelvic fin (Fore & Schmidt, 1973).

Partial genetic isolation occurs between Florida's Atlantic and Gulf Coast stocks (Tringali & Bert, 1996). Snook are protandric hermaphrodites: some males develop into females between 1 and 7 years of age, having a maximum 20-year lifespan. Females are generally larger than males of the same age, at the same time it is unusual to find females smaller than 500 mm in fork length. Snook growth rates are highly variable. For instance, Atlantic Coast fish grow more quickly and to a larger size than do fish on the Gulf Coast (Taylor *et al.* 2000).

Common snook form the basis of important fisheries throughout their range due to their sporting and culinary attributes (Tucker *et al.*, 1985; Matlock & Osburn, 1987). Numbers of common snook have declined over recent years due to shoreline development, fishing pressure, and loss of coastal habitats. As a result, common snook were designated as a game fish restricted to recreational harvest only. Depletion of some Florida stocks during the late 1970's and the early 1980's (Bruger & Haddad, 1986) resulted in common snook being declared a species of special concern, and they are now protected by strict regulations enacted by the Florida legislature.

The ultimate objective of hatchery-based production of common snook in Florida is the supply of high quality animals to restore declining stocks and enhance local populations. The quality of the juveniles, environmental conditions and releasing techniques are all involved in the success of restocking programs (Tsukamoto 1993). The objective of larval rearing is to mass-produce high-quality and healthy juvenile fish. The management of both the rearing environment and feeding regime are the most important aspects of this activity.

To improve larval rearing techniques, a good understanding of larval morphology, behaviour, live food and artificial diet requirements, and environmental conditions is fundamental (Liao et al., 2001).

Collection of data on the conditions required for spawning, larval rearing, and release of common snook into marine and freshwater systems began in 1974 at the Florida Game and Fresh Water Fish Commission (Ager *et al.*, 1978; Shafland and Koehl, 1980, Chapman et al., 1978). Information on the development of laboratory reared larvae and juveniles (Lau and Shafland, 1982) and on the lower lethal temperature (15°C) for juveniles (Howells *et al.*, 1990) was also documented. These studies described the basic common snook biology and the principles for captive rearing. Although research on this species in Florida and Texas was carried out during the 1970's, 1980's and 1990's, there are still a number of gaps in our understanding of the requirements for successful larval rearing and broodstock management.

1.1 Snook larval culture

Techniques of larviculture have gradually been developed from simply collecting the stocking material in the wild to using modern, advanced facilities for complete larviculture practices (Liao *et al.*, 2001). Common snook culture research in Florida, Texas, Mexico and Brazil has primarily relied on collection of fertilized eggs from mature wild fish and more recently at Mote Marine Laboratory on captive broodstock production. At this time, snook larviculture practices are still under-development when compared to many other marine fish species, such as red drum and cobia. This paper is focused on larval rearing during the first 14 days after hatching using wild, strip spawned eggs. The techniques investigated include the design of a larviculture system, diet requirements of larvae, and system management.

1.1.1 Importance of temperature on embryonic and larval development

Nearly every aspect of early fish development is affected by temperature (i.e., fertilization, hatching, first feeding) (Alderdice and Velsen, 1978; Heggberget and Wallace, 1984, Brännäns, 1987; Crisp, 1988; Kane, 1988; Jensen *et al.*, 1989; Beacham and Murray, 1990; Blaxter, 1992). Other aspects affected by temperature are the yolk conversion efficiency as demonstrated in salmonid embryos (Heming, 1982; Heming and Buddington, 1988; Marr, 1996; Peterson & Martin-Robichaud, 1995) and in stripped bass (Peterson *et al.*, 1996). Also larval size and fitness at the end of the endogenous feeding period are directly affected by temperature (Peterson *et al.*, 1977, 1996, Baynes and Howell, 1996). Therefore, temperature has a key controlling effect on metabolic processes through thermal dependence on enzymatic activity (Brett, 1970; Rombough, 1988; Blaxter, 1992).

1.1.2 Larval stocking densities

One of the key aspects of successful large-scale production is determining the optimum larval stocking densities. For several species of fish, such as sea bass (*Dicentrarchus labrax*) or sea-bream (*Sparus aurata*), optimum culture densities are well known. The optimal stocking density varies between species depending on the behavioural and physical characteristics (Tagawa *et al.*, 1997, 2004; Kaji *et al.*, 1999; Hernandez-Cruz *et al.*, 1994). Larval density studies for common snook have not been conducted, although some work has been done on fat snook (*Centropomus parallelus*) evaluating the effect of larval and juvenile densities on growth (Cerqueira *et al.*, 1995).

1.1.3 Prey density

Prey density (Werner and Blaxter, 1981) is one of the factors affecting feeding efficiency and consequently larval growth and survival under culture conditions. Enhancing feeding efficiency, at first feeding, can reduce the risk of starvation during the first days of development (Peña *et al.*, 2004). It has also been shown that foraging success increases with prey density (Wyatt, 1972; Laurence, 1974, 1978; Houde and Schekter, 1980; Munk and Kiørboe, 1985) until an asymptote is reached (Houde and Schekter, 1980; Klumpp and Von Westernhagen, 1986). Feeding levels (e.g. rotifer densities) must be tailored to the needs and consumption rates of the larvae at different ages so that food is not wasted, larvae are not underfed, and rearing water is not fouled. The usefulness of food to larvae at particular stages may be measured by food intake, growth and survival (Duray, *et al.*, 1996).

1.2 First feeding

One of the key restrictions in larval rearing is first feeding at early stages of development. This is a major bottleneck for larval culture, due primarily to their small size and often poorly developed digestive system (Person Le Ruyet, et al., 1993). Many marine fish larvae require motile prey organisms (Pedersen *et al.*, 1987; Pedersen and Hjelmeland, 1988). Visual skill is not only important for feeding but also for orientation, schooling and eluding predators (Blaxter, 1986, Batty 1987). Larval survival clearly depends on their ability to feed successfully (Heath, 1992). During the endo-exotrophic phase (Mani-Ponset *et al.*, 1996), larvae utilize nutrients from both yolk sac and their surrounding environment. This phase starts soon after hatching, especially in larvae with a small yolk sac (Calzada *et al.,* 1998). This first feeding phase is critical for larval survival; therefore, successful synchronization between exhaustion of endogenous reserves and first feeding must occur.

Larval mouth size at first feeding is also an important factor for larval survival. The mouth size of first-feeding larvae mechanically restricts the size of the food particles that can be ingested. In general, mouth size is correlated with body size, which in turn is influenced by egg diameter and the period of endogenous feeding (i.e., yolk sac consumption period). For example, Atlantic salmon eggs are usually at least four times larger than Gilthead sea bream eggs and consequently on hatching yield large salmon larvae with large yolk sac supplies (i.e., sufficient endogenous feed reserves for the first three weeks of their development). Whereas first-feeding Gilthead sea bream larvae are very small with limited yolk sac reserves, and consequently can only feed endogenously for about three days (Jones & Houde, 1981).

1.2.1 Background phytoplankton ('Green water')

Most marine fish larvae are visual feeders and feeding success of larvae at various developmental stages depends on the provision of suitable food, the rearing environment, and on the visibility and adequate density of the prey (Ina *et al.*, 1979, Hunter, 1980). Publications on the rearing of marine fish larvae indicate that phytoplankton cultures enhance survival rates (May, 1971; Al-Abdul-Elah, 1984; Hernandez-Cruz, *et al.*, 1994; Marliave, 1994). Furthermore, several papers have discussed the beneficial effect of adding microalgae to larval rearing tanks in order to improve larval growth and survival (Howell 1979; Scott & Middleton 1979; Jones & Houde, 1981; Bromley & Howell 1983; Vasquez- Yeomans, *et al.,* 1990; Naas, *et al.*, 1992; Hernandez-Cruz *et al.*, 1994; Marliave 1994; Tamaru, *et al.*, 1994). These papers discuss the effect of micro-algae on the nutritional and behavioural aspects of fish larvae. Some fish larvae take up substantial amounts of micro-algae during the initial days after hatching (Van der Meeren, 1991; Reitan, *et al.*, 1991) which maybe used as a food source. In recent years, the benefits of culturing larvae in `green water' is considered to be optical rather than nutritional to fish larvae (Marliave 1994).

1.2.2 Rotifers and their nutritional value

Live food organisms are an important food source for the first feeding of early larval stages. The most widely used starter live-food organism in fish larviculture is the marine rotifer *Brachionus plicatilis*. The successful development of commercial fish farms in the Mediterranean has been made possible by several improvements in production techniques for rotifers (Candreva *et al.*, 1996; Dehasque *et al.*, 1998). Rotifers are an ideal link in the food chain for different stages of fish and shrimp larvae. Rotatoria (=Rotifera) belong to the smallest metazoan of which over 1000 species have been described, 90% of which inhabit freshwater habitats. They seldom reach 2 mm in body length. Males have reduced sizes and are less developed than females; some measuring only 60 μm. The body of all species consists of a constant number of cells, with various *Brachionus* species containing approximately 1000 cells, which should not be considered as single identities, but as a plasma area. Growth of the animal is achieved by plasma volume increase and not by cell division. The epidermis contains a densely packed layer of keratin-like proteins and is called the lorica. The shape of the lorica and the profile of the spines and ornaments allow determination of different species and morphotypes. A rotifer body is differentiated into three distinct parts consisting of the head, trunk and foot. The head carries the rotatory organ or corona, which is easily recognized by its annular ciliation and is the characteristic that led to the name Rotatoria (bearing wheels). The retractable corona assures locomotion and a whirling water movement, for the uptake of small food particles (mainly algae and detritus). The trunk contains the digestive tract, the excretory system and the genital organs. A characteristic organ of rotifers is the mastax (a calcified apparatus in the mouth region) that is very effective in grinding ingested particles. The foot is a ring-type retractable structure without segmentation ending in one or four toes. *B. plicatilis,* a cosmopolitan inhabitant of inland saline and coastal brackish waters. It has a lorica length of 100 to 340 μm, with the lorica ending in 6 occipital spines (Fukusho, 1989).

The nutritional value of *B. plicatilis* is dependent on the nutritional value of its food source, which can influence its suitability as a starter feed for marine larvae, and is determined by the concentrations of highly unsaturated fatty acids ((n-3) HUFA), such as docosahexaenoic acid (DHA, 22:6(n-3)) and eicosapentaenoic acid (EPA, 20:5(n- 3)). Low dietary HUFA levels can lead to high mortality in fish larviculture. Koven *et al.* (1990) suggested that HUFAs function as essential components of bio-membranes, and that their levels in the tissue phospholipid fraction are associated with larval growth. Rainuzzo *et al.* (1997) emphasized the importance of DHA in the development of neural tissues such as brain and retina, considering that the larval head constitutes a significant part of the body mass, and that predatory fish larvae rely on vision to capture their food. Sorgeloos *et al.*, (1988) reported a strong correlation between dietary EPA content and survival, and between DHA and growth of Asian sea bass larvae. Watanabe (1993) concluded that DHA and EPA increased survival and growth of several marine fish larvae. At the same time, Kanazawa (1993) observed that high DHA levels increased the tolerance of red sea-bream larvae to various stressful conditions.

1.2.3 Copepods

The suitability of copepods as live prey for marine fish larvae is now well established, but their use in aquaculture remains sporadic. Although of lower nutritional value, the relative ease of production of rotifers and *Artemia* nauplii continues to ensure their predominance. Studies in the literature have highlighted differences in the levels and ratios of fatty acids, lipid classes and pigments between copepods and traditional live prey used in hatcheries. Such differences are important for fish larval nutrition, as previously mentioned. The consequences of poor nutrition during fish larval development can result in deformities or malpigmentation, and in some cases may be less obvious, such as effects on temperature tolerance or growth during later life stages. (Støttrup, 2000). Rearing the larvae of most marine fish species requires provision of live prey for variable periods from the onset of exogenous feeding. A common feature of these species is the production of small pelagic eggs. Larvae generally hatch at an early stage in their development of the digestive system as well as the development of organs critical for successful feeding, such as vision and motor development (Støttrup,. 2000). Effects such as tolerance to low temperatures during the juvenile stage have been shown to be related to the larval diet (enriched vs non-enriched *Artemia*) (Howell, 1994), which were not detectable during or at the end of the larval stage. Several studies have shown that rearing marine fish on natural zooplankton can ameliorate these nutritional deficiencies (Nellen, 1981)

1.2.4 Artificial microparticulate diets

A number of studies were carried out to find satisfactory, formulated diets that would substitute for natural live food (rotifers, *Artemia* sp.) in larval rearing of various fish species (Lazo, et al., 2000; Yufera, et al., 1999; Dabrowski, et al., 2003; Takeuchi, et al., 2003). Feeds used as first food during fish larval development must be fine-grained, acceptable, digestible and utilized for body protein/lipid synthesis by the larvae (Ostaszewska, *et al.*, 2005). They should also include the optimal composition of nutrients to achieve high survival and growth rate, and correct development (metamorphosis) of fish. Simultaneously with the efforts on feed formulation, studies of digestion physiology in the gastrointestinal system development in fish larvae must be performed (Ostaszewska, *et al.*, 2005). Ontogenesis, differentiation and development of functions of all organs are genetically determined. However, fish larvae are able to adapt, within some limits, to variable environmental and feeding conditions (Webb, 1999).

1.2.5 Snook aquaculture

Collection of data on the conditions required for spawning, larval rearing, and release into marine and freshwater systems began in 1974 at the Florida Game and Fresh Water Fish Commission (Ager *et al.*, 1978; Shafland and Koehl, 1980, Chapman et al., 1978). These studies provided information on the lower lethal temperature (15°C) for juveniles (Howells *et al.*, 1990) and preliminary developmental results for laboratory reared larvae and juveniles (Lau and Shafland, 1982). These studies described basic common snook biology and the principles for captive rearing. The early studies on this species in Florida and Texas conducted in the 1970's, 1980's and 1990's, were unable to identify the appropriate culture requirements to support captive spawning and larval rearing of common snook.

1.2.6 Objectives

The objectives of the series of experiments reported here were to improve larval survival of common snook during the first 14 days after hatching.

The main aims were to:

1. Investigate the influence of temperature on hatch rate
2. Establish the effect of egg stocking density on larval survival and growth
3. Determine the influence of flow rate on larval survival
4. Investigate the effect of the green water technique on larval survival and growth
5. Determine the influence of rotifer density on larval survival
6. Investigate effect of alternative live food species on larval survival and growth
7. Evaluate the acceptance of micro-diet feeding by larval snook

2. MATERIALS AND METHODS

All the larval rearing experiments reported here took place at the Mote Marine Laboratory facilities, located in Sarasota, Florida. During the first two years experiments were carried out in aquaculture systems located at Mote Marine Laboratory. The studies carried out in the later two years were conducted in the new aquaculture systems located at Mote Aquaculture Research Park (MAP).

Artificial seawater (Instant Ocean ®) was used at both locations; however, conditions at Mote Marine Laboratory were not ideal because the systems were located under a building that had poor ventilation, limited lighting, and a lack of temperature control. At MAP, experiments were conducted in a variety of tank systems equipped with state-of-art filtration, and in isolated and temperature controlled experimental rooms.

2.1 Live culture

The suitability of three live food types (microalgae, rotifers, copepods) were investigated in parallel. Each live culture was maintained in separate rooms using water from a different reservoir, in order to avoid any possibility of cross-contamination.

2.1.1 Microalgae

The main microalgae used in the snook trials was *Nannochloropsis occulata;* this is non motile, green coloured cell with no flagella. It is a small, elliptical cell, 4-6 μm in diameter, with few distinguishing features. The chloroplast usually occupies much of the cell. Cells tend to float in culture and stay in suspension without aeration. This organism is placed in a separate division from Nannochloris because of its lacks of chlorophyll b. These algae are a popular food source for rotifers and filter feeders.

N. occulata was used to feed the rotifer cultures and to create a green water environment in the larval systems. The procedure used for its culture was the classical batch culture method, which consists of inoculating culture tubes with low density of algae cells. After two weeks, test tube cultures were transferred into 250 ml flasks and later (1 week) into a larger 19 L carboy culture vessels. After a week, a 100-liter cone shape transparent tank was inoculated with a full carboy. The culture was kept running with four 200 L transparent fiberglass tanks during the experiment's duration to ensure reliable microalgae production. All cultures were exposed to 24 hour white light condition (1000 lux), water temperature was kept at 29 ºC, and had constant aeration.

N. oculata paste was also obtained from Reed Mariculture. This paste is a highly concentrated media (68 billion/ml) of *N. oculata* that was kept frozen until the day before it was used. The paste was used to reduce the time involved in batch culture of live algae and to test the difference between live and frozen paste algae as a food source for rotifers and for creating a green water environment.

2.1.2 Rotifers

Four different types of rotifers were used in the snook larval rearing trials. *Brachionus rotundiformis* or small (S-type) rotifers and *B. plicatilis* or large (L-type) rotifers, which can be clearly distinguished by their morphological characteristics: the lorica length of the L-type ranges from 130 to 340 µm (average 239 µm), and in the S-type ranges from 100 to 210 µm (average 160 µm). Moreover, the lorica of the S-type has pointed spines, while the L-type has obtuse angled spines. Two other types of rotifers were used. The SS type rotifer (Super small rotifers) ranges between 100-120 µm, which are preferred for the first feeding of fish larvae with small mouth openings (rabbitfish, groupers, and other fish with mouth openings less than 100 µm at first feeding). Those rotifers, however, are not genetically isolated from S- strains, but are smaller than common S-strains (Person Le Ruyet, *et al.*, 1993). The last strain was an SS rotifers from the University of Ghent, Belgium, that were genetically modified to resist warmer temperatures (above 30 ºC).

All the rotifers strains were cultured using a batch culture method. Batch cultivation, due to its simplicity, is probably the most common type of rotifer production in marine fish hatcheries (Fukusho, 1989; Nagata and Hirata, 1986; Snell, 1991). The culture strategy consists of either the maintenance of a constant culture volume with an increasing rotifer density or the maintenance of a constant rotifer density by increasing the culture volume. In batch culture, a total harvest of the rotifers is done with part of the rotifers being used as food for fish larvae and part used as inoculum for the next culture (Hirata, 1980; Lubzens, 1987). All the rotifers were fed *N. oculata*.

2.1.3 Copepods

The calanoid copepod *Acartia tonsa* (*Acanthacartia*) was cultured for some of the feeding experiments. This species was chosen due its small size (80-100μm), nutritional value and availability. The copepods were cultured at Florida State University (Tallahassee) and eggs were sent every two weeks on ice (4°C) in 100 ml flasks. Once in the lab, eggs were refrigerated at 4°C until they were needed. Copepod eggs were taken out the 100 ml flask and placed in a 500 ml transparent flask with seawater at 35 ppt and 28ºC under a 12 hours light:dark period. No aeration was needed during the 48 hours hatching period. After hatch, *A. tonsa* were fed to the snook larvae. Feeding densities varied depending on the experiment.

2.2 Larval rearing systems

All the experiments were conducted using two independent experimental systems: small and large microcosm systems.

2.2.1 Small microcosm system

The small microcosm system (System A) was a self-contained recirculating system (Rana, 1986), which allowed several different experiments to be run at the same time, with the appropriate replication.. The system was made of transparent plexiglass, and contained 48 2-L tanks (Figure 1), The system dimensions were 1.25 m in length by 70 cm in width by 20 cm in height. The individual tank dimensions (Figure 1) were 12x10x20 cm, with an opening to allow water exchange 17.5 cm from the bottom, which was covered by a mesh screen (75 μm). In addition to the larval tanks, the system included two sumps, a 140 liter sump (filled with biofiltration beads) and a 120 liter sump (with a fluidized bed and a carbon filter). Air stones, were placed in both sump and a pure oxygen ceramic stone was also in the second sump, in order to keep the dissolved oxygen levels between 8 to 10 mg/L. Flow rates within the larval tanks was individually regulated through a drip valve. Flow rates varied between tanks depending on the experimental requirements. The tank recirculating system was based on an overflowing system, with a drip valve on the inflow and a 75 μm mesh rectangular opening for tank outflow. For the first 3 days after hatching, a transparent plastic separator was placed close to the outflow to avoid egg and larvae impingement on the mesh to reduce possible mortalities. All the tanks outflows drain to a common canal through the UV light filter and into the sumps.

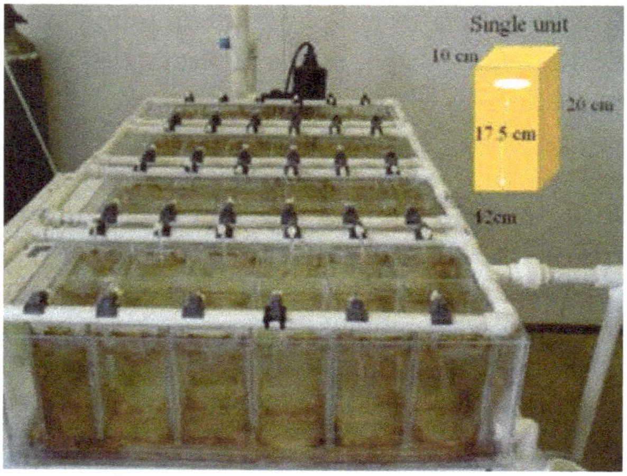

Fig. 1. System A (Microcosms) and individual tank dimensions

Daily 10% water exchange was conducted. Water quality was checked three times a day (every five hours) for temperature, salinity, dissolved oxygen and pH. Nitrite, ammonia and nitrate were checked on a weekly basis. During the feeding experiments, residual prey counts were taken prior to feeding. Tanks were fed 1, 2, 3 and 4 times a day depending on the experiment.

2.2.2 Large microcosm system

The large microcosm system (System B) was built in 2005 to provide additional replicated experimental tanks. The water volume in the System B tanks (6L) was 3 times the water volume of the System A tanks (2L) and was used to conduct simultaneous trials to compare the influence of increased water volume on larval survival. System B tanks were placed in a green bottom fiberglass raceway, where individuals tanks where maintained in a water bath (Figure 2). Twelve (6 l) tanks shared the same filtration system, which included a fluidized bed, a moving bed bioreactor, UV, and a protein skimmer. The system had two 300-litre sumps under the raceway, where the filtration system was set up. Air stones were placed in each tank to keep dissolved oxygen at desired levels. Water heaters were placed in the raceways to maintain constant temperature.

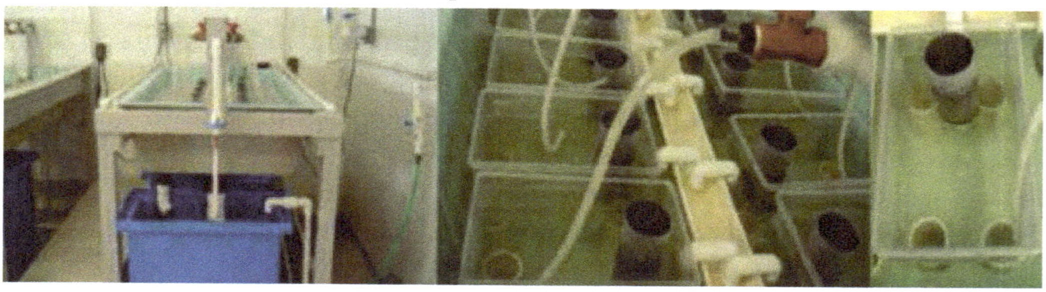

Fig. 2. Larval rearing System B. Full system and individual rearing tanks (left to right)

Inflow water was regulated individually per tank, and outflow water passed through a 75μm mesh standpipe, draining into a common drain channel (Figure 2, middle picture) leading to the first sump. Slight aeration was also supplied to the individual tanks. A 25% water exchange was carried out every week. Like in the microcosms, water quality (temperature, salinity, dissolved oxygen and pH) was checked three times a day (every five hours). Nitrite, ammonia and nitrate were monitored weekly.

2.3 Experimental methodology

After collecting common snook eggs in the field, eggs were fertilized and transported to the laboratory. Eggs were stocked in experimental systems within 2 to 3 hours after fertilization. Eggs were stocked at a salinity of 35 ppt, temperature of 28°C (except for the temperature experiments), 9-10 mg/l of dissolved oxygen and pH of 7.9. During the first two years, larvae were fed 3 days after hatching; during the last two years, feeding started 2 days after hatching. All specimens that survived past Day 14 after hatching were sacrificed, for total length and myomere height measurements.

2.3.1 Influence of temperature on hatch rate

Eggs were stocked in 2 L PVC floating containers, similar containers were at the same time floating in a 300 L raceway tank. Three raceway tanks were used, each one of them had a different temperature (23, 28 and 30°C). Each raceway tank had three 2 L PVC floating container and each container was stocked with 100 snook eggs per litre. Aeration was removed before stocking. Water quality parameters were maintained within acceptable values. Twenty hours after fertilization, containers were removed from the raceways and percent hatch was determined (see the formula below). This experiment was replicated four times to obtain reliable results.

Hatch rate=Total number hatched larvae/Total number of eggs stocked *100

2.3.2 Effect of egg stocking density on larval survival and growth

Nine tanks were stocked with three different egg densities (three replicates per density). The densities used were: high density (375 eggs/L), medium density (200 eggs/L) and low density (200 eggs/L). All tanks were fed S type rotifers three times a day at a concentration of 30 rot/ml. The tanks were harvested fourteen days after hatch (DAH) and larvae were counted and measured (total length and myomere height). This experiment was run in both systems A and B.

2.3.3 Influence of flow rate on larval survival

Common snook eggs were stocked at a density of 200 eggs/L and exposed to three different flow rates (no flow, slow flow (10 ml/min) and high flow (30 ml/min)).

Each flow treatment had 3 replicates per day. All tanks were fed three times a day with SS type rotifer at a concentration of 30 rotifers/ml. System A tanks (27 tanks) were stocked, nine per flow treatment. Larvae were harvested on 3, 6 and 10 DAH to determine survival.

2.3.4 Effect of background phytoplankton in the water (Green water technique) on larval survival and growth

Snook were stocked at 200 eggs/L in ten system A tanks and the flow rate was set at 15 ml/min. Five tanks were stocked with *Nannochloropsis oculata* at 1000/ml (green water) and five tanks without (clear water). All tanks were fed SS type rotifers three times a day with a concentration of 30 rot/ml. The experiment was terminated at 14 DAH and all larvae were counted to establish larval survival and measured (total length and myomere height) for growth.

2.3.5 Influence of rotifer density on larval survival

The effect of rotifer density on larval survival was evaluated in 15 System A and System B tanks. Three SS type rotifers densities were used with: 5 rotifers/ml, 15 rotifers/ml and 30 rotifers/ml. All the tanks were stocked at the same time with 200 eggs/L. Residual rotifer counts were taken before every feeding and feeding amounts were adjusted, in order to maintain the same rotifer concentration throughout the experiment. All larvae were harvested 14 DAH and counted to calculate survival.

2.3.6 Effect of alternative live food species on larval survival and growth

To investigate the effect of alternative live food species on larval survival and growth, four diets were given to the larvae: a) 100% rotifers, b) 75% rotifers and 25% copepods, c) 50% rotifers and 50% copepods, and d) 75% copepods and 25% rotifers. All the tanks were stocked at the same time and fed 30 prey/ml; the number of each prey (rotifer or copepod) was determined by the above-mentioned percentages. Twelve tanks were used in System A and System B for this experiment and there were three tanks per experiment diet. Flow rate was maintained at 10 ml/min during the 14 day trial. Total length and myomere height were taken at 14 DAH and stomach contents were examined.

2.3.7 Acceptance of micro-diet feeding by larval snook

A total of 21 (2L PVC) floating tanks were placed in the raceway and each tank was stocked with 300 eggs/L. The experiment was run twice for a period of 7 days, from day 2 after hatching till day 8 after hatching. Seven tanks were fed one of the following three diets: a) a SS type strain (150 µm) rotifer, at a density of 30 rot/ml, fed three times a day, b) a 100 µm artificial micro-diet, fed twice a day, c) a 150 µm artificial micro-diet fed twice a day. Each day three tanks were taken out and larvae stomach contents were examined under a microscope.

3. RESULTS

3.1 Influence of temperature on hatch rate

The influence of incubation temperature on hatch rate was investigated to establish the optimal temperature for incubating snook eggs. Embryos exposed to 23ºC water had the lowest mean hatch rate of 5.9% (range 0.8 to 11.4%) (Figure 3). The highest hatch rates of 23.5% occurred at 28ºC (range 21.7% to 27%). No significant difference in hatching rate ($p > 0.05$) was found between the 28ºC and 30°C treatment, but a significant difference ($p < 0.05$) was found when 28ºC was compared to 26ºC and 23°C.

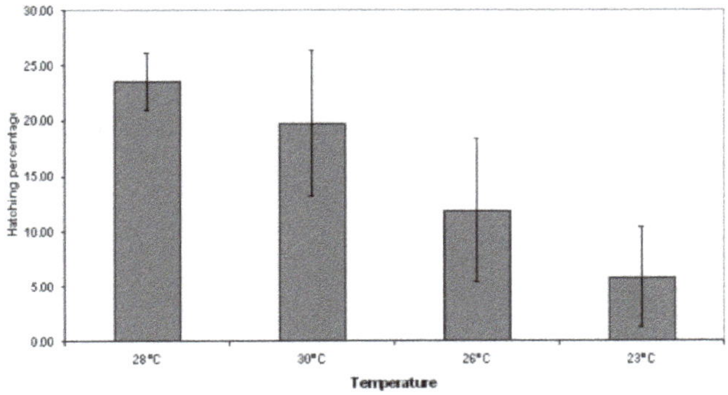

Fig. 3. Influence of temperature on percent hatch in snook larvae.

3.2 Effect of egg stocking density on larval survival

The effect of egg stocking density on larval survival at 14 DAH is presented in (Figure 4). At a stocking density of 375 eggs/l (High density) per tank, the mean survival from three tanks was 0.6%. Tanks stocked with a medium density (200 eggs/l) had a similar mean survival percentage with 0.5%. The highest survival was obtained at the low density stocking (100 eggs/l) with a mean overall 1.2% survival by 14 DAH (Figure 4). No significant difference ($p > 0.05$) was obtained between the high and medium density treatments, but a significant difference ($p < 0.05$) was obtained between the low density treatment and the other two.

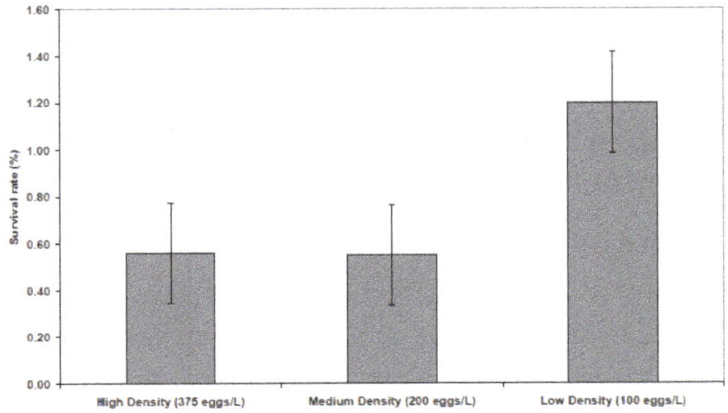

Fig. 4. Larval survival at three stocking densities at 14 DAH.

3.3 Influence of flow rate on larval survival

Snook larvae were exposed to three different flow rates. Those fish exposed to no flow during the first 10 days (Figure 5) had a mean survival rate of 45% by 3 DAH, survival then decreased to 19%by day 6. At 10 DAH, larval mean survival was only 5% from the initial stocking density. The slow flow (10 ml/min) treatment had a mean survival of 40% at 3 DAH, decreasing 3 days later to a 32% survival. Finally a mean survival of 21% was observed by 10 DAH (Figure 5). The third experiment had the lowest values in terms of survival. At high flow (30 ml/min), survival at 3 DAH was 16%, decreasing on Day 6 to a mean of 10% survival and by day 10 after hatch larval mean survival was 9% (Figure 5). No significant difference ($p>0.05$) was observed between the no flow and slow flow treatment at 3 DAH, but a significant difference ($p< 0.05$) in survival was found between all the treatments 6 and 10 DAH.

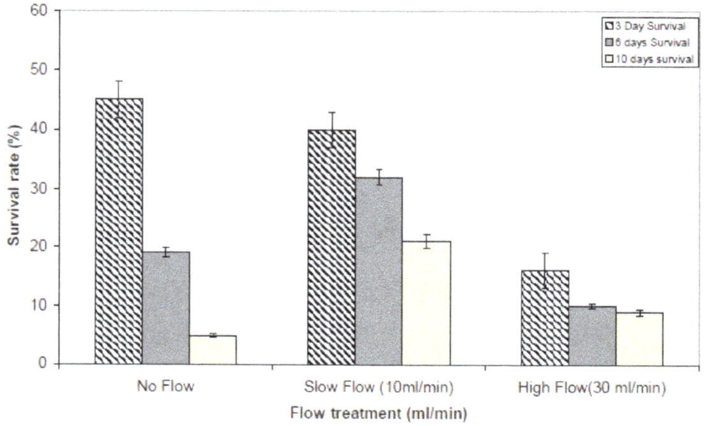

Fig. 5. Snook larval survival under three types of flow conditions.

3.4 Effect of background phytoplankton in the water (Green water technique) on larval survival and growth

Larvae were stocked in water with no algae and in water with algae (*N. occulata*) at a concentration of 1000/ml. Larvae tanks without algae are referred to as 'clear water' (Figure 6). The survival of larvae was significantly ($p<0.05$) influenced by the presence of *N. occulata* in the rearing water. The mean survival at 14 days post hatching in clear water was significantly lower at 0.17%, compared with 0.55% survival in green water tanks (Figure 6).

Length and myomere height in larvae from the clear water tanks had an average standard length (SL) of 3.20 mm and average myomere height of 0.70 mm (Figure 7); on the other hand, larvae in tanks subjected to green water technique had an average length of 3.34 mm and average myomere height of 0.73 mm. Although standard length and myomere height in the green water tanks were higher, no significant difference was found ($p= 0.053$).

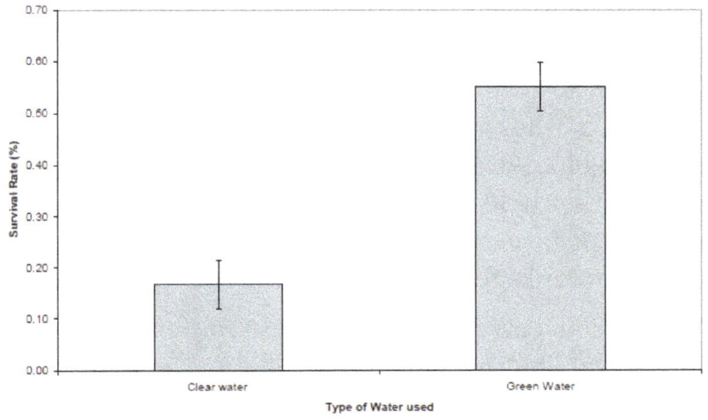

Fig. 6. 14 DAH snook larval survival from tanks with *N. oculata* (Green water) and without phytoplankton (clear water)

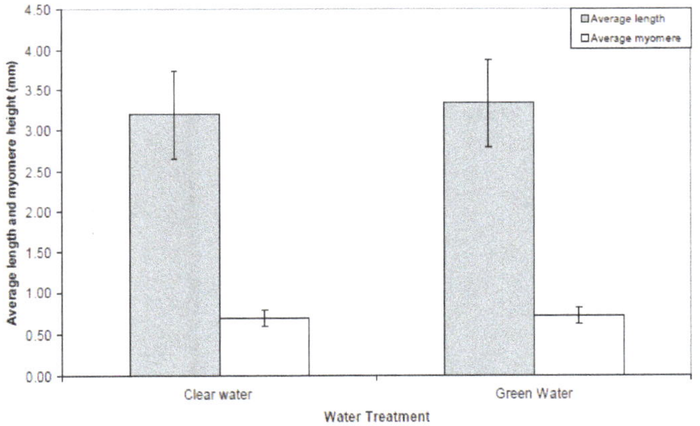

Fig. 7. Average snook larval length (SL) and myomere height from green water and clear water tank

3.5 Influence of rotifer density on larval survival

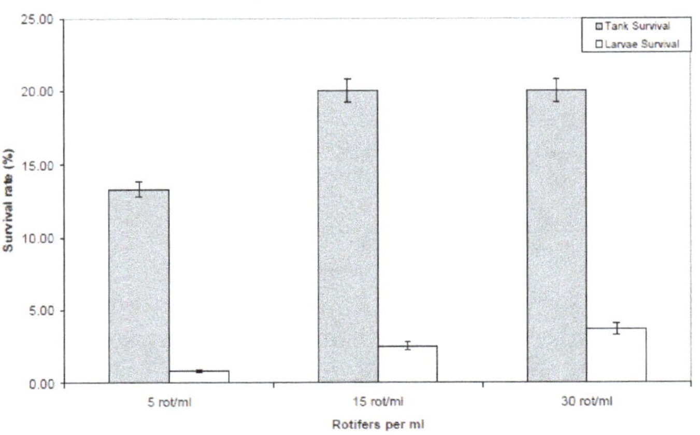

Fig. 8. Effect of 3 different rotifer densities on larval survival

Three different rotifers densities were evaluated to determine the effect on growth and survival of common snook larvae (Figure 8). After 14 days, larvae fed 5 rotifer/ml diet had an average survival rate of 0.86% and only 13% of all tanks stocked had live larvae. All the tanks where common snook larvae were fed 15 rotifers/ml had a mean larvae survival of 2.5%, and 20% of tank of all tanks stocked had live larvae. In the third treatment with 30 rotifers/ml, 20% of the tanks had live larvae and a total average of 3.67% larvae survival. No significant differences was found in larval survival between tanks fed 15 and 30 rotifer/ml (p=0.053).

Increasing rotifer concentrations from 5 to 30 rotifers/ml did not significantly (p>0.05) affect the larval length or myomere height (Figure 9). Larvae fed with 15 rotifers per ml had the highest mean length (SL) at 3.58 mm, followed by the other two treatments (3.48 and 3.49 mm in average respectively). Mean myomere height results were 0.74, 0.78 and 0.69 mm respectively.

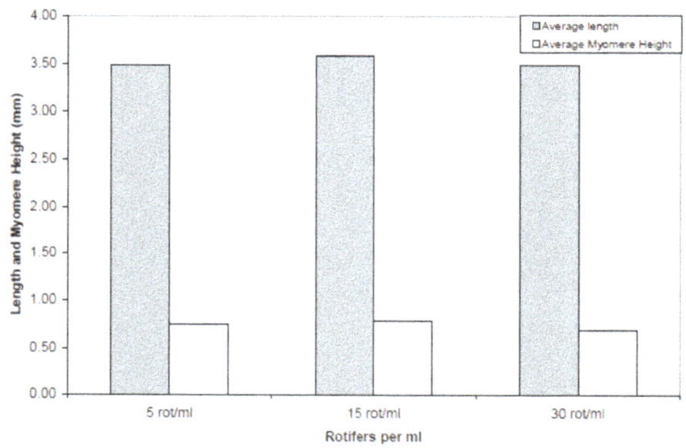

Fig. 9. Snook larval length (SL) and myomere height at three rotifers densities.

3.6 Effect of alternative live food species on larval survival and growth

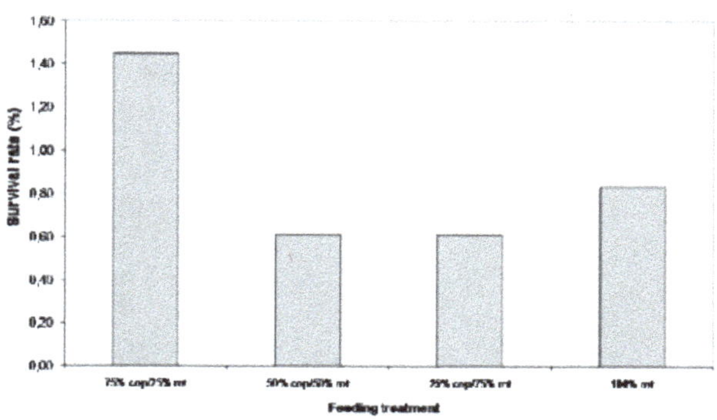

Fig. 10. Snook larval survival after 14 days associated with four diets that combined copepods and rotifers at different percentages. Feeding treatment 75/25 consisted of 75% copepods and 25% rotifers, 50/50 consisted of 50% copepods and 50% rotifers, and the 25/75 consisted of 25% copepods and a 75% rotifers.

Four different diets were used in this experiment: 3 using copepods (*Acartia tonsa*) and SS type rotifers, one with 100% rotifers. The diet with 75% copepods and 25% rotifers (Figure 10) had the highest mean survival (1.44%), the second highest average larval survival occurred with the diet that had 100% rotifers (0.83%). The other two diets (50% rotifers and 50% copepods, and 25% copepods and 75% rotifers) had similar results with 0.61% mean survival. A significant difference was found between the 75/25 copepod/rotifer diet and the other three diets ($p > 0.05$). No significant difference was found between the 50/50 and the 25/75 diets ($p < 0.05$).

The influence of live food combinations on larval growth is presented in Figure 11. Larvae at the 75/25 (copepod/rotifer) and the 50/50 tanks had the average highest length (4.70 mm), followed by the 25/75 (copepod/rotifer) with an average length (SL) of 4.32 mm, which was very similar to the 100% rotifer diet (average length (SL) of 4.28 mm). No significant difference was found between the first two treatments ($p > 0.05$) or between the other two (25/75 and 100 rot), but a significant difference was found between the first two and the last two ($p < 0.005$). Myomere height was similar on all the diets ranging from 1 mm to 1.04 mm.

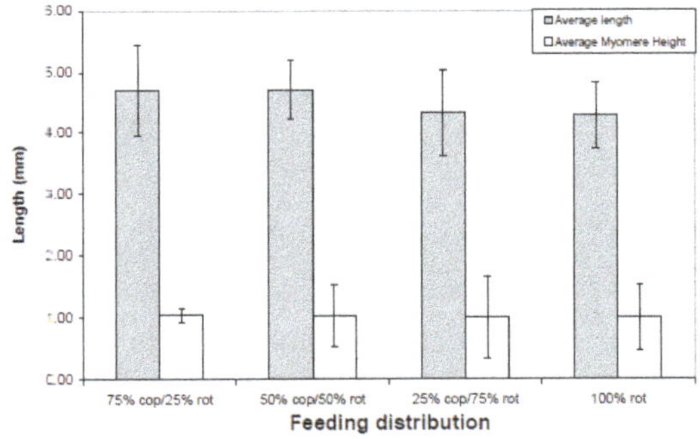

Fig. 11. Average length (SL) and myomere height from larvae exposed to four diets.

The stomach contents of 14 day old larvae were examined for food items (Figure 12). In the 75/25 treatment, 30.8% of the larvae had both rotifers and copepods in the guts, in the 50/50 diet, 30% of the larvae also had both rotifers and copepods in the guts, and in the 25/75 diet 18 % of the larvae had both rotifers and copepods in the guts. The rest of the larvae only had rotifers.

3.7 Acceptance of micro-diet feeding by larval snook

The stomach contents from larvae fed micro-diets were examined in this experiment. In tanks fed with rotifers, the percent of larvae with rotifers in the stomach after 7 days ranged from 10 to 37 (Figure 13). In tanks where 100 µm micro-diet food was offered, percent larvae with dry diet ranged from 6 to 28%. Finally, in tanks where the 150 µm micro-diet was offered the percent of larvae with dry food ranged from 7 to 27%.

4. DISCUSSION

4.1 Importance of temperature during incubation

Fish are affected by many intrinsic and extrinsic factors. These factors can affect developmental controls resulting in phenotype alterations (Johnston *et al.*,1996; Adriaens & Verraes, 2002). Among the abiotic factors, temperature has the most significant effect on development and growth (Blaxter, 1992; Kamler, 1992a; Hochachka & Somero, 2001), influencing developmental timing and formation and function of key tissues and structures (Kamler, 1992b; Fuiman *et al.*, 1998; Koumoundouros *et al.*, 1999) and the synchronization of these continuous developmental paths (Kovac, 2002). Temperature can also have a direct effect on physiology through its effects on enzyme reaction rates (Hochachka and Somero 2001).

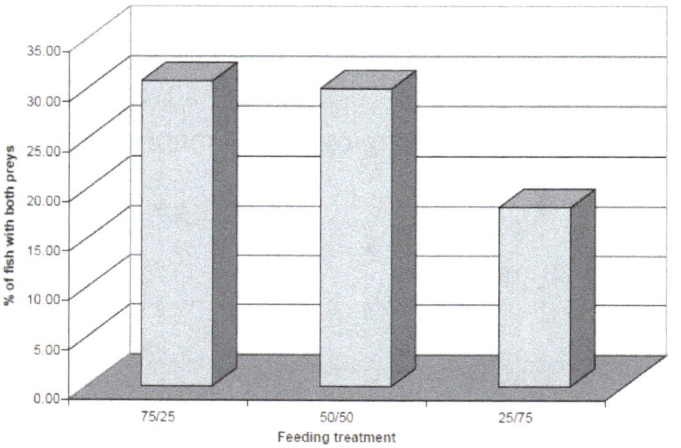

Fig. 12. Percent of rotifers and copepods found in the snook larvae stomach contents from four diets.

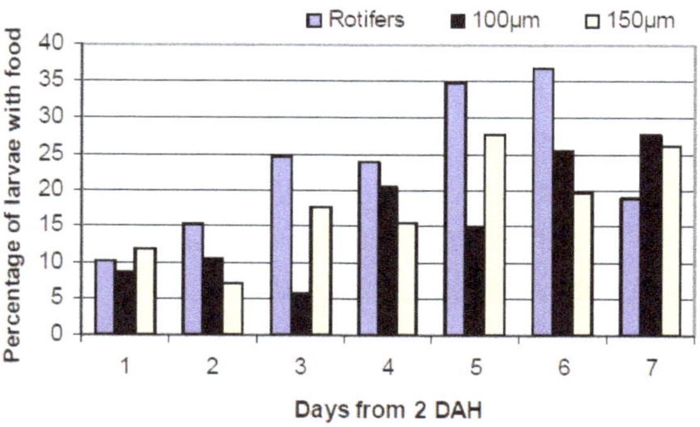

Fig. 13. Percent of larvae with food in the digestive system from the three diets (SS type rotifers, 100 and 150µm micro-diets).

Hatching rate of common snook varied with temperature in this study. Eggs incubated at 28ºC showed the best hatching rates; however, no growth trials were conducted. These results confirm Rideout *et al.*, (2004) findings of a direct influence of temperature during the incubation period on larval growth, which also has a direct effect on hatch rates (Pepin *et al.*, 1997). It has also been argued that survival may be proportional to larval size, since larval mortality rates have been shown to be inversely proportional to growth rates (Pepin, 1991).

4.2 The effect of stocking density on larval survival

Egg stocking density is a key factor in larviculture; without an optimal stocking density, overall survival can be affected. The common snook stocking density experiments results showed the lower stocking density (100 eggs/l) to be the one with the higher survival. This finding agrees with Hernandez-Cruz *et al.*, (1999), who obtained low survival in red porgy when eggs were stocked in high densities. On the other hand, Tagawa *et al.*, (1997) found higher survival at higher rearing densities in Japanese flounder and this finding was confirmed in three other marine teleosts (Kaji *et al.*, 1999). Tagawa *et al.*, (2004). The higher survival at high densities was attributed to substances (proteins) secreted by larvae that were beneficial for their survival. Although more experiments need to be done on common snook density, the results clearly showed low survival at high densities, which might be due to the a high level of competition for prey or over-crowding the environment and reduced water quality conditions.

4.3 Influence of 'green water' on larval survival

Food availability is a key factor during first feeding and the consumption rate is dependent on availability. Additionally, the developmental stage of the individual affects the consumption rate (Houde and Schekter, 1980; Kentouri, 1985). Based on the previous findings, this study examined survival of larvae in green water and in clear systems. Final results showed a significant difference in survival between the two treatments, where larvae grown in a green water (*N. occulata* had ingested the algae and had higher survival than those reared in clear water. This result agrees with previously published data (Papandroulakis *et al.*, 2002; Divanach *et al.*, 1998; Oie *et al.*, 1997; Holmejford *et al.*, 1993). Also, it has been reported in cod (Van der Meeren, 1991) and halibut (Reitan, *et al.*, 1991). This may support the idea that microalgae are used as a direct food source at the start of feeding confirming the important role that phytoplankton has during the early stages of several species. Another explanation for the increased survival in green water is the role of phytoplankton in stabilization and improvement of the rearing medium and its direct (Moffatt, 1981; Reitan *et al.*, 1993; Van der Meeren, 1991) or indirect nutritional effect (Tamaru *et al.*, 1993). Phytoplankton has been reported as being a protective agent, antagonistic towards pathogenic bacteria (Kennedy et al., 1998; Støttrup *et al.*, 1995). Skjermo & Vadstein (1999) noticed that microflora in rearing tanks of *Hippoglossus hippoglossus* were more stable in the presence of phytoplankton, increasing the total bacteria population by about 45%. The same authors noticed also that bacteria in larval gut were similar to the ones in rearing water, being mostly species with low growth rate. In addition, Nicolas *et al.*, (1989) showed that stomach microflora affects survival during early stages of marine larvae. Other studies have hypothesized that green water produces a background effect that allows the fish to better locate its prey (Marliave, 1994). This effect has been documented to improved the larval rearing in a number of fish species, including dolphinfish *Coryphaena hippurus* (Ostrowski, 1989), yellow perch *Perca flavescens* (Hinshaw, 1985), walleye *Stizostedion vitreum* (Corraza & Nickum, 1981; Colesante, 1989), white bass *Morone chryops* (Denson & Smith, 1996), grouper *Epinephelus suillus* (Duray *et al.*, 1996), Dover sole *Solea solea* (Dendrinos *et al.* 1984), barramundi *Lates calcarifer* (Pearce 1991); and red porgy *Pagurs pagrus* (Rotllant *et al.* 2003).

4.4 Importance of rotifer density in common snook larviculture

Prey density is an important factor in successful larval rearing (Fushimi, 1983). Low density can cause larval starvation or nutritional deficiencies leading to high mortalities. High densities can deteriorate water quality and lead to system fouling, decreased oxygen levels, and increased ammonia levels. Finding the right prey density is crucial to avoid these problems and also to reduce costs associated with live food production.

During the rotifer density experiments, both 15 and 30 rotifers/ml treatment resulted in high survival. Although 30 rotifers/ml had the highest larval survival, the difference between the two treatments was not significant; therefore, the higher cost to produce 30 rot/ml treatment indicates that 15 rot/ml is the optimal density for the 2 and 6 L tank systems. Recent attempts to calculate optimal prey density for cod larvae (*Gadus morhua*) found that survival reaches a maximum level and then begins to decrease if prey densities are further increased (Puvanendran and Brown, 1999). The decrease in larval survival when higher prey densities are used may be a result of poor water quality due to the release of metabolites by the prey (Houde, 1975) or it may be related to a reduction in the ability of the larvae to capture prey; what Laurel et al. (2001) term a "confusion effect". Optimal rotifers densities differ between species, for instance the black sea bream needs 1-3 rotifers/ml (Kafuku & Ikenoue, 1983), and the red sea bream needs between 3-10 rotifers/ml (Fushimi, 1983). In the case of common snook, results showed that 5 rotifers/ml was inadequate to meet their food demand. Further experiments need to be carried out to find out the optimal density, this time focusing between 10 to 20 rotifers per ml.

4.5 Alternative prey for common snook larvae

The suitable size of prey for fish larvae varies with larval mouth size (Shirota, 1970), and fish larvae select larger prey size as they grow (Ivlev, 1961). Although many researchers have reported larval rearing trials with marine fishes, only a few studies have been conducted to compare the appropriate rotifer size among fish species and among different growth stages (Oozeki et al., 1992, in Hagiwara *et al.*, 2001). Four strains of rotifers and one copepod species were tested to find the optimal prey type for common snook larvae; prey that will suit the physical needs as the larvae develops and grows in size. Previous work done on snook larval culture used L type rotifers and had little success. The experiments run with the L type rotifers in these experimental trials, showed extremely low survival and DHA analysis of larvae showed a steep decrease in DHA from 1 to 6 DAH. DHA values in 6 DAH larvae were below 1% of total lipid concentration (Yanes-Roca et al. 2009). In both experimental and production systems mass mortality regularly occurred between 5-6 DAH (Yanes-Roca et al. 2009) and 75-85% of all stocked tanks did not have live larvae after 6 DAH. These results along with the finding that only 5% of the larvae fed L type rotifers had food in their stomachs lead to the conclusion that the snook larvae were dying of starvation. This is likely due to the L type rotifer prey size, which was larger than the snook larvae mouth gape. Common snook larval rearing was successful when the larvae were fed copepods grown naturally in outdoor ponds (Lau & Shafland 1982), confirming the prey size hypothesis. Experimental trials comparing survival and growth with several strains of rotifers and copepods (*Acartia tonsa*) also showed an increase in snook larval survival. These findings agree with Doi *et al.*, (1997a,b) and Toledo *et al.*, (1997) found nauplii of copepods to be effective when fed to red spotted grouper, *Epinephelus coioides* and with Støttrup *et al.*, (1997) who found an increase in larval survival when rotifer feeding was supplemented with *Tisbe* sp copepod.

4.6 Acceptance of micro-diets by larval marine fish

Production of marine fish juveniles in commercial hatcheries still depends on the supply of live prey, such as rotifers and *Artemia*. Artificial micro-diet substitution for live prey is crucial to lower production costs and sustain production of high and constant quality juveniles (Cahu & Zambino-Infante, 2001). The use of micro-diets for common snook larvae during the pre-weaning period was also tested. Since the first rearing of plaice (*Pleuronectes platessa*) larvae to metamorphosis using an artificial diets (Adron *et al.*, 1974), many trials have been conducted, with different degrees of success, to utilize artificial diets in larval rearing of species, such as seabass *Dicentrarchus labrax* (Gatesoupe *et al.*, 1977; Cahu & Infante, 1994; Kolkovski *et al.*, 1997), sole *Solea vulgaris* (Gatesoupe *et al.*, 1977), Atlantic silverside *Menidia menidia* (Seidel *et al.*, 1980), red seabream *Chrysophrys major* and Ayu *Plecoglossus altivelis* (Kanazawa *et al.*, 1982). In all cases, poor results were obtained when live food was replaced completely by micro-diets. However, during the last decade, the pre- weaning period has been greatly reduced in many species, such as European sea bass *Dicentrarchus labrax* (Person Le Ruyet *et al.*, 1993, Zambonino-Infante *et al.* 1997). Cahu *et al.*, (1998) reported that 35% of European sea bass larvae were fed exclusively compound diet from mouth opening. In other marine species, some survival was obtained when fed compound diet from mouth opening, such as sea bream *Sparus aurata* (Fernandez-Diaz & Yufera, 1997) and red sea-bream *Pagrus major* (Takeuchi *et al.*, 1998).

Common snook larvae were offered two sizes of micro-diets (100μm, 150μm) and rotifers for 7 days, from 2 to 9 DAH. Although there was higher percentage of larvae with food in their stomachs in tanks fed SS type rotifers, a significant number of larvae offered micro-diet had food in their stomachs. These results were observed in pike-perch (Ostaszewska *et al.*, 2005), where micro-diets were readily accepted, digested and absorbed as well as in the Japanese eel (Pedersen *et al.*, 2003), and the gilthead sea-bream (Salhi *et al.*, 1997). Earlier studies suggested that co-feeding with live food improved yellow perch growth and assimilation of artificial diets (Kolkovski *et al.*,1997), a method that could be applied to common snook, assuming that digestive enzymes of live food organisms supported digestive processes in fish larvae (Boulhic & Gabaudan 1992; Jones *et al.*, 1993). However, some publications have reported contradictory results (Cahu & Zambino-Infante, 1997; Kolkovski *et al.*, 1993).

Future research on micro-diets is needed to evaluate the effect on survival or growth. Results indicate that snook larvae seem to accept the artificial diet and based on the already mentioned literature findings on several marine larval species survival and growth, it appears that the common snook larvae pre-weaning period could be reduced.

5. CONCLUSIONS

Although common snook larval survival has improved, mortality is still high and more improvements in rearing techniques are needed to increase survival. Many factors could be responsible for the low survival rates observed, including the introduction of bacteria along with the live food (i.e., rotifers). Rotifers are major carriers of bacteria (Muroga and Yasunobu, 1987; Munro *et al.*, 1993, 1994). Studies to evaluate the effect of bacteria on snook larval culture may help the overall survival. In these trials, two different experimental tank sizes were used (2 L and 6 L tanks) and survival was better in the larger tanks. More research is needed to determine the optimal tank size to examine growth and survival in larval snook. After four years of research on the snook larval rearing techniques, positive improvements have been made and the critical bottlenecks and potential solutions have been identified. Findings such as optimal rearing temperature (28°C), appropriate flow and water management (green technique) are basic for future research. Other critical results were: finding appropriate prey size (SS rotifers and copepods), optimal stocking and prey densities and the acceptance of microdiets prior to weaning.

REFERENCES

1. Adriaens, D. & Verraes, W. (2002). An empirical approach to study the relation between ontogeny, size and age using geometric morphologies. In: Aerts, P., D'Aout, K., Herrel, A. & VanDamme, R, (Ed.), *Functional and Ecological Vertebrate Morphology* , pp. 293–324. Maastricht: Shaker Publishing.]
2. Adron, J.W.; Blair, A. & Cowey, C.B. (1974). Rearing of plaice (*Pleuronectes platessa*) larvae to metamorphosis using an artificial diet. *Fishery Bulletin* 72, 353–357
3. Ager, L.A.; Hammond, D. E. & Ware. F. Artificial spawning of snook. (1978). Proceedings of the annual conference South-eastern Association Game Fish Commissioners 30, 158-166p.
4. Al-Abdul-Elah K.M. (1984). Procedures and problems of marine fish hatcheries with special reference to Kuwait. Master of Science Dissertation, University of Stirling.
5. Alderdice, D.F. & Velsen, F.P.J. (1978). Relationship between temperature and incubation time for eggs of chinook salmon (*Oncorhynchus tshawytscha*). *Journal of the Fisheries Research Board of Canada* 35 , 69–75
6. Batty, R.S. (1987). Effect of light intensity on activity and food-searching of larval herring (*Clupea harengus*): a laboratory study. *Marine Biology* 94, 323-327.
7. Baynes, S.M. & Howell, B.R. (1996). The influence of egg size and incubation temperature on the condition of the *Solea solea* (L.) larvae at hatching and first feeding. *Journal of Experimental Marine Biology and Ecology* 199, 59–77
8. Beacham, T.D. & Murray, C.B. (1990). Temperature, egg size, and development of embryos and alevins of five species of Pacific salmon: a comparative analysis. *Transactions of the American Fisheries Society* 119, 927–945
9. Blaxter, J.H.S. (1992). The effect of temperature on larval fishes. *Netherlands Journal of Zoology* 42, 336–357
10. Blaxter, J.H.S. (1986). Development of sense organs and behavior of teleost larvae with special reference to feeding and predator avoidance. *Transactions of the American Fisheries Society* 115, 98-114.
11. Boulhic, M & Gabaudan J. (1992). Histological study of the organogenesis of the digestive system and swim bladder of the Dover sole, (*Solea solea*) (Linnaeus1758). *Aquaculture* 102, 373-396.
12. Brannas, E. (1987). Influence of photoperiod and temperature on hatching and emergence of Baltic salmon (*Salmo salar* L.). *Canadian Journal of Zoology* 65, 1503–1508.

13. Brett, J.R. (1970) 3. Temperature. 3.3. Animals. 3.3.2. Fishes. In: Kinne O., (Ed.) *Marine Ecology* , pp. 515–573 London: Wiley Interscience]

14. Bromley P.J & Howell B.R. (1983). Factors influencing the survival and growth of turbot larvae, *Scophthalmus maximus* L. during the change from live to compound feeds. *Aquaculture* 31, 31-40.

15. Bruger, G.E. & Haddad, K.D. (1986). Management of Tarpon, bonefish, and snook in Florida. *Multi-jurisdcitional Management of Marine Fisheries: Marine Recreational Fisheries* 11, 53-57.(Abstract)

16. Calzada, A.; Medina, A. & Gonzalez de Canales, M.L. (1998). Fine structure of the intestine development in cultured sea bream larvae. *Journal of Fish Biology* 53, 340-365.

17. Candreva, P.; Dhert, P.; Novelli, A. & Brissi, D. (1996). Potential gains through alimentation nutrition improvements in the hatchery. In: Chatain, B., Sargalia, M., Sweetman, J., Lavens, P. Eds. . 388 pp, 149–159. Oostende, Belgium.

18. Cahu. C. & Zambonino Infante J. (2001) Substitution of live food by formulated diets in marine fish larvae. *Aquaculture* 200, 161-180.

19. Cahu, C.L.; Zambonino Infante, J.L.; Escaffre, A.M.; Bergot, P. & Kaushik, S. (1998). Preliminary results on sea bass *Dicentrarchus labrax* larvae rearing with compound diet from first feeding. Comparison with carp *Cyprinus carpio* larvae. *Aquaculture* 169, 1–7

20. Cahu, C.L. & Infante, J.L.Z. (1994). Early weaning of sea bass (*Dicentrarchus labrax*) larvae with a compound diet: effect on digestive enzymes. *Comparative Biochemistry and Physiology* 109A, 213–222

21. Cerqueira, V.R.; Macchiavello, J.A.G. & Brugger, A.M. (1995). Producao de alevinos de robalo, *Centropomus parallellus*, (Poey 1890) atraves de larvicultura intensivaem laboratorio. 191-197.

22. Chapman, P.; Horel, G.; Fish. W., Jones. K. & Spicola. J. (1978). Artificial culture of snook, rookery Bay, 1977. Florida Game and Fresh Water Fish Commission.

23. Crisp, D.T. (1988) Prediction from temperature of eyeing, hatching and 'swim-up' times for salmonid embryos. *Freshwater Biology* 19, 41–48.

24. Colesante R.T. (1989). Improved survival of walleye fry during the first 30 days of intensive rearing on brine shrimp and zooplankton. *Progressive Fish-Culturist* 51, 109-111.

25. Corraza L. & Nickum J.G. (1981). Positive phototaxis during initial feeding stages of walleye larvae. Rapports et Proces-Verbaux Reunions du Conseil International pour l' Exploration de la Mer 178, 492-494p.

26. Dabrowski K; Lee K.-J & Rinchard J. (2003). The smallest vertebrate, teleost fish, can utilize synthetic dipeptide based diets. *Journal of Nutrition* 133, 4225-4229.

27. Dehasque, M.; De Wolf, T.; Candreva, P.; Coutteau, P. & Sorgeloos, P. (1998). Control of bacterial input through the live food in marine fish hatcheries. In: Grizel, H., Kestemont, P. , Aquaculture and Water: Fish Culture, Shellfish Culture and Water Usage. pp. 66–67. Oostende.: European. Aquaculture Society.

28. Denson M.R. & Smith T.I.J. (1996). Larval rearing and weaning techniques for white bass *Morone chrysops*. *Journal of World Aquaculture Society* 27, 194-201.

29. Dendrinos P.; Dewan S. & Thorpe J.P. (1984). Improvement in the feeding efficiency of larval, post larval and juvenile Dover sole (*Solea solea* L.) by the use of staining to improve the visibility of Artemia used as food. *Aquaculture* 38, 137-144.

30. Divanach, P.; Papandroulakis, N. & Kentouri, M. (1998). Bases de techniques de l'evage larvaire de la daurade *Sparus aurata* . 107–114p.

31. Doi, M.; Ohno, A.; Taki, Y.; Singhagraiwan, T. & Kohno, H. (1997a). Nauplii of the calanoid copepod, *Acartia injiensis* as an initial food organism for larval red snapper, *Lutjanus argentimaculatus*. *Suisan Zoshoku* 45, 31–40

32. Doi, M.; Toledo, J.D.; Golez, M.S.N.; Santos, M.D.L. & Ohno, A. (1997b). Preliminary investigation of feeding performance of larvae of early red-spotted grouper, *Epinephelus coioides*, reared with mixed zooplankton. *Hydrobiologia* 358, 259–263

33. Duray M.N; Estudillo C.B & Alpasan L.G. (1996) The effect of background color and rotifer density on rotifer intake, growth, and survival of the grouper (*Epinephelus suillus*) larvae. *Aquaculture* 146, 217-224.

34. Fernandez-Diaz, C. & Yufera, M. (1997). Detecting growth in gilthead seabream *Sparus aurata* L. larvae fed microcapsules. *Aquaculture* 153, 93–102.

35. Fore, P.L. & T.W. Schmidt (1973). Biology of juvenile and adult snook, *Centropomus undecimalis*, in the ten Thousand Islands, Florida. Atlanta, Georgia.

36. Fuiman, L.A.; Poling, K.R. & Higgs, D.M. (1998). Quantifying developmental progress for comparative studies of larval fishes. *Copeia* 199, 602–611.

37. Fushimi, T. (1983). III-5 ingestion by fish larave and juveniles. In: Koseishya-Koseikaku *The rotifers, Brachiomus plicatilis- Biology and mass culture*, pp. 69-93. Tokyo: Japanese Society of Science and Fisheries.

38. Fukusho, K. (1989). Biology and mass production of the rotifer, *Brachionus plicatilis*. *International Journal of Aquaculture and Fish Technology* 1, 292–299

39. Gatesoupe, F.J.; Girin, M. & Luquet, P. (1977). Recherche d'une alimentation artificielle adapte´e a` l'e´levage des stades larvaires des poissons. II- Application a` l'e´levage larvaire du bar et de la sole 4, 59–66p.

40. Hagiwara A.; Gallardo W.G.; Assavaaree M.; Kotani T & De Araujo A.B. (2001). Live food production in Japan: recent progress and future aspects. *Aquaculture* 200, 111-127.

41. Heath, M.R. (1992). Field investigation of the early life history stages of marine fish. *Advances in Marine Biology* 28, 2-174.

42. Heggberget, T.G. & Wallace, J.C. (1984) . Incubation of the eggs of Atlantic salmon, *Salmo salar*, at low temperatures. *Canadian Journal of Fisheries and Aquatic Sciences* 41 , 389–391.

43. Heming, T.A. (1982) Effects of temperature on utilisation of yolk by chinook salmon (*Oncorhynchus tshawytscha*) eggs and alevins. *Canadian Journal of Fisheries and Aquatic Sciences* 39, 184–190

44. Heming, T.A. & Buddington, R.K. (1988) Yolk absorption in embryonic and larval fishes. In: S. Hoar and D.J. Randall, (Eds.) *In: Fish Physiology* , pp. 407–446 San Diego.: Academic Press]

45. Hernandez-Cruz C.M.; Salhi M.; Fernandez-Palacios H & Izquierdo M.S. (1994) Improvements in the culture of *Sparus aurata* L. larvae in relation to the use of antibiotics, phytoplankton and rearing system. *Aquaculture* 124 , 269-274. Hernández Cruz, C.M.; M Salhi, M.; Bessonart, M.S.; Izquierdo, M.M.; González, H. & Fernández Palacios. (1999). Rearing techniques for red porgy (*Pagrus pagrus*) during larval development. Aquaculture 179:489-497.

46. Hinshaw J.M. (1985). Effects of illumination and prey contact on survival and growth of yellow perch, *Perca flavescens*. *Transactions of the American Philological Association* 114, 540-545.

47. Hirata, H. (1980). Culture methods of the marine rotifer, *Brachionus plicatilis*. *Mini Reviews and Data File of Fishreries Research* 1, 27–46.

48. Hochachka, P.W. & Somero G. N. (2001). *Biochemical Adaptation*. Oxford: Oxford University Press.

49. Holmejford, I.; Gulbrandsen, J.; Lein, I..; Refstie, T.; Leger, P.; Harboe, T.; Huse, I.; Sorgeloos, P.; Bolla, S., Olsen, I.; Reitan, K.I.; Vadstein, O.; Oie, G. & Danielsberg, A. (1993) An intensive approach to atlantic halibut fry production. *Journal of the World Aquaculture Society* 24 , 275–284

50. Houde, E.D. (1975). Effects of stocking density and food density on survival, growth and yield of laboratory reared larvae of sea bream *Archosargus rhomboidalis* (L.) (Sparidae). *Journal of Fish Biology* 7, 115– 127

51. Houde, E.D. & Schekter, R.C. (1980) Feeding by marine fish larvae: developmental and functional responses. *Environ. Biol. Fishes* 5 4,
52. Howell B.R. (1979). Experiments on the rearing of larval turbot, *Scophthalmus maximus* L. *Aquaculture* 18, 215-225
53. Howell B.R. (1994). Fitness of hatchery-reared fish for survival in the sea. *Aquaculture and Fisheries management* 25, 3-17.
54. Howells, R.G.; Sonski, A.J.; Shafland, P.L. & Hilton B.D. (1990). Lower temperature tolerance of snook, *Centropomus undecimalis. Northeast Gulf Science* 11, 155-158.
55. Hunter, J.R. (1980). The feeding behavior and ecology of marine fish larvae. Fish Behaviour and its Use in the Capture and Culture of Fishes. International Center for Living Aquatic Resources. Management 512p.
56. Ina, K., Ryogi, Y. & Higashi, K. (1979). Color sensitivity of red sea bream, Pagrus major. *Bulletin of the Japanese Society of Sciencie Fisheries.* 45(l), l-5
57. Ivlev, V.S. (1961). Experimental Ecology of Feeding of Fishes. New Haven, CT: Yale University Press.
58. Jensen, A.J., Johnsen, B.O. & Saksgĺrd, L. (1989) . Temperature requirements in Atlantic salmon (*Salmosalar*), brown trout (*Salmo trutta*), and Arctic char (*Salvelinus alpinus*) from hatching to initial feeding compared with geographic distribution. *Canadian Journal of Fisheries and Aquatic Sciences* 46, 786–789
59. Johnston, I.A.; Vieira, V.L.A. & Hill, J. (1996). Temperature and ontogeny in ectotherms: muscle phenotype in fish. In: Johnston, I. A. & Bennett, A. F. *Phenotypic and Evolutionary Adaptations of Organisms to Temperature*, pp. 153–181. Cambridge: Cambridge University Press.]
60. Jones D.A.; Kamarudin M.S & LeVay L. (1993). The potential for replacement of live feeds in larval culture. *Journal of the World Aquaculture Society* 24, 199-210.
61. Jones A & Houde E. D., (1981). Mass rearing of fish fry for aquaculture. Realism in aquaculture: achievements, constrains and perspectives. 351-374p. European Aquaculture Society
62. Kafuku, T. & Ikenoue H. (1983). Modern methods of aquaculture in Japan. In: Anonymous *Developments in aquaculture and fisheries sciences*, pp. 216 Amsterdam: Tokyo and Elsevier]
63. Kane, T.R. (1988) Relationship of temperature and time of initial feeding of Atlantic salmon. *The Progressive Fish-Culturist* 50, 93–97
64. Kaji, T.; Tanaka, M. & Tagawa, M. (1999) Laboratory study of density-dependent survival after handling in yolksac larvae of tunas and a grouper. *Fishery Science* 65, 482– 483.

65. Kamler, E. (1992a). *Early Life History of Fish: An Energetics Approach*, London: Chapman & Hall.

66. Kamler, E. (1992b). Mixed feeding period. In: Anonymous *Early Life History of Fish*, pp. 176- 181. London: Chapman & Hall

67. Kanazawa, A.; Teshima, S.; Inamori, S.; Sumida, S.; and Iwashita, T. (1982). Rearing of larval red sea bream and ayu with artificial diets. *Memoirs of the Faculty of Fisheries, Kagoshima University* 31, 185–192

68. Kanazawa, A. (1993). Essential phospholipids of fish and crustaceans. In: Kaushik, S.J., Luquet, P. *Fish nutrition in practice*, INRA edn. pp. 519-530. Paris, Les Colloques.

69. Kennedy, S.B..; Tucker, J.W..; Neidig, C.L..; Vermeer, G.K., Cooper, V.R..; Jarrell, J.L. & Sennet, D.G. (1998). Bacterial management strategies for stock enhancement of warm water marine fish: a case study with common snook *Centropomus undecimalis*. *Bulletin of Marine Science* 62, 573–58.

70. Kentouri, M. (1985). Comportement larvaire de 4 Sparides mediterraneens en elevage: *Sparus aurata, Diplodus sargus, Lithognathus mormyrus, Puntazzo puntazzo* Poissons teleosteens . Universite de Sciences et Techniques du Languedoc.

71. Klumpp, D.W. & Von Westernhagen, H. (1986) Nitrogen balance in marine fish larvae: Influence of developmental stage and prey density. *Marine Biology* 93, 189–200

72. Kolkovski S, Koven, W.M & Tandler A. (1997). The mode of action of Artemia in enhancing utilization of microdiet by gilthead seabream *Sparus aurata* larvae. *Aquaculture* 155, 193-205.

73. Koumoundouros, G.; Divanach, P. & Kentouri, M. (1999). Ontogeny and allometric plasticity of *Dentex dentex* (*Osteichthyes: Sparidae*) in rearing conditions. *Marine Biology* 135, 561–572

74. Kovac, V. (2002). Synchrony and heterochrony in ontogeny (of fish). *Journal of Theoretical Biology* 217, 499–507

75. Koven, W.M..; Tandler, A.; Kissil, G.W.; Sklan, D.; Friezlander, O. & Harel, M. (1990). The effect of dietary n-3 polyunsaturated fatty acids on growth, survival and swimbladder development in *Sparus aurata* larvae. *Aquaculture* 91 131–141.

76. Lau, S.R. & Shafland P. L. (1982). Larval development of snook *Centropomus undecimalis* (Pisces: *Centropomidae*). *Copeia* 618-627.

77. Laurence, G.C. (1978). Comparative growth, respiration and delayed feeding abilities of larval cod (*Gadus morhua*) and haddock (*Melanogrammus aeglefinus*) as influenced by temperature during laboratory studies. *Marine Biology* 50, 1– 7

78. Laurence, G.C. (1974). Growth and survival of haddock (*Melanogrammus aeglefinus*) larvae in relation to planktonic prey concentration. *Journal of the Fisheries Research Board of Canada* 31, 1415– 1419

79. Laurel, B.J.; Brown, J.A. & Anderson, R. (2001). Behaviour, growth and survival of redfish larvae in relation to prey availability. *Journal of Fish Biology* 59, 884– 901

80. Lazo J.P.; Dinis M.T.; Holt G.J.; Faulk C & Arnold C.R. (2000). Co-feeding microparticulate diets with algae: toward eliminating the need of zooplankton at first feeding in larval red drum (*Sciaenops ocellatus*). *Aquaculture* 188,

81. Liao, I.C.; Su, H.M. & Chang, E.Y. (2001). Techniques in finfish larviculture in Taiwan. *Aquaculture* 200, 1-31.

82. Lubzens, E. (1987). Raising rotifers for use in aquaculture. *Hydrobiologia* 147, 245–255.

83. May R.C. (1971). An annotated bibliography of attempts to rear the larvae of marine fishes in the laboratory.

84. Mani-Ponset, L.; Guyot, E.; Diaz, J.P. & Connes, R. (1996). Utilization of yolk reserves during post-embryonic development in three teleostean species: the sea bream, (*Sparus aurata*), the sea bass, (*Dicentrarchus labrax*) and the pike-perch (*Stizostedion lucioperca*). *Marine Biology* 126, 539-547.

85. Marliave J.B. (1994). Green water: optical rather than nutritional effect. AZA Conference Proceedings, 266-269p.

86. Marr, D.H.A. (1996) Influence of temperature on the efficiency of growth of salmonid embryos. *Nature* 212, 957–959

87. Matlock, G.C. & Osburn, H.R. (1987).Demise of the snook fishery in Texas, USA. *Northeast Gulf Science* 9, 53-58.(Abstract)

88. Moffatt, N.M. (1981). Survival & growth of northern anchovy larvae on low zooplankton densities as affected by the presence of a *Chlorella* bloom. In: Lasker, R., Sherman, K. *The Early Life History of Fish: Recent Studies*, pp. 475–480

89. Munk, P. & Kiørboe, T. (1985). Feeding behaviour and swimming activity of larval herring (*Clupea harengus*) in relation to density of copepod nauplii. *Marine Ecology Progress Series* 24, 15– 21

90. Munro, P.D., Barbour, A. & Birkbeck, T.H. (1994). Comparison of the gut bacterial flora of starting feeding larval turbot reared under different conditions. *Journal of Food Protection*. 40, 288

91. Munro, P.D.; Birkbeck, T.H. & Barbour, A. (1993). Bacterial flora of rotifers *Brachionus plicatilis*: evidence for a major location on the external surface and methods for reducing the rotifer bacterial load. In Reinertsen, H., Dahle, L.A.,

Jorgensen, L., Tvinnereim, K. *Fish Farming Technology*, pp. 93–100. Balkema, Rotterdam:

92. Muroga, K. & Yasunobu, H. (1987). Uptake of bacteria by rotifer. *Bulletin of the Japanese Society of Scientific. Fisheries* 53, 2091
93. Naas K.E, NÒss T & Harboe T. (1992). Enhanced first feeding of halibut larvae (*Hippoglossus hippoglossus* L.) in green water. *Aquaculture* 105, 143-156.
94. Nagata, W.D. & Hirata, H. (1986). Mariculture in Japan: past, present, and future prospective. *Mini Reviews and Data File of Fisheries Research* 4, 1–38.
95. Nellen W. (1981). Live animal food for larval rearing in aquaculture.- non-Artemia organisms Realism in Aquaculture: Achievements, Constrains, Perspectives 215- 260. European Aquaculture Society.
96. Nicolas JL, Robic E & Ansquer D (1989). Bacterial flora associated with a tropical chain consisting of microalgae, rotifers and turbot larvae: influence of bacteria on larval survival. Aquaculture 83: 237–248
97. Oie, G., Makridis, P.; Reitan, K.I. & Olsen, Y. (1997). Protein and carbon utilization of rotifers *Brachionus plicatilis* in first feeding of turbot larvae *Scophthalmus maximus* L. *Aquaculture* 153, 103–122.
98. Oozeki, Y.; Hagiwara, A.; Eda, H. & Lee, C.S. (1992). Developmental changes of food selectivity of striped mullet *Mugil cephalus* during the Larval Stage. *Nippon Suisan Gakkaishi* 58, 1381
99. Ostaszewska T; Dabrowski, K.; Czuminska, K. & Olech W. (2005). Rearing pike-perch larvae using formulated diets- first success with starter feeds. *Aquaculture research* 36, 1167-1176.
100. Ostrowski A.C. (1989). Effect of rearing tank background color on early survival of dolphin larvae. *Progressive Fish- Culturist* 51, 161-63.
101. Papandroulakis N.; Divanach P & Kentouri M (2002). Enhanced biological performance of intensive sea bream *Sparus aurata* larviculture in the presence of phytoplankton with long photophase. *Aquaculture* 204, 45-63.
102. Pearce, D. (1991). Measuring Sustainable Development. Blueprint 2. Earthscan, London Pedersen B.H.; Ueberschar B & Kurokawa T. (2003). Digestive response and rates of growth in pre-leptocephalus larvae of Japanese eel *Anguila japonica* reared on artificial diets. *Aquaculture* 215, 321-338.
103. Pedersen, B.H. & Hjelmeland, K. (1988). Fate of typsin and assimilation efficiency in larval herring *Clupea harengus* following digestion of copepods. *Marine Biology* 97, 467–476.
104. Pedersen, B.H.; Nilssen, E.M. & Hjelmeland, K. (1987). Variations in the content of trypsin and trypsinogen in larval herring *Clupea harengus* digesting copepod nauplii. *Marine Biology* 94, 171–18

105. Peña R.; Dumas S.; Saldivar-Lucio R.; Garcia G.; Trasvin a A & Hernandez-Ceballos D. (2004) The effects of light intensity on first feeding of the spotted sand bass *Paralabrax maculatofasciatus* Steindachner larvae. *Aquaculture Research* 35, 345-349.

106. Pepin, P. (1991). The effect of temperature and size on development and mortality rates of the pelagic early life history stages of marine fish. *Canadian Journal of Fisheries and Aquatic Sciences* 48, 503–518

107. Pepin, P.; Orr, D.C. & Anderson, J.T. (1997). Time to hatch and larval size in relation to temperature and egg size in Atlantic cod (*Gadus morhua*). *Canadian Journal of Fisheries and Aquatic Sciences* 54 (Suppl. 1), 2–10

108. Peterson, R.H.; Martin-Robichaud, D.J. & Berge, Ĺ. (1996) Influence of temperature and salinity on length and yolk utilization of striped bass larvae. *Aquaculture International* 89–103, 4.

109. Peterson, R.H. & Martin-Robichaud, D.J. (1995). Yolk utilization by Atlantic salmon (*Salmo salar* L.) alevins in response to temperature and substrate. *Aquacultural Engineering* 14, 85–99

110. Person Le Ruyet, J.; Alexandre, J.C.; Thebaud, L. & Mugnier, C. (1993). Marine fish larvae feeding: formulated diets or live preys?. *Journal of the World Aquaculture. Society*. 24, 211–224

111. Peterson, R.H.; Spinney, H.C.E. & Sreedharan, A. (1977) . Development of Atlantic salmon (*Salmo salar*) eggs and alevins under varied temperature regimes. *Journal of the Fisheries Research Board of Canada* 34, 31–43

112. Puvanendran, V. & Brown, J.A. (1999). Foraging, growth and survival of Atlantic cod larvae reared in different prey concentrations. *Aquaculture* 175, 77–92

113. Rainuzzo, J.R.; Reitan, K.I. & Olsen, Y. (1997). The significance of lipids at early stages of marine fish: a review. *Aquaculture* 155 , 103-115.

114. Rana, K, (1986). Parental influences on egg quality, fry production and fry performance in *Oreochromis niloticus* (Linnaeus) and *O. mossambicus* (Peters). PhD thesis, University of Stirling, Scotland.

115. Reitan K.I.; Bolla S,& Olsen Y. (1991). Ingestion and assimilation of microalgae in yolk sac larvae of halibut, *Hippoglossus hippoglossus* (L.). Larvi '91, Fish and Crustacean Larviculture Symposium 15, 332-334.p. European Aquaculture Society

116. Reitan, K.I.; Rainuzzo, J.R.; Oie, G. & Olsen, Y. (1993). Nutritional effects of algal addition in first-feeding of turbot *Scophthalmus maximus* L. larvae. *Aquaculture* 118, 257–275.

117. Rideout, R.M.; Trippel, E.A. & Litvak, M.K. (2004). Paternal effects on haddock early life history traits. *Journal of Fish Biology* 64, 695–701

118. Rombough, P.J. (1988). Respiratory gas exchange, aerobic metabolism, and effects of hypoxia during early life. In: Hoar S and D.J. Randall, (Ed.) *Fish Physiology* , pp. 59– 61 San Diego: . Academic Press]

119. Rotllant J.; Tort L.; Montero D.; Pavlidis M.; Martinez M.; Wendelaar Bonga S.E & Balm P.H.M. (2003). Background colour influence on the stress response in cultured red porgy *Pagrus pagrus*. *Aquaculture* 223, 129-139.

120. Salhi M.; Izquierdo, M.S.; Hernandez de la Cruz C. M.; Socorro J & Fernandez-Palacios H. (1997). The improved incorporation of polyunsaturated fatty acids and changes in liver structure in larval gilthead seabream fed on microdiets. *Journal of Fish Biology* 51, 869–879

121. Scott A.P & Middleton C. (1979). Unicellular algae as a food for turbot (*Scophthalmus maximus,* L.) larvae the importance of dietary long-chain polyunsaturated fatty acids. *Aquaculture* 18, 227-240.

122. Seidel, C.R.; Schauer, P.S.; Katayama, T. & Simpson, K.L. (1980). Culture of Atlantic silverside fed on artificial diets and brine shrimp nauplii. *Bulletin of the Japanese Society of Scientific Fisheries* 46 , 237–245

123. Shafland, P.L.& Koehl D.H. (1980). Laboratory rearing of the common snook. Proceedings of the annual Conference South-eastern Association of Fisheries and Wildlife Agencies 33 (425-431).

124. Skjermo, J. & Vadstein, O. (1999).Techniques for microbial control in the intensive rearing of marine larvae. *Aquaculture* 177, 333–343

125. Snell, T.W. (1991).Improving the design of mass culture systems for the rotifer, *Brachionus plicatilis*. In. Fulks, W, Main, K.L. Rotifer and Microalgae Culture Systems, 61–71.p.

126. Sorgeloos P.; Léger PH & Lavens P (1988). Improved larval rearing of European and Asian seabass, seabream, mahimahi, siganid and milkfish using enrichment diets for *Brachionus* and *Artemia*. *World Aquaculture* 19, 78-79.

127. Shirota, A. (1970). Studies on the mouth size of fish larvae. *Nippon Suisan Gakkaishi* 36, 353–368

128. Støttrup, J.G. (2000). The elusive copepods: their production and suitability in marine aquaculture. *Aquaculture Research,* 31, 703-711.

129. Støttrup J. G & Norsker N.H (1997). Production and use of copepods in marine fish larviculture. *Aquaculture* 155, 231-247.

130. Støttrup, J.G.; Gravningen, K. & Norsker, N.H. (1995). The role of different algae in the growth and survival of turbot larvae *Scophthalmus maximus* L. in intensive rearing systems. 201, 173–186p.

131. Tagawa M.; Kaji, T.; Kinoshita, M. & Tanaka, M. (2004). Effect of stocking density and addition of proteins on larval survival in Japanese flounder, *Paralichthys olivaceus*. *Aquaculture* 230, 517-525.

132. Tagawa M,; Kaji, T.; Kinoshita, M. & Tanaka, M. (1997). Effect of thyroid hormone supplementation to eggs on the larval survival of Japanese flounder. edn. p. 48p.

133. Takeuchi T.; Wang Q.; Furuita H.; Hirota T.; Ishida S & Hayasawa H. (2003). Development of microparticle diets for Japanese flounder *Paralichthys olivaceus* larvae. *Fisheries Science* 69, 547-554.

134. Takeuchi, T.; Ohkuma, N.; Ishida, S.; Ishizuka, W.; Tomita, M.; Hayasawa, H. & Miyakawa, H., (1998). Development of micro-particle diet for marine fish larvae 1st International Symposium on Nutrition and Feeding of Fish, Abstract only 193p.

135. Tamaru C.S.; Murashige R & Lee C.S. (1994). The paradox of using background phytoplankton during the larval culture of striped mullet, *Mugil cephalus* L. *Aquaculture* 119, 167-174.

136. Tamaru C.S.; Lee C, & Ako H. (1993). Improving the larval rearing of stripped mullet (*Mugil cephalus*) by manipulating quantity and quality of the rotifer. *Brachiounus plicatillis*. Proceedings of a US-Asia Workshop

137. Taylor R. G.; Whittington J. A.; Grier, H.J. & Crabtree R. E. (2000). Age, growth, maturation, and protandric sex reversal in common snook, *Centropomus undecimalis*, from the east and west coast of South Florida. *Fisheries Bulletin* 98, 612-624.(Abstract).

138. Toledo, J.D.; Golez, S.N.; Doi, M. & Ohno, A. (1997). Food selection of early grouper, *Epinephelus coioides*, larvae reared by the semi-intensive method. *Suisan Zoshoku* 45, 327–337

139. Tringali, M.D. & T. M. Bert (1996). The genetic stock structure of common snook (*Centropomus undecimalis*). *Canadian Journal of Fishery and Aquatic Science.* 53, 974- 984.(Abstract)

140. Tsukamoto, K. (1993). Quality of fish for release. In: Kitajima, C. *Healthy Fry for Release, and Their Production Technique*, pp. 102–113 Koseisya Koseikaku, Tokyo.

141. Tucker, J.W.J.; Landau, M.P. & Falkner, B.E. (1985). Culinary value and composition of wild and captive common snook, *Centropomus undecimalis*. *Florida Scientist* 49, 196- 200.(Abstract)

142. Van Der Meeren, T. (1991). Algae as first food for cod larvae *Gadus morhua* L. : filter feeding or ingestion by accident? *Journal of Fish Biology* 39, 225–237

143. Vasquez-Yeomans L.& Carrillo-Barrios-Gomez E & Sosa-Cordero E. (1990).The effect of the nanoflagellate *Tetraselmis suecica* on the growth and survival of grunion, *Leuresthes tenuis*, larvae. *Environmental Biology of Fisheries* 29, 193-200.

144. Watanabe, T. (1993). Importance of docosahexaenoic acid in marine larval fish. *Journal of World Aquaculture Society* 24, 495-501.

145. Webb J.F. (1999). Larvae in fish development and evolution. In: B.K. Hall & M.H. Wake. *The Origin and Evolution of Larval Forms* , pp. 109-158. San Diego, CA, USA: Academic Press]

146. Werner R.G & Blaxter J.H.S (1981) The effect of prey density on mortality, growth and food consumption in larval herring (*Clupea harengus L.*). *Rapports et Proces-verbaux du Conseil international pour l'Exploration de la Mer* 178, 405-408.

147. Wyatt, T. (1972). Some effects of food density on the growth and behaviour of plaice larvae. *Marine Biology* 14, 210–216

148. Yanes-Roca, C.; Rhody, N.; Nystrom, M.; Main, K. (2009). Effects of Fatty acid composition and spawning season patterns on egg quality and larval survival in common snook (*Centropomus undecimalis*). Aquaculture 287, pp 335-340

149. Yufera M.; Pascual E & Fernandez-Diaz C. (1999). A highly efficient microencapsulated food for rearing early larvae of marine fish. *Aquaculture* 177, 249-256.

150. Zambonino Infante, J.L.; Cahu, C.L. & Peres, A. (1997) Partial substitution of di- and tripeptides for native proteins in sea bass diet improves *Dicentrarchus labrax* larval development. *Journal of Nutrition* 127, 608–614.

11 ADVANCES IN DOMESTICATION AND CULTURE TECHNIQUES FOR CRAYFISH PROCAMBARUS ACANTHOPHORUS

1. INTRODUCTION

1.1 Ecological importance

The crayfish are a group of crustaceans that habit in different environments in the world, both in lotic systems as lentic, in addition to caverns, which makes them cosmopolitan organisms with a wide range of tolerance to environmental conditions. Over 600 crayfish species are known to exist in the worldwide, with at least 100 species in Australia and about 300 in the Americas (Holdich, 1993), mostly (85%) in North and Central America (Rojas, 1998). Species in Mexico include one in the *Orconectes* genus, 10 in the *Cambarellus* genus and 44 in the *Procambarus* genus, the latter also distributed in Belize, Honduras and the United States (Villalobos, 1948; Hobbs, 1984; Rojas, 1998; López, 2006). The genus *Procambarus* habits in temporary water bodies, during the dry season can be seen in small holes in the soil, similar to the anteaters, which conduct to tunnels and chambers with sufficient moisture for the crayfish to survive to the drying (López, 2008). The crayfish have been adapted in various ways, according to environmental conditions that occur in the places they want to colonize. The first adaptation is their ability to spend a lot of time, even months, faced with the lack of water and breathe atmospheric oxygen (Huner, 1995) in some cavemen environments it has been recorded that these organisms exhibit a diminution in their effective breathing rate as a response to the decrease in the concentration of oxygen, and undersupply of food (Mejía, 2010). Despite their abundance, less than a dozen crayfish species are cultivated worldwide and only two species constitute sizable commercial fisheries (Huner, 1994). Crayfish have a high potential for use in aquaculture systems because they are at the bottom of the trophic chain, feeding largely on carrion and detritus, for that they are therefore considered fundamental for maintaining ecological balance in natural ecosystems (Rojas, 1998), and is possible their maintained in control conditions. The global diversity of crayfish allows to establish a productive activity associated with diverse environments, even in reduced environments with eutrophication, sulfate-reducing bacteria and that there is a proliferation of algae (Sánchez et al., 2009).

1.2 Aquaculture importance

Crayfish like other decapods crustaceans, have some biological characteristics that make them potentially important species for aquaculture, among which include: adaptation to conditions of captivity and handling; accept artificial feeds of different origins (shrimp, fish aquatic plants, vegetables), can even be fed diets with vegetable protein (75%); have a relatively short life cycle (two years or less). First studies indicates that these organisms can breed in captivity at early age (about four months), and first spawning had high survival rates (> 75%), with reproduction all year, and some females had more than one spawning per year.

The physiological characteristics of crayfish allow them to adapt to extreme climatic variations, diversifying their potential habitats, ensuring reproduction and contributing to progeny survival under adverse conditions. This occurs under natural or artificial conditions, making them promising organisms for use in aquaculture systems (Gutiérrez- Yurrita, 1994; Rodríguez-Almaraz & Mendoza-Alfaro, 1999). Of the crayfish species which can be cultivated in subtropical environments, *Procambarus acanthophorus* stands out for its biological attributes, such as high number of progeny per spawn, resistance to a wide range of environmental and water quality conditions, and successful performance in captivity (Arrignon, 1985; Cervantes, 2008; Cervantes-Santiago et al., 2010a).

1.2 Advances in laboratory research for aquaculture facilities

1.3.1 Environmental requirements of the species

Advances in laboratory research indicate that crayfish *P. acanthophorus* has a high potential for being used in aquaculture, for that reason different trials were done to determinate the best biotechnology for semi intensive and intensive culture conditions in monoculture and polyculture facilities.

Studies under laboratory conditions had showed that crayfish *P. acanthophorus* (Villalobos, 1948; 1993), can be quickly adapted to conditions of captivity, despite coming from natural environments. In the study the organisms were maintained at an average temperature of 26 ± 2^{0}C and photoperiods between 12 and 14 h light: dark, oxygen concentrations from 0.5 to 5 mgL^{-1}, indicating that they may like other crustaceans endure low contents O_2. The pH tolerance of the species ranges from 6 to 9, the ammonium concentration is about 0.5 mlL-1, and hardness greater than 200 mlL-1 as calcium carbonates.

1.3.2 Feeding

Even though it is known that crayfish accept balanced food for aquaculture species as shrimp. Laboratory experiment with 20 formulated diets containing different protein (200, 250, 300, 350 and 400 gkg-1) and lipid (60, 80, 100 and 120 gkg-1) levels (Table 1) on growth and survival in juvenile crayfish (*P. acanthophorus*) during 12-week nutritional trial, indicate that the protein requirement for young *P. acanthophorus* is in a range between 210 and 280 gkg-1, without observing a specific requirement of lipids, the results also suggest that in culture, it is possible to use foods with a maximum of 279 gkg-1 of protein and 60 gkg-1 lipid for better growth in crayfish, which can use up to 75% protein of vegetal source and only 25% from animal source, with growth performance and uptake efficient (Cervantes, 2006) (Table 1).

The study results indicate that nutritional diets can be used with protein content between 211 and 232 gkg-1 to feed growing crayfish in order to minimize feed costs, indicating that these organisms consume protein from vegetal source and assimilated efficiently (1.09:1 FCR) regardless of sex. This was verified by the assessment of carcass composition of crayfish fed with 20 experimental diets and by sex, where females were found to store more lipids (%) and body caloric energy (MJ/100 g) without significant differences in relation with the values reported for males. Other important information obtained during the nutritional study was the detection of ovigerous females in treatments 200/120, 250/60 and 400/120, but could not be attributed to the generation of egg protein or lipids tested, which suggests that in general all the diets allowed the bodies to cover their energy requirement for basic functions, but also could reproduce and promote sexual maturation.

The crayfish survival ranged from 66% to 86%, without differences between treatments. Because of the lowest survival recorded in the juvenile fed with the 250/12 gkg-1 diet, although this treatment yielded the most efficient parameters (WG, SGR and DWG), a correlation analysis was performed between weight gain and survival to determine a possible influence of mortality on growth. The weight gain survival ratio was not significant ($r2 > 0.037$; $P > 0.1365$), indicating that crayfish growth was only affected for the experimental diets.

Table 1. Formulation and proximate composition of experimental diets containing different protein/lipid ratios fed *P. acanthophorus*.

Ingredients (g kg^{-1})	200 / 60	200 / 80	200 / 100	200 / 120	250 / 60	250 / 80	250 / 100	200 / 120	300 / 60	300 / 80	300 / 100	300 / 120	350 / 60	350 / 80	350 / 100	350 / 120	400 / 60	400 / 80	400 / 100	400 / 120
Fish meal[1]	77.1	77.1	77.1	77.1	96.3	96.3	96.3	96.3	115.6	115.6	115.6	115.6	134.8	134.8	134.8	134.8	154.1	154.1	154.1	154.1
Soya meal[2]	327.9	327.9	327.9	327.9	409.8	409.8	409.8	409.8	491.8	491.8	491.8	491.8	573.8	573.8	573.8	573.8	667	681	695.1	702.6
Wheat bran[3]	58.1	58.1	58.1	58.1	72.6	72.6	72.6	72.6	87.2	87.2	87.2	87.2	101.7	101.7	101.7	101.7	88.3	53.5	18.6	0.0
Fish oil[4]	22.3	32.3	42.3	52.3	20.4	30.4	40.4	50.4	18.4	28.4	38.4	48.4	16.5	26.5	36.5	46.5	14.6	24.6	34.6	44.6
Corn oil[5]	20.8	30.8	40.8	50.8	18.6	28.6	38.6	48.6	16.3	26.3	36.3	46.3	14	24	34	44	12.4	23.3	34.2	44.7
Corn starch	431.3	411.3	391.3	371.3	319.28	299.8	279.8	259.8	208.3	188.3	168.3	148.3	96.7	76.7	56.7	36.7	1..1	1.0	0.9	0.0
Vitamin and mineral premix[6]	30.0	30.0	30.0	30.0	30.0	30.0	30.0	30.0	30.0	30.0	30.0	30.0	30.0	30.0	30.0	30.0	30.0	30.0	30.0	30.0
Carboxy-methylcellulose	15.0	15.0	15.0	15.0	15.0	15.0	15.0	15.0	15.0	15.0	15.0	15.0	15.0	15.0	15.0	15.0	15.0	15.0	15.0	15.0
Soya lecithin	5.0	5.0	5.0	5.0	5.0	5.0	5.0	5.0	5.0	5.0	5.0	5.0	5.0	5.0	5.0	5.0	5.0	5.0	5.0	5.0
Pigment	10.0	10.0	10.0	10.0	10.0	10.0	10.0	10.0	10.0	10.0	10.0	10.0	10.0	10.0	10.0	10.0	10.0	10.0	10.0	10.0
Vitamin C[7]	2.5	2.5	2.5	2.5	2.5	2.5	2.5	2.5	2.5	2.5	2.5	2.5	2.5	2.5	2.5	2.5	2.5	2.5	2.5	2.5
Proximate composition (g·kg^{-1} dry base)																				
Moisture	63.1	54.9	63.6	59.4	66.2	61.3	59.4	52.3	92.5	82.4	82.9	79.1	62.7	57.4	75.5	53.7	66.5	75.2	67.4	56.1
Crude protein	217.6	219.6	220.0	227.7	276.9	271.9	271.2	268.2	317.9	316.9	315.4	319.4	377.2	373.0	376.5	381.5	434.4	424.5	425.9	420.9
Ether extract	68.3	95	116.9	134.8	67.5	91.1	114.7	136	62.6	81.4	98.4	112.8	63.4	84.1	110.7	135.1	62.3	86.4	104.9	126.1
Ash	42.3	46.0	44.8	45.9	56.4	55.3	54.8	54.6	65.5	63.9	64.5	64.6	74.1	77.3	74.5	77.3	83.5	83.5	82.1	99.4
Crude fiber	10.7	11.9	11.0	12.3	10.8	16.1	20.4	20.9	25.8	22.9	21.0	20.1	29.3	31.6	34.1	40.8	34.8	36.5	30.4	28.2
NFE	598.0	572.6	543.7	519.9	522.2	504.3	479.5	468.0	435.7	432.5	417.8	404.0	393.3	376.6	328.7	311.6	318.5	293.9	289.3	269.3
Energy (Mj 100g^{-1})	18.92	18.83	20.01	17.24	18.42	18.67	18.75	17.24	18.29	18.25	18.46	19.00	18.54	18.92	19.17	18.83	18.08	18.04	17.58	18.46

1.3.3 Life cycle and reproduction

A vital aspect to consider when determining an organism culture potential, is its reproductive capacity under controlled conditions, which in turn depends on its ability to adapt to the culture system, feed and water quality. Factors reported to significantly affect crayfish reproductive capacity include water temperature, photoperiod, and sex ratio (Yeh & Rouse, 1995; Carmona-Osalde et al., 2002; 2004a, 2004b). The results in laboratory conditions demonstrate that the crayfish *P. acanthophorus* is a candidate for aquaculture production in a closed cycle since it effectively reproduces in captivity. During the time that organisms captured from the wild remained in captivity, it was observed mating and reproduction, from which ovigerous females were obtained, indicating that breeding in captivity could be obtained. Under the study conditions, P. *acanthophorus* exhibited the peak of the reproductive activity during November and December, when average water temperature was 25 °C. The lowest reproductive activity occurred in February and March, when wide variations in water temperature (25 ± 5°C) may have affected organism metabolism and consequently their reproductive cycle (Figure 1).

In a similar researches, Rodríguez-Serna et al. (2000) reported that reproduction in *P. llamasi* occurs year around, although, in contrast to *P. acanthophorus*, this species has three spawning peaks between November and June. Its maximum activity is in May and June and its minimum in August and October, when temperatures above 26.7 °C negatively affect its reproductive efficiency. Temperature is clearly a limiting factor for reproduction in *P. llamasi*, with optimum spawning at 21 °C, although breeder length and degree of female sexual maturity can also affect spawning (Carmona-Osalde et al., 2004a).

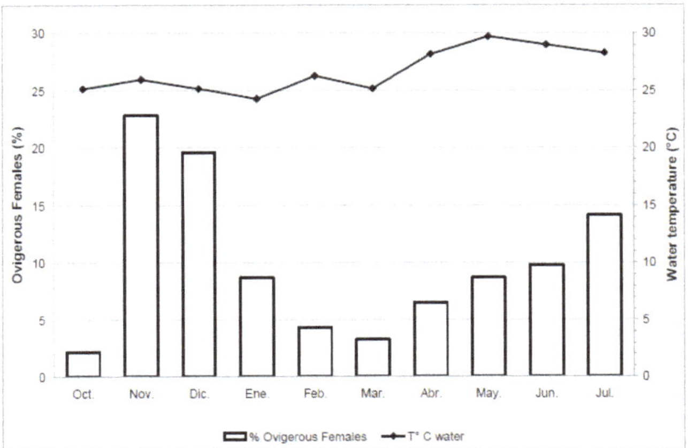

Fig. 1. Proportion of ovigerous females versus water temperature (°C) during a ten-month period.

In the study, a total of 92 ovigerous females were recorded from the 192 placed in the reproduction tanks during the study. ANOVA showed no significant variation for the number of ovigerous females at the three sex ratios used in the treatments, with 30 ovigerous females at 1:1, 40 at 1:3 and 22 at 1:5 male: female ratios, respectively (Figure 2).

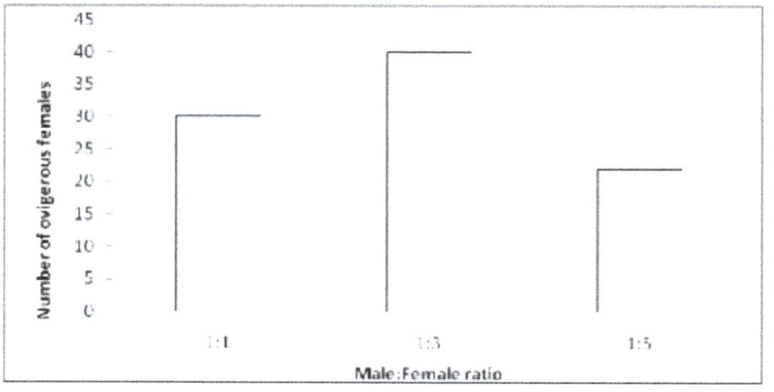

Fig. 2. Number of ovigerous females per sex ratio treatment

The results indicate that a higher quantity of gravid females can be obtained if sex ratio of 1:3 (male: female) is using during reproduction, which indicates that the ratio male: female is a variable that can affect the mass production juveniles. This is higher than the 1M: 1F ratio reported for optimum reproductive efficiency in peneid shrimp (Martínez, 1999), but lower than the 1M: 5F ratio reported for *Cherax quadricarinatus* (Yeh & Rouse, 1995). In the crayfish *Astacus astacus*, a 1M: 3F ratio is inefficient since all females do not reproduce at this ratio, for that reason a 1M: 2F ratio is recommended for best results (Taugbol & Skurdal, 1990). In *P. llamasi*, higher densities resulted in higher mating and ovigerous female rates during reproduction, meaning density significantly influenced female sexual maturation and possibly also male maturation (Carmona-Osalde, et al., 2004a).

Other interesting result is that mating was first observed when females reached an average length of 30 mm, although the first viable spawns did not occur until females reached 37 mm total length. This is therefore considered the female length at first sexual maturity for *P. acanthophorus*. This is similar to *P. llamasi* females, which exhibit initial reproductive behavior at 30 mm but have viable spawns only between 40 and 60 mm length. This indicates that crayfish mature and reproduce at an early age, an advantage for production under controlled conditions.

Aspects of reproductive biology such as fertility vary by species, ranging from as little as five eggs in *Astacus pachypus* up to 960 in *Cherax destructor* (Lee & Wickins, 1992). Egg counts for *P. acanthophorus* females in the present study ranged from 77 to 467 per female, with an average of 240.9. The ranges are similar with reported average fertility ranges for other crayfish species: 100-700 eggs in *P. clarkii* (Lee & Wickins, 1992; McClain & Romaire, 2007); 200-700 eggs in *P. llamasi* (Rodríguez-Serna et al., 2000); 300-400 eggs in *P. zonangulus* (Reynolds, 2002); and 323 eggs in *P. leniusculus* (Celada et al., 2005).

Egg count per female in crayfish depends on organism age and length. For instance, egg production in *A. leptodactylus* varies according to female length, with organisms measuring 47 to 76 mm producing between 200 and 400 eggs (average = 305.9), and 72 mm long females producing a average maximum of 588 eggs (Köksal, 1988; Mustafa et al., 2004). In the present study, total female length in *P. acanthophorus* had a linear, positive and significant ($p \leq 0.01$) relationship (r2=0.654) with egg count. This is similar to the positive relationship between cephalothorax length and egg count reported for *Austropotamobius pallipes* where females with a 25 mm minimum carapace length produced a maximum of 80 eggs (Brewis & Bowler, 1985). The number of eggs per spawn can also depend on egg size. The crab *Cambaroides japonicus* produces only 22 to 75 eggs per spawn but these are 2.13 to 2.50 mm in diameter; this is significantly larger than mean egg size in other crustaceans and may contribute to this species high survival rate (Nakata & Goshima, 2004).

Independent of egg count, not all eggs hatch. This can be caused by biotic and/ or abiotic factors that lead to losses during embryo growth. Temperature was the main factor causing egg loss in the present study since at higher temperatures fungi began to grow on the egg masses, affecting water quality in the hatchery units. Rodríguez-Serna et al. (2000) reported a similar incident in *P. llamasi* which seriously affected egg survival. Water quality and stability are clearly key elements during crayfish incubation since this is apparently a phase when pathogenic microorganisms attack eggs. Maximum egg viability in the present study was 97.4% in 56 mm long females. Embryo survival varied according to female length but did not exhibit a clear pattern; it was highest (42%) in 56-60 mm females and almost absent in 66-70 mm females. This decrease may be the result of female age since egg quality, and therefore viability, generally decreases as age increases, although this cannot be emphatically stated in the present case because the breeders were collected from the wild and their ages were therefore not exactly known. Two suggestions arise from the above results for commercial production of *P. acanthophorus*. First, breeders should be between 41 and 60 mm in length to ensure the highest possible egg and viable progeny counts. Second, adequate female nutritional condition and genetic quality need to be ensured since these are expressed in progeny quality and survival, perhaps by maintaining well-fed breeder stocks and employing constant selection to improve genetic quality.

1.3.4 Physicochemical parameters

Water chemical and quality parameters during the crayfish reproduction trial were within tolerance ranges for organisms of the same sex (Malone & Burden, 1988; McClain & Romaire, 2007; Cervantes-Santiago et al., 2010b): temperature, 23.8±2.2 °C; dissolved oxygen, 5.7±0.18 mg L-1; total hardness, 110 mg L-1CaCO3; pH,

8.67±0.13; ammonium, 0.18±0.10 mg L-1; N-nitrite, 0.25±0.20 mg L-1; and N-nitrate, 32.5±20.6 mg L-1.

1.3.5 Fertility

The ratio between total female length and egg counts (r2=0.6541) was positive, linear and significant (p≤0.01), defined by the equation y= 8.4126X - 216.4313. Average ovigerous female length was 54.4 mm (max = 71 mm; min = 37 mm) and average egg count per female was 240.9 (±S.D. 93.08) (max = 467; min = 77) (Figure 3).

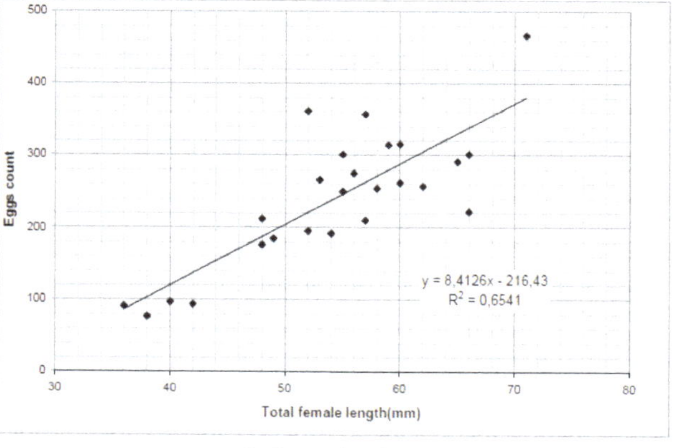

Fig. 3. Ratio of egg count to total female length in *Procambarus acanthophorus* during study.

1.3.6 Egg viability

Average egg viability was 29.1% (± S.D. 31.7; n=66), with a maximum of 97.4% and a minimum of zero. Overall, females between 41 and 60 mm had the highest egg viability (97.4%). Those within the 46-50 mm size had an average viability greater than 40%, while those in the 66-70 mm had the lowest (2.9%) (Figure 4).

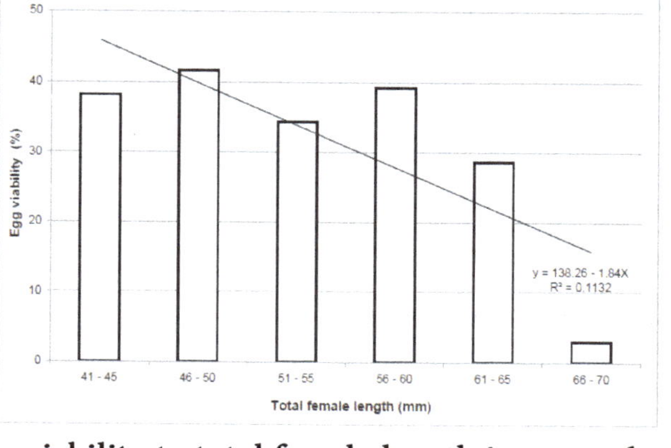

Fig. 4. Ratio of egg viability to total female length in *Procambarus acanthophorus* during study.

Water temperature probably affected egg viability since when it surpassed 30 °C, the eggs detached from the female's abdomen and were soon infected by fungi.

1.3.7 Female length at first sexual maturity

The present reproductive biology results for the crayfish *Procambarus acanthophorus* show it to be apt for use in aquaculture systems. Based on specimens grown in captivity, it is known that it reproduces year around and females are sexually mature at 37 mm length (13 weeks) (Cervantes, 2008). This allows for constant reproduction and implementation of a continuous breeder replacement program to ensure high quality and genetically variable spawns. Under the study conditions, spawn viability was variable (29.1- 97.4%), although this could probably be kept above 44% by controlling environmental parameters (e.g. keeping water temperature below 30 °C) and thus providing a regular supply of juveniles for culture. In addition, the species did not exhibit aggressiveness or territoriality during the study. These results, in conjunction with previous studies indicating these species preference for feeds containing vegetable protein sources (Cervantes et al., 2007), confirm that the crayfish *P. acanthophorus* is a candidate for culture under controlled conditions, be it as a preservation strategy for commercial purposes.

1.3.8 Embryology development

Incubation of crayfish *P. acanthophorus* eggs has a duration period of three to four weeks, depending on temperature and embryonic development inside the egg. At the time of hatching, the larvae present physical characteristics and eating behavior similar to an adult (Cervantes, 2006; 2008). A detailed description of embryonic development until juvenile stage, suggests that the development lasts 21 to 27 days on average, where it can identify nine embryonic stages, four post-embryonic and one juvenile. Also was observed that embryonic development, presents 11 color changes of the eggs, which are however asynchronous the same ovigerous mass, so it is considered that the coloring of eggs during embryonic development is not a clear indicator the stage of development.

Under the laboratory conditions, fourteen embryonic development stages were identified for *P. acanthophorus* using the structure descriptions and nomenclature of Anderson (1982). The females produced fertile eggs which exhibited nine embryonic stages, with an average total elapsed time of 15±3 days. After embryonic stages, crayfish had four post- embryonic stages and a final juvenile stage, which lasted an average of 10 days. Total elapsed time of development from fertilization through juvenile stage was on average 25 ± 3 days. Egg diameter and later embryo and juvenile length increased constantly until reaching a final length 600 times larger than initial diameter of a recently fertilized egg) (Figures 5 to 18).

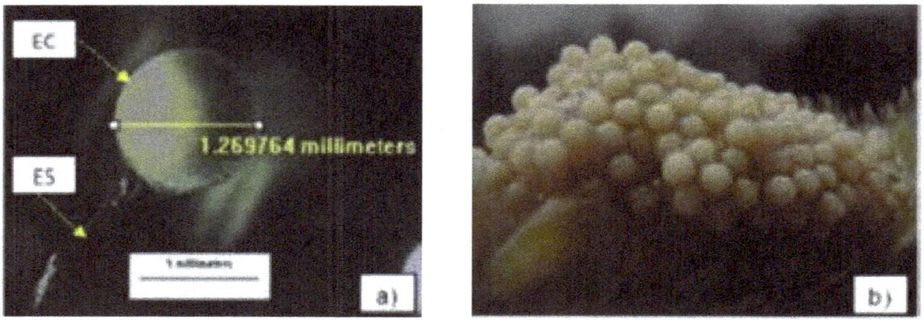

Fig. 5. Stage 1 (days 1-2; 0-11% development). - a) The recently fertilized eggs have a spherical shape and large quantities of mucus are present evidence of recent spawning. b) Egg mass has uniform light beige color which corresponds to the vitellus. By day 2 the vitellus has divided and small drops appear, probably the beginning of scission. It is observed the egg capsule (EC) and egg stolon (ES).

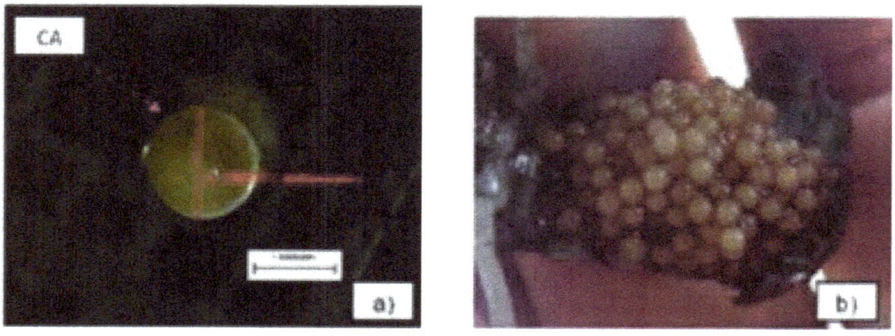

Fig. 6. Stage 2 (days 3-4; 11-22%). - a) Cellular division continues and the vitellus is completely divided into small drops. A region of greater cell accumulation (CA) is observed which corresponds to the zone where the blastopore will form. b) The egg mass begins to change color (yellow-olive green).

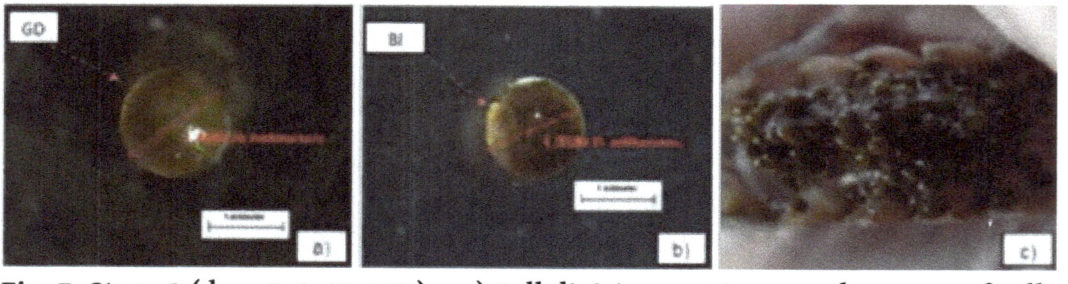

Fig. 7. Stage 3 (days 5-6; 22-33%). - a) Cell division continues and a group of cells begins to form on the egg ventral surface, corresponding to the germinal disk (GD). Starting on day 6, the cell layer begins to expand and form a depression, corresponding to the gastrula. b) The gastrula's ventral plate sinks in to form a groove, which is how the blastopore (Bl) appears; the forward portion of the caudal papilla starts to develop. The blastopore then closes and the rear portion of the caudal papilla appears. d) The egg mass changes in color from dark beige to a translucent olive green.

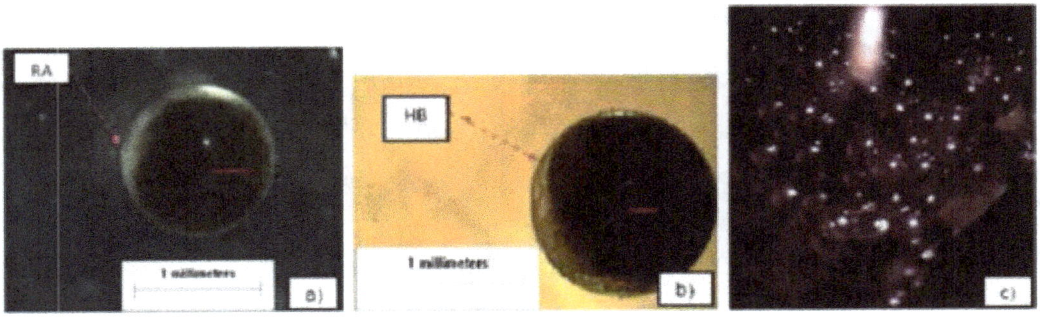

Fig. 8. Stage 4 (days 7-8; 33-44%). - a) An embryo with rudimentary anterior appendages (RA) is evident. Outlines of the frontal lobules and antennules, antennae and mandible can be distinguished. b) On day 8 the heart beat (HB) and embryo contractions can be. c) The egg mass slowly changes in color from olive green to brown.

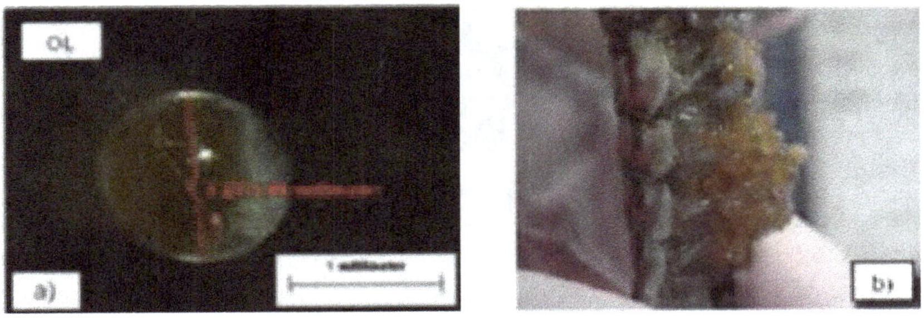

Fig. 9. Stage 5 (days 9-10; 44-55%).- a) Primordial eyes or ocular lobules (OL) appear as two elevations in front of the body, a transversal groove appears in the vitellus, crossing the middle of the egg, and a long, thin, anterior-curved caudal papilla is present. The embryo contracts more frequently and the vitellus is clearly visible. b) The egg mass has a translucent, light yellow color and the ocular spots appear as black dots.

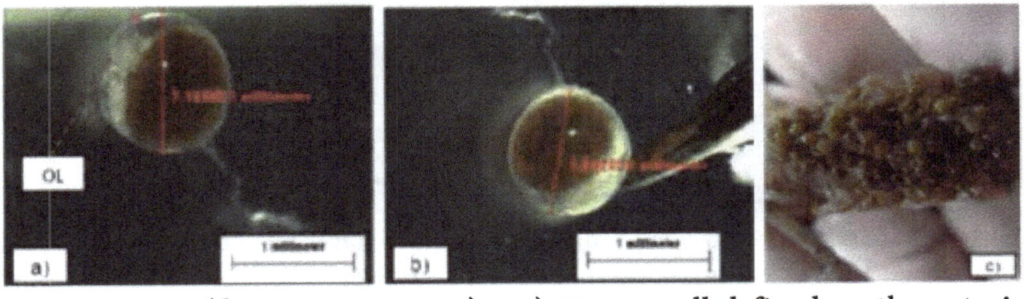

Fig. 10. Stage 6 (days 11-12; 55-66%). - a) OL are well-defined on the anterior body, while the abdominal somites and periopods remain rudimentary, although the chelae are visible. b) The caudal papilla is folded after and covered by the periopods, almost reaching the head. c) The egg mass is heterogeneously colored, with tones varying from light brown, to olive green and khaki.

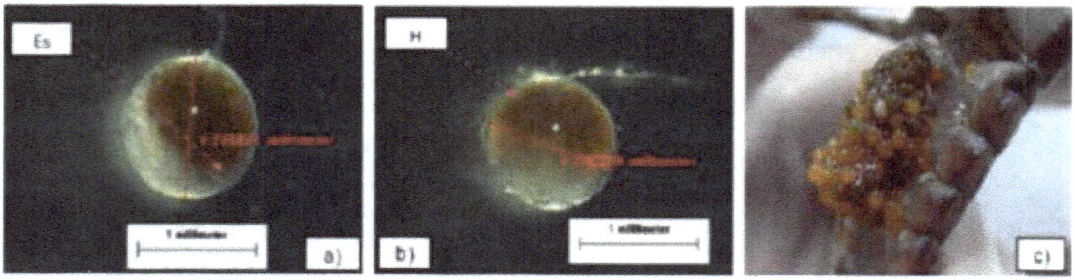

Fig. 11. Stage 7 (days 13-14; 66-77%). - a) Eyespots (Es) are visible, and deep grooves cross through the vitellus along the dorsal medial line. b) The heart (H) can be seen to beat strongly and regularly, Embryo interior is clear and more complex, the periopods are elongated and thin, and a small rostrum appears between the eyes. c) Egg mass coloring is heterogeneous, varying from olive green to bright orange.

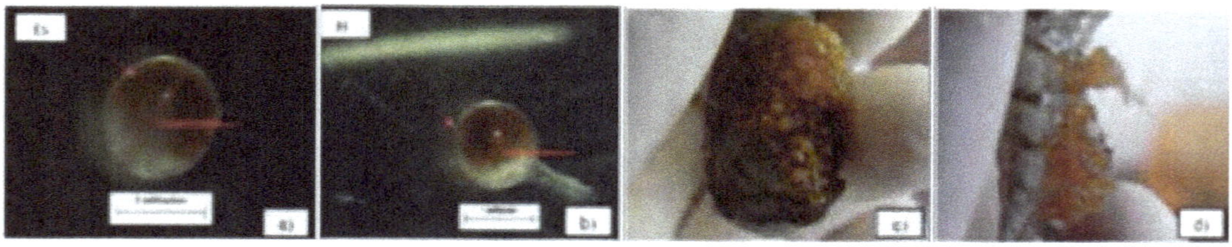

Fig. 12. Stage 8 (days 15-17; 77-88%). - a) The embryo occupies approximately three quarters of the egg ventral surface. b) The thoracic appendages are more developed and the chelae are totally formed. The eyes are sessile and elongate. c) The egg mass has taken on a translucent bright orange color. d) The embryo and eyes are clearly visible.

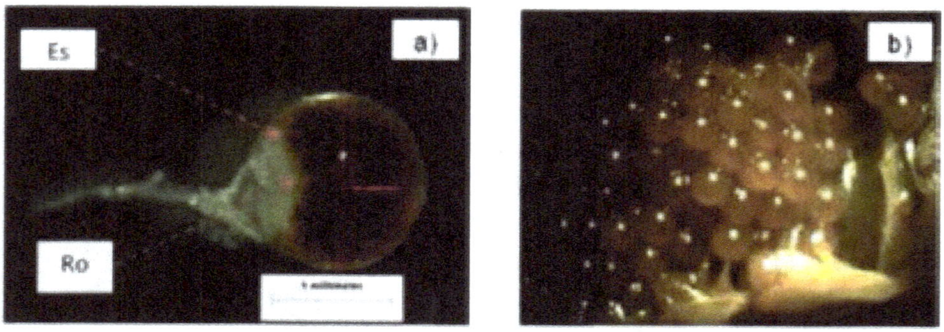

Fig. 13. Stage 9 (day18; 88-99%). - a) Shortly before hatching, the embryo appears compressed inside the egg such that the appendages seem flat and overlapped; there is no space remaining in the chorion. The chelae have grown in front of the eye base, the rostrum (Ro) is visible between the eyes and a groove sagittally crosses half the embryo. b) Egg mass color is bright yellow, rudimentary appendages are visible on the translucent embryos and the eyes are clearly identifiable.

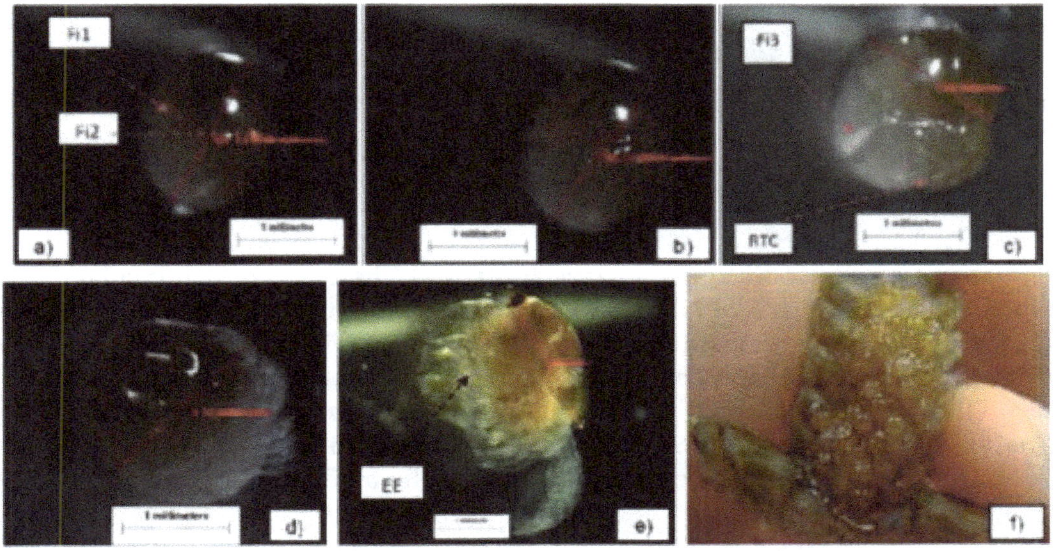

Fig. 14. Stage 10. Hatching (day 18; Development 100%). - The hatching process lasts an average of 10 to 15 min, from when the chorion breaks to when the embryo is completely free. a, b) It begins with the first fissure (Fi1) in the surface of the chorion barely visible, fissure (Fi2) of the chorion covers the folds of the eyes and maxillae; c) notable fissure (Fi3) accompanied by total rupture of the chorion (TRC). d) Breaking the chorion, the periopods are the first to get out. e) Expulsion of the embryo (EE) to the outside of the chorion. f) External appearance of hatchings, births shown asynchronism. This is considered a critical phase in embryo survival since exiting the chorion requires considerable energy expenditure.

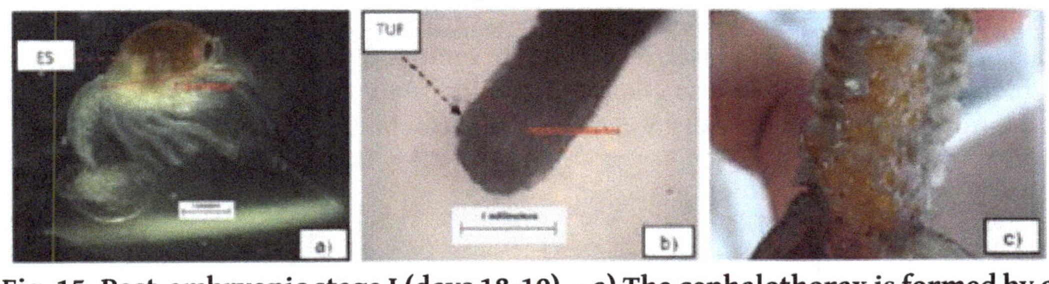

Fig. 15. Post-embryonic stage I (days 18-19). - a) The cephalothorax is formed by a yellow, elongated dorsal hump containing the remaining vitellus, which supplies nutrients to the organism during the following post-embryonic stages; the rest of the body is translucent. The eyes are round and sessile (ES) and contain dark pigment on about one quarter of the overall surface. The antennae and antennules are caudally curved and have sensory villi. b) The partially developed telsons and uropods are fused together in the membranous ligaments that attach the organisms to the chorion interior. c) The hatched organisms remain attached to the mother's pleura.

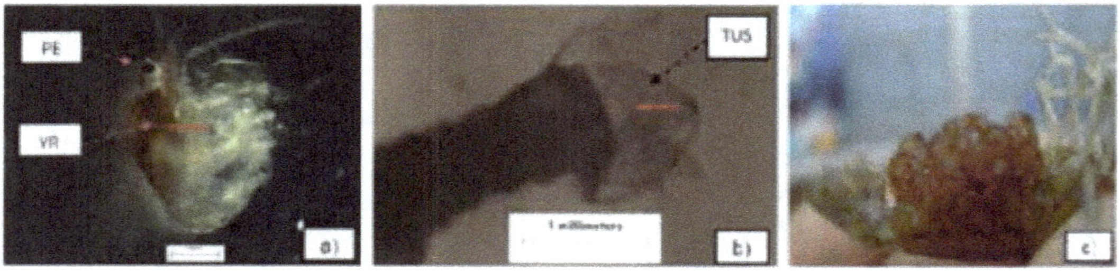

Fig. 16. Post-embryonic stage II (days 20-21).- a) The eyes are pedunculate (PE) with dark pigment, the dorsal hump which corresponds to the yolk remaining (YR) that nourishes the organism is smaller than in the previous stage (almost half its original size) and the cephalothorax has almost reached it final anatomy. Red dots begin to cover the entire body, the beginning of chromatophore pigmentation. b) The telson and uropods (TUS) appear to be separate with bristles at the ends. c) The organisms remain attached to the mother's pleura.

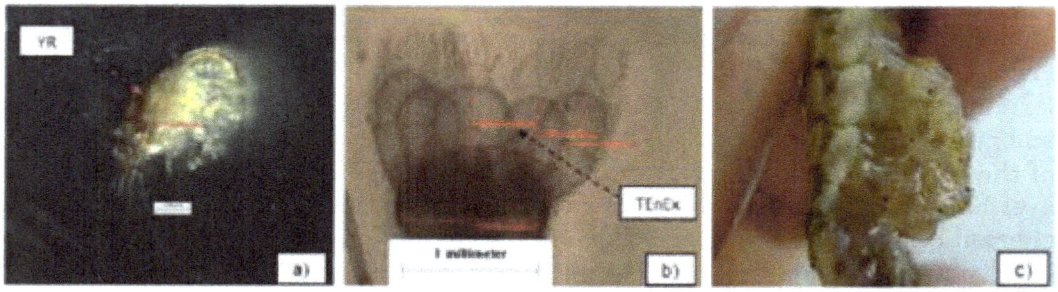

Fig. 17. Post-embryonic stage III (days 22-23). - a) The number and size of chromatophores increases over the entire body, but peduncles eyes not have dark pigment. Yolk Reserves (YR) are still present in the vitellus but almost exhausted, no exogenous feeding activity is observed. b) Telson and uropods are larger-well defined, this last are divided in endopodites and exopodites (TEnEx) which are still immobile and short but well-defined. c) Independent locomotion does not yet occur and the organisms remain attached to the mother's periopods.

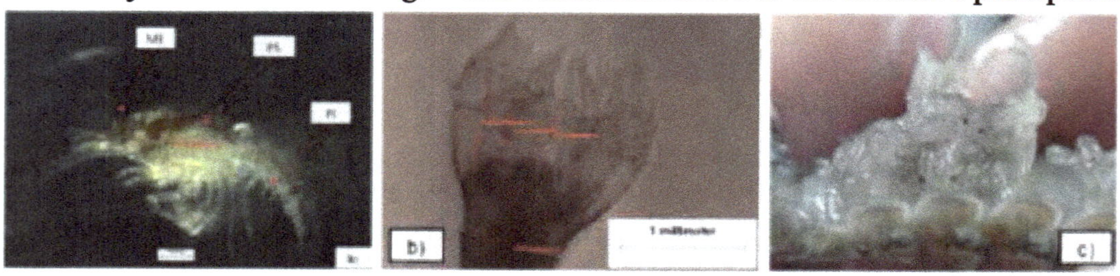

Fig. 18. Post-embryonic stage IV (days 24-25).- a) Fully development eyes or mature eyes (ME), pigmentation spots (PS) dispersed throughout the body, pleopods (Pl) are visible; the vitellus is exhausted but exogenous feeding has not begun. b) Telson elongated and uropods compound (protop, endopod, exopod) with bristles and sensorial filaments at the ends; but independent movement does not yet occur. Three pairs of pleopods (Pl) are visible on the abdominal somites, and c) the organisms remain attached to the mother's pleopods.

Descriptions were based on the external morphological changes observed in the embryos during ontogeny because the most important ontogonic events during development in *P. acanthophorus* occur in the embryos while still in the chorion. These define the different stages and morphological changes in ways similar to those reported by Montemayor et al. (2010) in *P. regiomontanus* (Villalobos, 1954) (Cambarideae family); Sandeman & Sandeman (1991) in *C. destructor*, and García-Guerrero et al. (2003) in *C. quadricarinatus* (both of the Parastacidae family).

Embryonic development in some crustaceans is highly dependent on water temperature (Bottrell, 1975; Herzig, 1983). In the present study, the crayfish *P. acanthophorus* embryos developed at an average temperature of 23.8 ± 2.2 °C during January-March. This coincides with Cervantes (2008), who reported that this species can reproduce year round under laboratory conditions as long as average temperature is kept at 25 °C. Auvergne (1982) stated that optimum temperature for each life stage in crustaceans is species dependent; for instance, *Astacus astacus* (Linnaeus, 1758) has a range of between 18 to 20 °C whereas *Procambarus clarkii* (Girard, 1852) requires a range of 22 to 26 °C. In further examples, Sandeman & Sandeman (1991) reported satisfactory development in *C. destructor* eggs incubated at 19 °C; García-Guerrero et al. (2003) described successful embryo development in *C. quadricarinatus* at 26°C; and García-Guerrero & Hendrickx (2009) reported proper development in fertilized *Macrobranchium americanum* eggs at 24 °C.

As embryo development progressed, egg length increased; initial egg diameter was 1.3 mm and total length of juvenile organisms was 6±1 mm. This development trajectory differs from the 16 stages (22 days) reported for *P. regiomontanus*, with eight embryonic stages, eight post-embryonic stages and an average juvenile size of 2 cm (Montemayor et al., 2010). However, both these *Procambarus* species have embryonic development periods near the 30- day average for cambarid crustaceans. *Procambarus clarkii* completes its embryonic development in an average of three to four weeks (McClain & Romaire, 2007), and *P. llamasi* completes it in 27 to 30 days (Rodríguez-Serna et al., 2000). These contrast with the longer development periods of other species. *C. quadricarinatus* has a 42-day development period with ten embryonic stages, two post-embryonic stages and a juvenile stage (García-Guerrero et al., 2003), while C. destructor has a 40-day development period with an unknown number of embryonic stages, at least two post-embryonic stages and a juvenile stage (Sandeman & Sandeman, 1991). In stark contrast to all the above species, *Austropotamobius pallipes* (Lereboullet, 1858), a cold water species, requires over seven months to complete embryo development (Holdich & Lowery, 1988).

During the embryonic development trial, selected live eggs were kept in Petri dishes to evaluate survival. Artificially incubated fertile eggs were found to remain viable after development stage eight. This means that egg lots could be artificially incubated to reduce development period and synchronize times in mass production settings and/or if only small lots of reproductive-age females are available. The present study is the first report of embryonic development and artificial incubation of eggs in *P. acanthophorus*. It constitutes a significant contribution to the biology of this and other decapods crustaceans with potential for use in alternative, sustainable aquaculture systems.

1.3.9 Culture

1.3.9.1 Optimum density for growth

Advances in knowledge for the commercial culture of crayfish have shown that in this specie territorial habits are not present, allowing stocking densities as high as 100 orgm-2 without affecting the growth density and survival, which is considered an advantage in production. In addition, laboratory studies indicate that it is possible to keep under recirculation systems the biculture of crayfish and tilapia at densities between 30-50 orgm-2, where the main species is tilapia, and crayfish support sustainable use of water.

1.3.9.1.1 Response parameters

A experimental trial was done when different densities of culture where evaluated, survival (%), showed no significant differences between treatments and remained in a range between 67 and 85%, with the increase of survival were crayfish were kept at densities of 50 and 90 orgm-2, while the lowest occurred where the crayfish were kept at a density of 60 orgm-2, without a positive relation between survival and density (Table 2).

Table 2. Response parameters ± s.d of the crayfish was culture at different densities.

Treat.	SUP. (%)	IW (g)	FW (g)	IWG (gday-1)	WG (%)	SGR (%day-1)	IFC (gday-1)	FCR
T1	85.2±17	1.0±0.3	5.6±1	0.1±0	450±58.9	1.0±0.1	0.2±0	3.1±0.1
T2	67.5±16	1.0±0.3	5.6±0.6	0.1±0	507±207	1.1±0.2	0.2±0	3.4±0.7
T3	72.9±40	1.0±0.1	5.6±0.6	0.1±0	474±33.2	1.1±0	0.2±0	3.3±0.2
T4	80.4±10	1.0±0.3	5.3±1.2	0.1±0	463±166	1.0±0.2	0.2±0	3.5±0.8
T5	84.4±14	1.1±0.2	5.3±0.7	0.1±0	412±85.8	1.0±0.1	0.2±0	3.5±0.4
T6	75.0±30	0.9±0.2	5.3±0.6	0.05±0	504±62.1	1.1±0.1	0.2±0	3.5±0.1
T1=50, T2=60, T3=70, T4=80, T5=90, T6=100 (org/m2)								
Treat.= Treatments (org/m2); SUP= Survival; IW= Initial Weight; FW= Final Weight; IWG= Individual weight gain; SGR= Specific growth rate; IFC= Individual food consumed and FCR= Feed Conversion rate								

With respect to growth, the final weight (FW g), weight gain (WG%) and specific growth rate (SGR%/day) for all treatments was similar. The organisms in treatment

1, had a greater weight gain, during the first 30 days of culture maintained a growth rate similar to that of crayfish of the other treatments, then increased its rate of growth, although not differ significantly from other organisms under study, it was considered that the density did not significantly affect growth or survival of the crayfish. The juvenile kept it in treatment 5 and 6 had the lowest growth without significant differences between treatments (95% confidence).

1.3.9.2 Biculture

Actually research on the feasibility of polyculture systems including fish and crustaceans, has produced inconsistent conclusions, for that reason is difficult to determinate the potential of polyculture in recirculation system for sustainable aquaculture. In a study with red claw crayfish *C. quadricarinatus* and tilapia *Oreochromis niloticus* in a polyculture system in earthen ponds, Brummett & Alon (1994) reported positive results for the crayfish growth and survival; whereas using the same species combination Rouse & Kahn (1998) reported that competition for feed and space between species negatively affected survival in *C. quadricarinatus*. An alternative to the above system is the use of recirculation systems in crustacean/fish polyculture because water quality and feed supply can be controlled, and shelters can be provided for crustaceans to prevent territorial competition, allowing in both species a survive and grow properly (Karplus et al., 2001). Polyculture is particularly appealing since it makes extremely efficient use of resources and can increase production. It can be quite viable for producers as long as appropriate species are identified in terms of biology and market demand. Using a polyculture with common carp, grass carp, silver carp, tilapia, mullet and Malaysian prawn, Cohen et al. (1983) reported efficient water use and an increase in production from 3.5 to 11 tons/ha/year. In a polyculture system for carp (*Cyprinus carpio*) and crayfish (*Cambarellus montezumae*) growth in artificial ponds in which feeding was focused mainly on the carp, Auró et al. (2000) reported that these species could coexist and use the food in the system, making it a viable system.

Competition is increasingly high for good quality water sources for productive activities such as aquaculture. There is also a need to optimize space during production and to promote productive and sustainable alternative activities in rural areas. Under these conditions, the most adequate option for polyculture systems is to use recirculation systems, ensuring that the species in the system do not compete for resources and have established market niches (Kazmierczak & Caffey, 1995). With the objective to determine the feasibility of a crayfish *P. acanthophorus*/ tilapia *O. niloticus* polyculture and monoculture using a water recirculation system, as aquaculture sustainable alternative, a experimental research in outdoor facilities was conducted. Six plastic tanks (3 m diameter x 1.2 m depth) in a recirculation system with a biological and sand filter were used. During 90-day experimental period, three treatments were evaluated with two replicates per treatment in a completely random design. T1: crayfish monoculture (1.02±0.2 g); T2 polyculture: crayfish (1.04±0.2 g) and tilapia (2.99±0.1 g); T3: tilapia monoculture (3.45±0.6 g).

Survival in the T2 crayfish was significantly lower (34.7%) compared to that in the T1 crayfish (72%). In contrast, the tilapia in both T2 and T3 had similar survival (>95%) and growth rates (83-86 g) with no apparent effect from the presence of the crayfish (Table 3).

Table 3. Growth and efficiency parameters in crayfish and tilapia monocultures and crayfish/tilapia polyculture in a water recirculation system1.

Treatment		S (%)	IW (g)	FW (g)	IWG (g)	WG (%)	SGR (%)
T1	Crayfish	72±12.7a	1.02±0.2a	4.8±0.4a	0.042±0a	373.5±55a	68.4±3.3a
T2	Crayfish	34.7±20.5b	1.04±0.1a	3.9±0.3b	0.033±0b	282.6±73b	59.9±3.3b
	Tilapia	98.7±0a	2.99±0.1a	88.5±9.9a	0.95±0.1a	2861.2±26a	194.2±5.3a
T3	Tilapia	93.7±9.9a	3.45±0.6a	84.3±10.7a	0.89±0.1a	2344.9±71a	192.1±5.6a
[1]Values in the same column with the same superscript are not statistically different ($p>0.05$)							
S% = survival rate; IW = initial weight; FW = final weight; IWG = individual weight gain; WG% = percentage weight gain; SGR = specific growth rate.							

The lower crayfish survival rate in T2 had no apparent effect on growth, as might be expected due to the density effect, since IWG and WF in T2 (0.033±0g; 3.9±0.3g, respectively) were significantly lower than in T1 (0.042±0g; 4.8±0.4g, respectively). Tilapia growth in T2 and T3 followed a steeply-sloped exponential curve whereas crayfish growth was constant but with a lesser slope, reflecting their lower growth rate (Figure 19).

Water quality parameter values during the trial were within the ranges tolerated by tilapia fingerlings and crayfish culture. Dissolved oxygen (DO) concentration was 3.8 mgL-1 throughout the experimental period, lower than the 5 mg/L recommended for optimum growth in *P. acanthophorus* (Cervantes-Santiago et al., 2007). This level coincides with the >3 mgL-1 DO level recommended for *P. clarkii* (Huner, 1994), and suggests that *P. acanthophorus* can adapt to environmental variations during cultivation. Like crayfish, tilapia can also tolerate low DO levels (<2 mgL-1), although levels greater than 3 mgL-1 are recommended for good growth (El-Sayed & Abdel-Fattah, 2006). This DO level also favors proper functioning of the biological filter in the recirculation system and prevents the death of nitrifying bacteria (Yousef et al., 2003). Increased temperature during the experimental period improved growth in the tilapia and crayfish, although efficient growth in *P. acanthophorus* is reported to occur at temperatures <28°C. Optimum growth in tilapia occurs at 28°C, even though the species can tolerate a range of 15-35°C. Apparently, environmental parameters are no impediment to polyculture of

Nile tilapia and *P. acanthophorus*. Higher temperatures (23-33°C) have also been reported to increase growth in a polyculture of *C. quadricarinatus* and tilapia (Rouse & Kahn, 1998). In addition, the temperature tolerance exhibited by *P. acanthophorus* coincides with overall temperature tolerance (10-38 °C) among *Procambarus* genus crayfish (Holdich, 2002). Of course, individual species have specific optimum temperature ranges for growth; for instance, *P. clarkii* prefer temperatures from 22- 30°C (Holdich, 2002) with optimal levels around to 20-25°C (Huner & Gaude, 2001). This tolerance for a wide range of environmental temperatures highlights the potential for crayfish cultivation in commercial systems.

Water pH levels were adequate for proper growth in both tilapia and crayfish, although both can tolerate pH from 3.5 to 12, another advantage for polyculture of these species (Huner, 1994; El-Sayed & Abdel-Fattah, 2006). Ammonium, N-nitrite and N-nitrate (mgL-1) values were below sublethal and lethal levels for the two cultured species (Huner, 1994; El-Sayed & Abdel-Fattah, 2006; Cervantes, 2008), indicating proper biological filter functioning.

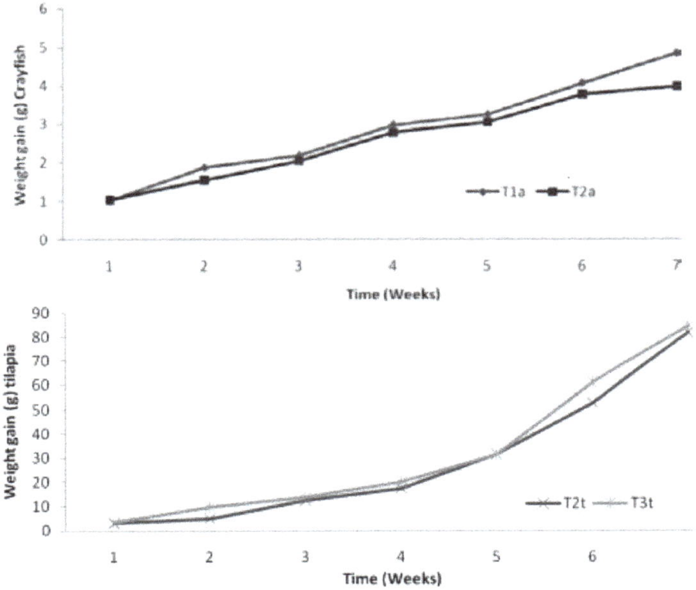

T1: Crayfish, T2: Crayfish/Tilapia, T3: Tilapia

Fig. 19. Weight gain in crayfish and tilapia monocultures and crayfish/tilapia polyculture in a water recirculation system.

This is the first report of a crayfish *P. acanthophorus* / Nile tilapia *O. niloticus* polyculture. The feasibility of this species pair in polyculture as a productive alternative when a recirculation system to increase water efficiency use was evaluated, because both tilapia and crayfish are readily available, and the *Procambarus* genus is diverse around the world, similar than tilapia. The polyculture system was successful in that both species developed properly during the experimental period, although survival among the *P. acanthophorus* (37.4%) was significantly lower than in the crayfish monoculture (72%), and crayfish IWG and FW were lower in the polyculture than in the monoculture. Presence of the tilapia in the same tanks was apparently the main potential cause of this overall lower growth performance in the polyculture since all other variables were within proper ranges for crayfish. There may have been interspecies competition for space and/or feed, or predation by the tilapia of the crayfish during molting partially due to a substantial size difference between species. The latter possibility was not evaluated in the present study since the same initial stocking sizes were used in all treatments: average initial stocking sizes were 1.03±0.77 cm (crayfish) and 3.15±1.04 cm (tilapia), giving a clear growth advantage to the tilapia. The tilapia did not necessarily need to be antagonistic for them to prey on the much smaller crayfish in T2. Different types and quantities of shelters could help to protect the crayfish from the larger tilapia, and manipulating initial stocking size of the different species might help to reduce predation.

Auró et al. (2000) highlighted the feasibility of crayfish/fish polyculture in a study using carp *Cyprinus carpio* and crayfish *Cambarellus montezumae* in artificial ponds. The species coexisted and had enough food resources at densities of up to 50 Org m-3 for both species under good water quality conditions. The crayfish *P. acanthophorus* has excellent potential for use in aquaculture systems, because it can be culture to high densities, tolerates handling, adapts to variable environmental conditions and accepts different kinds of artificial diets in captivity, although it does not reach sizes as large as other crustaceans, such as *C. quadricarinatus* (Cervantes, 2008; Cruz-Ordoñez, 2009). Its use in sustainable rural aquaculture production systems is promising under monoculture, and possibly under polyculture conditions after further research into optimum initial stocking size, densities and crayfish shelter type and quantity. Greater production of alternative protein sources using aquaculture in rural areas is an important step towards increasing food availability and diversity.

In addition to the above, the biculture system may be associated to aquaponic production of herbs (cilantro, basil and aquaponic green fodder) with satisfactory results.

1.3.9.3 Genetic improvement

An experiment was designed to estimate genetic variability of growth as heritability (h2) of the weight at different ages. For which was captured 2135 organisms (4.1g ± 1.79) from its natural medium (G0) of which 10% heavier were selected (i = 1,755) for each sex: 140 females (5.62g ± 1.97) and 48 males (6.02g ± 1.9) to form the progenitors of the line selection (LS), while the control line (LC) consisted of organisms take it at random. These organisms were maintained for reproduction in two rectangular fiberglass tanks of 2.4 m2 and 0.15 m deep (one tank per line), and a density of 49 orgm-2, and relation females: males (3:1). Biweekly organisms were reviewed to identify gravid females From these organisms 30 full- sib families per line (LC and LS) were obtained (F1), and grown individually for five months, tracking them individually in a recirculation system consisting of 60 rectangular plastic tubs (54 cm x 37 cm x 22 cm) distributed in three levels with mechanical and biological filtration under laboratory conditions and fed 2 times day with balanced feed for shrimp (35% protein). Once the organisms in lines arrived at three months old, and due to differential mortality in the families and in order to reduce the effect of the environment, the density was standardized to 11 organisms per family (55 orgm-2) to continue growth in the same. Heritability for growth in broad sense (h2) and in each age, for both, the control (LC) and selection line (LS) was estimated from variance components (ANOVA method REML) using a full-sib design from the formulas described by Roff (1997). Growth between the lines in F1 was also compared for each age.

Table 4 shows estimates of heritability from full-sib design for F1. The value of h2 estimates for LC was 0.48 initially and decreased until the fourth month of age, not to suffer variations due to the rearrangement of the population in the third month compared to LS which began with a value of 1.1 reducing the third month (0.27) have to rise again in the fourth month (0.58). Due to the differential survival and in order to reduce the effect of common environment which is reported in previous studies with crustaceans (Benzie et al., 1997; Hetzel et al., 2000; Pérez-Rostro & Ibarra, 2003), the density of families in the third month were standardized, leaving 11 organisms per family in F1.

Table 4. Estimates of heritability (h2) ± E.S in F1 at the different ages (months), in selection line (LS) and control line (LC)

Age (month)	Heritabilty estimate (h^2) ± E.S. LC	LS
1	0.48 ± 0.11	1.10 ± 0.14
2	0.08 ± 0.04	0.61 ± 0.12
3	0.23 ± 0.07	0.27 ± 0.08
4	0.20 ± 0.10	0.58 ± 0.14
5	0.27 ± 0.11	0.34 ± 0.12

Estimates of heritability (combined males and females) in F1 at the end of the trial was similar for both lines (0.27 ± 0.11 and 0.34 ± 0.12, LC and LS, respectively), coinciding with results obtained by Pérez-Rostro & Ibarra (2003) in white shrimp *Litopenaeus vannamei*, where values of 0.20 were obtained at 17 weeks of culture as well and those of Cameron et al. (2004) in the red claw crayfish *C. quadricarinatus*, where a decrease in the value of heritability of 0.38 for the first year to 0.13 after 4 years.

When growth gain between the lines was compared, it was seen that LS was significantly heavier (9.6%) than the control line (Figure 20).

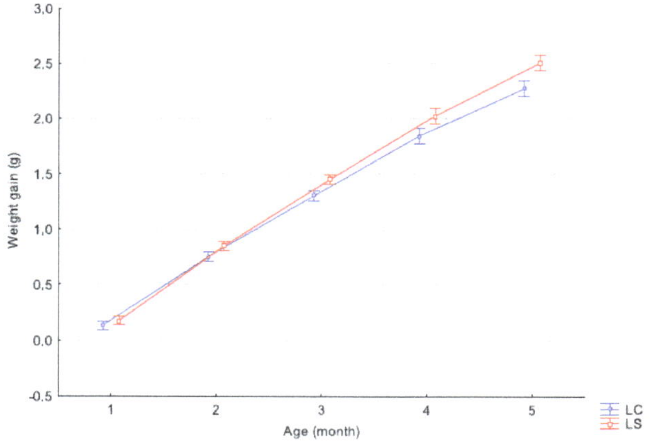

Fig. 20. Weight gain between selection line (LS) and control line (LC) during the five months of culture.

The values of heritability indicate that the species has a positive response to selection, so they can continue to implement a screening program to improve the growth of the species and thus promote commercial cultivation.

REFERENCES

1. Anderson, D. (1982). Embryology. In: D. Bliss and L. Abele (eds.), Embryology and Genetics. (Academic Press, New York). *The Biology of Crustacea* 1:1-42.
2. Arrignon, J. (1985). *Cría del cangrejo de Río*. Editorial Acribia. Zaragoza, España. 222 Pp.
3. Auró, A.A.; Fragoso, C.; Ocampo, C.M.; Sumano, L.H. & Osorio, D.S. (2000). Evaluación del crecimiento de carpas (*Cyprinus carpio* var. *rubrofuscus*) y acociles (*Cambarellus moctezumae*) en bicultivo, alimentados con ensilado de cerdaza empastillado, en un embalse artificial y en tanques de fibra de vidrio. http://albeitar.portalveterinaria.com. México, 418/634 pp 5.
4. Auvergne, A. (1982). *El cangrejo de río. Cría y explotación*. Editorial Mundi-Prensa. Spain. 119 pp.
5. Benzie, J.A.H; Kenway, M. & Trott, L. (1997). Estimates for the heritability of size in juvenile *Penaeus monodon* prawns from half-sib matings. *Aquaculture* 152: 49-53.

6. Bottrell, H.H. (1975). Generation time, length of life, instar duration and frequency of moulting, and their relationship to temperature in eight species of Cladocera from the River Thames. Reading. *Oecology* (Berlin), 19:129-140.

7. Brewis, J.M.; Bowler K. (1985). A study of reproductive females of the freshwater cray⇒sh *Austropotamobius pallipes*. *Hydrobiology Journal* 121:145–149.

8. Brummett, R.E. & Alon, N.C. (1994). Polyculture of nile tilapia (*Oreochromis-niloticus*) and australian red claw crayfish (*Cherax-quadricarinatus*) in earthen ponds. *Aquaculture* 122: 47-54

9. Cameron, P.M.; Jones, C.M. & Shanks, A.S. (2004). Selection for increased weight at 9 months in redclaw crayfish (*Cherax quadricarinatus*). *Aquaculture* 237: 131-140.

10. Carmona-Osalde, C.; Rodríguez-Serna, M.; Olvera-Novoa, M.A.; Gutiérrez-Yurrita, P.J. (2004a). Gonadal development, spawning, growth and survival of the cray⇒sh *Procambarus llamasi* at three different water temperatures. *Aquaculture* 232: 305–316

11. Carmona-Osalde, C.; Rodríguez-Serna, M.; Olvera-Novoa, M.A.; Gutiérrez-Yurrita, P.J. (2004b). Effect of density and sex ratio on gonad development and spawning in the cray⇒sh *Procambarus llamasi*. *Aquaculture* 236: 331–339.

12. Carmona-Osalde, C.; Rodríguez-Serna, M.; Olvera-Novoa, M.A. (2002). The in⇓uence of the absence of light on the onset of ⇒rst maturity and egg laying in the cray⇒sh *Procambarus (Austrocambarus) llamasi* (Villalobos, 1955). *Aquaculture* 212: 289–298 Celada, J.D.; Antolín, J.I.; Carral, J.M.; Sáez-Royuela, M.; Rodríguez, R. (2005). Successful sex ratio of 1M: 4F in the astacid cray⇒sh *Pacifastacus lenisculus* Dana under captive breeding conditions. *Aquaculture* 244: 89–95.

13. Cervantes, S.E. (2006). *Efecto de la sustitución de harina de pescado por harina de soya, en dietas para el crecimiento y supervivencia del acocil Procambarus* (*Astrocambarus*) *acanthophorus*, Villalobos, 1948, bajo condiciones de laboratorio. Tesis profesional de licenciatura. Instituto Tecnológico de Boca del Río. Boca del Río, Veracruz México. 55 pp.

14. Cervantes, S.E. (2008). *Relación proteína-lípidos y fuentes alternas de proteína en dietas para el aprovechamiento sustentable del acocil Procambarus (Austrocambarus) acanthophorus*. Thesis for degree of Master in Sciences in Aquaculture. Instituto Tecnológico de Boca del Río, Veracruz, México. 93 pp.

15. Cervantes-Santiago, E.; Hernández-Vergara, M.P.; Pérez-Rostro, C.I. (2007). Sustitución de harina de pescado por harina de soya en dietas para crecimiento y supervivencia del acocil *Procambarus (Austrocambarus) acanthophorus*, en condiciones de laboratorio. Memorias en extenso del *Simposium Internacional*

Aquamar 2007, y VI Feria Internacional de Acuacultura y Pesca. Boca del Rio, Veracruz, México. Impulso Tecnológico, Edición especial, pp. 107–114. ISBN:1405-0323.

16. Cervantes-Santiago, E.; Hernández-Vergara, M.P.; Pérez-Rostro, C.I.; Olvera-Novoa, M.A. (2010a). Protein/lipids ratios for efficient growth and survival in the crayfish *Procambarus (austrocambarus) acanthophorus* under controlled conditions. *Aquaculture Research* 41(9): 62-71.
17. Cervantes-Santiago, E.; Hernández-Vergara, M.P.; Pérez-Rostro, C.I.; Olvera-Novoa, M.A. (2010b). Reproductive performance of the crayfish *Procambarus (austrocambarus) acanthophorus* Villalobos 1948 under controlled conditions. *Aquaculture* 308 (1-2): 66-70.
18. Cohen D., Z. Ra'anan & Barnes A. (1983). Production of the freshwater prawn *Macrobrachiurn rosenbergii* in Israel. Integration into fish polyculture systems. *Aquaculture* 3(1): 67-76
19. Cruz-Ordoñez, S.B. (2009). *Cultivo piloto del acocil Procambarus (Austrocambarus) acanthophorus en bicultivo con tilapia Oreochromis niloticus en un sistema de recirculación.* Thesis for degree of Master in Sciences: Aquaculture. Instituto Tecnológico de Boca del Río, Veracruz México. 78 pp.
20. El-Sayed & Abdel-Fattah, M. (2006). *Tilapia culture.* CABI Publishing. Oceanography Department, Faculty of Science, Alexandria University. Alexandria, Egypt. pp. 139- 145.
21. García-Guerrero, M.U. & Hendrickx, M.E. (2009). External description of the embryonic development of the prawn, *Macrobrachium americanum* Bate, 1868 (decapoda, palaemonidae) based on the staging method. *Crustaceana* 82 (11): 1413-1422.
22. García-Guerrero, M.U.; Hendrickx, M.E. & Villareal, C.H. (2003). Description of the embryonic development of *Cherax quadricarinatus* Von Martens, 1868 (Decapoda Parastacidae) based on the staging method. *Crustaceana* 76 (3): 269-280.
23. Gutiérrez-Yurrita, P.J. (1994). *Estudios fisioecológicos sobre algunos aspectos del metabolismo energético de Procambarus bouvieri y Procambarus digueti* (Crustacea: Decapoda: Cambaridae). Master in Science Thesis. Faculty of Sciences, Universidad Nacional Autónoma de México. México. 180 pp.
24. Herzig, A. (1983). The ecological significance of the relationship between temperature and duration of embryonic development in planktonic freshwater copepods. *Hydrobiología* 100: 65-91.
25. Hetzel, J.S.D.; Crocos, J.P.; Davis, P.G.; Moore, S.S.; Preston, C.N. (2000). Response

to selection and heritability for growth in the Kuruma prawn, *Penaeus japonicus. Aquaculture* 181: 215-223.

26. Hobbs, H.H. Jr. (1984). On the distribution of the crayfish genus *Procambarus* (Decapoda: Cambaridae). *Journal of Crustacean Biology*, 4: 12-24.
27. Holdich, D.M. & Lowery, R.S. (1988). *Freshwater crayfish, biology, management and exploitation*. Timber Press. 480 pp.
28. Holdich, D.M. (1993). A review of astaciculture freshwater crayfish farming *Department of Life Science, Univm~iiyoJ Noiiingham, Nottingham NG7 ZRn, U.K. Aquat. Living Kesour* 6:307-317.
29. Holdich, D.M. (2002). *Biology of freshwater crayfish*. Blackwell Science. EUA pp. 702
30. Huner, J.V. (1994). *Freshwater Crayfish Aquaculture in North America, Europe, and Australia*. Food Products Press. USA. pp. 311
31. Huner, J.V. (1995). Ecological observations of red swamp crayfish *Procambarus clarkii* (Girard, 1852) and White crayfish *Procambarus zanangulus*, Hobbs & Hobbs 1990 as regards their cultivation in eathen ponds. *Freshwater Crayfish*, 10: 456-468 p.
32. Huner, J.V. & Gaudé III., A.P. (2001). Hatcheries for red swamp crawfish. *Crawfish Tales* 8 (1): 23-25.
33. Karplus, I.; Harpaz, S.; Hulata, G.; Segev, R. & Barki, A. (2001). Culture of the Australian red-claw crayfish (*Cherax quadricarinatus*) in Israel - IV. Crayfish incorporation into intensive tilapia production units. *Journal of Aquaculture-Bamidgeh* 53 (1): 23-33.
34. Kazmierczak, Jr. & Caffey, R.F. (1995). Management ability and the economics of recirculating aquaculture production systems. *Marine Resources Economies*. 10, 187–209
35. Köksal, G. (1988). *Astacus leptodactylus* in Europe. In:Holdich, D.M., Lowery, J.S. (Eds.), Freshwater Cray⇒sh: Biology, Management and Exploitation. Croom Helm, London, pp. 365–400.
36. Lee, D.O'C. & Wickins, J.F. (1992). *Crustacean Farming*. Blackwell Scient, Oxford. 392 pp.
37. López, M.M. (2006). Diversificación de *Procambarus (Villalobosus)* Hobbs, 1972. (Crustacea: Decapoda: Cambaridae). Tesis doctoral. Posgrado en Ciencias del Mar y Limnologia. UNAM. 225-228 p

38. Lopéz, M.M. (2008). *Sistemática de los cangrejos de río de México.* In: Álvarez F., Rodríguez- Almaraz G. A. (1ª Ed.) Crustáceos de México. Estado Actual de su Conocimiento. Universidad Autónoma de Nuevo León. PromeP, Mexico:115-165

39. Malone, R.F.; Burden, D.G. (1988). Design of recirculating soft craw⇒sh shedding systems. Louisiana Sea Grant College Program. 73 pp.

40. Martínez, C.L.R. (1999). Cultivo de camarones peneidos, principios y prácticas. AGT Editor, S.A., México. 282 pp.

41. McClain, W.R. & Romaire, R.P. (2007). Procambarid craw⇒sh life history and biology. *Southern Regional Aquaculture Center.* Publication No. 2403. EUA. 6 pp.

42. Mejía, O.L.M. (2010). Pautas de comportamiento alimentario y adaptaciones progresivas en los apéndices auxiliares en *Agostocaris bozanici* y *Barbouria yanezi* (Crustacea: Decapoda: Caridea: Agostocarididae, Hyppolitidae). *Revista mexicana de biodiversidad.* S193-S201. 81 p

43. Montemayor, L.J.; Mendoza, A.R.; Aguilera C. & Rodríguez A.G. (2010). Influencia de la alimentación, sobre la reproducción y crecimiento del cangrejo de rio regio *Procambarus regiomontanus*, especie en peligro de extinción. July- September 2010. *Ciencia* UANL 13(3) 276-286.

44. Mustafa, H.M.; Barım, Ö.; Türkgülü, I.; Harlıoglu, A.G. (2004). Potential fecundity of an introduced population, Keban Dam Lake, Elazığ,Turkey, of freshwater cray⇒sh, *Astacus leptodactylus leptodactylus* (Esch, 1852). *Aquaculture* 230: 189–195

45. Nakata, K. & Goshima, S. (2004). Fecundity of the Japanese cray⇒sh, *Cambaroides japonicus*: ovary formation, egg number and egg size. *Aquaculture* 242: 335–343

46. Pérez-Rostro, C.I. & Ibarra A.M. (2003). Quantitative genetic parameter estimates for size and growth rate traits in Pacific White shrimp, *Penaeus vannamei* (Boone 1931) when reared indoors. Aquaculture Research 34: 1-11.

47. Reynolds, J.D. (2002). *Growth and reproduction.* In: Holdich, D.M. (Ed.), Biology of Freshwater Cray⇒sh. Blackwell, USA, pp. 152–184.

48. Rodríguez-Almaraz, G.A. & Mendoza-Alfaro, R. (1999). *Crustáceos nativos de agua dulce*: Conocimiento y Utilización. 3ª Reunión de Redes de Acuacultura, Cuernavaca, Morelos. Pp. 181-191.

49. Rodríguez-Serna, M.; Carmona-Osalde, C.; Olvera-Novoa, M.A.; Arredondo-Figueroa, J.L. (2000). Fecundity, egg development and growth of juvenile cray⇒sh *Procambarus (Austrocambarus) llamasi* (Villalobos 1955) under laboratory conditions. *Aquaculture Research* 31: 173–179

50. Roff, D.A. (1997). Evolutionary quantitative genetics. Editorial Ghapman y Hall. Printed in USA. 41-42 p.

51. Rojas, P.Y.R. (1998). Revisión taxonómica de ocho especies del genero *Procambarus* (Crustacea: Decapada: Cambaridae) del centro de Veracruz, México. Tesis de licenciatura, Facultad de Ciencias, UNAM. México. 157 PP.

52. Rouse, D.B. & Kahn, B.M. (1998). Production of Australian Red Claw *Cherax quadricarinatus* in Polyculture with Nile Tilapia *Oreochromis niloticus. Journal of the World Aquaculture Society* 29(3): 340-344

53. Sánchez, M.P.E.; Romero, J.J.; Negrete, R.P. (2009). Aprovechamiento de los ambientes reducidos para la producción de organismos acuáticos susceptibles a cultivo, para el consumo humano. *Vet. Mex.*, 40 (1): 55-67 p.

54. Sandeman, R. & Sandeman, D. (1991). Stages in development of the embryo of the freshwater crayfish *Cherax destructor. Roux Archives of Developmental Biology* 200: 27-37.

55. Taugbol, T.; Skurdal, J. (1990). Effect of density on brood size in noble cray⇒sh *Astacus astacus* L., subjected to indoor rearing conditions. *Aquaculture Research* 21: 17–23

56. Villalobos, F.A. (1948). Estudios de los cambaridos mexicanos VII. Descripción de una Nueva Especie del Género *Procambarus. Procambarus acanthophorus* n. sp. Anales del Instituto de Biología de la Universidad Nacional Autónoma de México 19: 175-182.

57. Villalobos-Hiriart, J.L.; Cantú Díaz-Barriga, A. y Lira-Fernández, E. (1993). Los crustáceos de agua dulce de México. *Revista de la Sociedad Mexicana de Historia Natural* 44: 267-290

58. Yeh, H.S. & Rouse, D.B. (1995). Effects of water temperature, density, and sex ratio on the spawning rate of red claw cray⇒sh *Cherax quadricarinatus* (von Martens). *Journal of the World Aquaculture Society* 26: 160–164.

59. Yousef, S.; Al-Hafedh, A.A. & Alfaque, M.A. (2003). Performance of plastic biofilter media with different configuration in a water recirculation system for the culture of Nile tilapia (*Oreochromis niloticus*). *Aquacultural Engineering* 29: 139–154.

12 CULTIVATING FRESHWATER PEARL MUSSEL FROM GLOCHIDIA TO ADULT IN VITRO

1. INTRODUCTION

Freshwater pearl mussel culture involves three stages: glochidia, juveniles, and adults. Juvenile freshwater mussels have been effectively raised in a controlled environment by affixing glochidia to fish until they developed into the early juvenile stage, as documented by several researchers (Fukuhara et al., 1990; Panha, 1992; Buddensiek, 1995; Uthaiwan et al., 2003; Hanlon & Neves, 2006). Sterilized artificial media can be used for culturing glochidia to skip the parasitic stage. The development of this technique is documented in a series of reports by various authors. Furthermore, certain species of freshwater mussel glochidia raised in artificial environments have shown the ability to mature into adulthood, stimulate gonadal development to sexual maturity, and develop marsupia. These species include Hyriopsis (Limnoscapha) myersiana (Kovitvadhi et al., 2006, 2008), Chamberlainia hainesiana (Kovitvadhi et al., submitted), and Hyriopsis (Hyriopsis) bialatus (Kovitvadhi & Kovitvadhi., in preparation). The author will detail the process of preparing and cultivating glochidia in artificial media, along with all procedures for raising early juveniles to adults in this chapter. The document will cover water quality and appropriate food for raising from early juvenile to adult stages, as well as morphological development seen using light and scanning electron microscopes.

2. CULTURING GLOCHIDIA IN SYNTHETIC MEDIA

Glochidia were raised in synthetic media using the methods outlined by Kovitvadhi (2000), Uthaiwan et al. (2002), and Kovitvadhi et al. (2006) until they developed into 0-day-old juveniles. The specifics of each stage of glochidia culture in artificial media were as follows:

2.1 Formulation of synthetic media

The composition of artificial media for culture of glochidia is shown in Table 1. Artificial media were based on those improved by Keller & Zam (1990), which consisted of a modification from the formulae of Isom & Hudson (1982). The main differences concerned the composition of the commercial media M199. While the protein source was exclusively horse serum in Keller & Zam (1990) and common carp, *Cyprinus carprio* fish plasma was used as an alternative support to the medium for cultured glochidia of *H. (L.) myersiana* (Uthaiwan et al., 2002; Kovitvadhi et al., 2006, 2007, 2008, 2009), *Hyriopsis (Hyriopsis) bialatus* (Areekijeree et al., 2006; Supannapong et al., 2008; Srakaew et al., 2010; Chumnanpuen et al., 2011; Kovitvadhi & Kovitvadhi., preparation), *Chamberlainia hainesiana* (Kovitvadhi et al., submitted) and *Anodonta cygnea* (Lima et al., 2006). Glochidia could transform into juvenile in the media containing common carp fish plasma as protein source. These glochidia were completely transformed within 8-11 days with a survival rate up to 93% except 34% of *A. cygnea*. All surviving larvae transformed into the juvenile stage except *A. cygnea* <34%. For these reasons, composition of artificial medium (Table 1) was according to Uthaiwan et al. (2002), based on Isom & Hudson (1982) and Keller & Zam (1990), was proposed for culture of glochidia.

Table 1. Composition of artificial medium for culture of glochidia

Composition of media	The ratio of artificial medium
M199	2
Common carp fish plasma	1
Antibiotics and antimycotic	0.5

2.2 Preparation of glochidia media

2.2.1 M199 preparation

Dissolved one packet of M199 powder (Gibco, No. 6231100-035) in 1 liter volume of sterile distilled water and added 2 g of NaHCO3. Thereafter, M199 was filtered through 0.45 and 0.20 µm filter paper, respectively and kept at 4 °C.

2.2.2 Fish plasma preparation

The common carp was anesthetized with 50 mg/l of quinaldine. Fish blood was obtained from the caudal vein in the tail region using a syringe needle size 18 coated with sodium heparin at a concentration of 1000 units per milliliter. The blood sample was placed in a sterile plastic test tube and centrifuged at 1000 and 3000 rpm for 10 minutes each. The clear yellow plasma part was isolated and transferred to a new test tube, then centrifuged at 3000 rpm for 10 minutes. The plasma was separated and filtered using 0.45 µm and 0.20 µm filter paper, then stored at temperatures between -10 and -20 °C.

2.2.3 Antibiotics and antimycotic preparation

The composition of antibiotics and antimycotic chemicals (Isom & Hudson, 1982) is shown in Table 2.

Table 2. Combination of antibiotics and antimycotic for culture of glochidia.

Compound	Concentration (µg / ml)
Antibiotics	
Carbenicillin	100
Gentamicin sulfate	100
Rifampin	100
Antimycotic	
Amphotericin B	5

2.3 Glochidia media preparation

Medium199 (see Section 2.2.1), fish plasma (see Section 2.2.2) and antibiotics/ antimycotic (see Section 2.2.3) were mixed in the ratio 2:1:0.5, respectively (Table 1). The artificial media were divided into sterile plastic test tubes and kept at -10 to -20 °C for stocking culture media.

2.4 Glochidia preparation

Adult freshwater mussels were gathered from their natural habitat. They were sexed by seeing sperm and eggs in fluid extracted from the gonads using a sterile syringe. Thirty adult mussels, consisting of fifteen females and fifteen males, were co-cultured in a cylindrical net cage of 50 cm in diameter and 50 cm in height within an earthen pond to produce mature glochidia. They were permitted to consume natural food without restriction. After 1-2 weeks, all females were seen using tongs to partially open the shell, revealing the marsupial color, which signifies the larval stage development. During the immature stage, glochidia were yellow, transitioning to a partly brown color at the beginning of adulthood. Only fully brown marsupia containing gravid mussels were chosen to study the robust and appropriate glochidia for cultivation in artificial environments. Afterward, the outer shell of a pregnant mussel with fully developed brown marsupia was rinsed with tap water and then sterile tap water. The glochidia were aspirated using a sterile 1 ml syringe and then released into a depression or well slide containing sterilized distilled water. The glochidia were examined using a light microscope at a magnification of ×400. If their shells closed regularly, they were sucked in as indicated before. The shells were cleaned by spraying sterilized distilled water to remove tissue residues, mucus, and glochidia shell pieces. Thorough cleansing and enhanced glochidia were utilized for cultivation. Cultivate glochidia from pregnant mussels in artificial media within 5 hours of harvesting.

2.5 Glochidia culture

Around 5000–6000 glochidia were placed into a 90×15 mm culture dish with 10 cc of artificial media from Table 1. The culture pans were placed in a low-temperature incubator set at 25 degrees Celsius with 5% carbon dioxide. During the culture time, the culture medium was replaced with new media. 4 ml of sterilized distilled water was applied to the culture dish to induce transformation when the mantle (Fig. 1A) was noticed the day before the glochidia transformed into juveniles. Observation of juvenile transformation was conducted using a light microscope (×400) to track the movement of the juvenile foot (Fig. 1B) and assess the success of glochidia transformation into the juvenile stage based on juvenile movement.

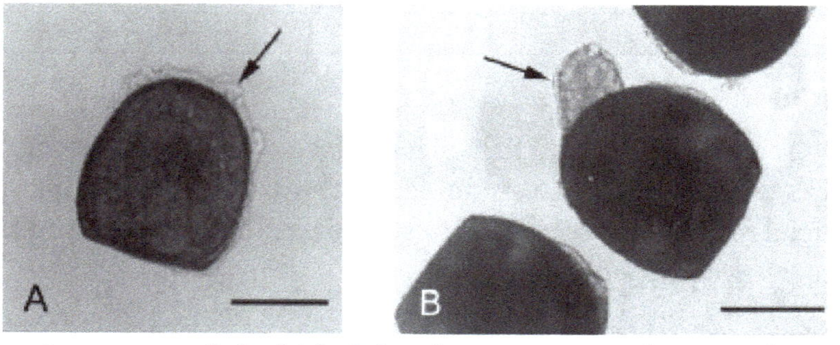

Fig. 1. Light microscope of glochidial development to early juvenile. A; Glochidial development with the mantle edge bordering shell outside (arrow). B; Early juvenile with a foot (arrow). Bar = 100 µm (Uthaiwan et.al., 2001).

3. CHOOSING PHYTOPLANKTON SPECIES AS NOURISHMENT FOR YOUNG ORGANISMS

Phytoplankton is an essential food source for several species of freshwater mussel juveniles, as demonstrated by several studies (Hudson & Isom, 1984; Gatenby et al., 1996; Gatenby et al., 1997; ÓBeirn et al., 1998; Uthaiwan et al., 2001). Kovitvadhi et al. (2000) found that phytoplankton made up 99% of the gastrointestinal tract content of adult freshwater pearl mussels in their native habitat. Adult phytoplankton from the gut were cultured and chosen to feed the juveniles. Harvesting mussels of various sizes from their natural environment, containing phytoplankton in their digestive system, then processing them for cultivation in sterilized water with f/2 media (Guillard & Ryther, 1962) by cutting and sucking. The cells were incubated under light conditions of at least 10,000 Lux for 12 hours and in an atmosphere containing approximately 3% carbon dioxide. Phytoplankton intended for cultivation were regularly sub-cultured and cleansed every 2-10 days using the streak plating technique as described by Hoshaw & Rosowski in 1973. Streak plates were exposed to light until individual colonies of phytoplankton emerged, which could take 10-30 days. Subsequently, the phytoplankton were placed in a slant tube. When phytoplankton were needed to feed juveniles, they were multiplied in a 1-liter bottle using the same specific formula.

Table 3. Characteristic of phytoplankton isolation from gastrointestinal tract of adult freshwater mussel in natural habitat and *in vitro* digestibility for carbohydrate (µg maltose mg plankton-1) and protein (µg DL-alanine equivalent mg plankton-1) of ten algal species at seven days old, using crude enzyme extracts from 30-day-old juveniles. 1 Kovitvadhi et al. (2006), 2 Supannapong et al. (2008).

Phytoplankton	Size[1] (µm)	Filtration time (h)[1]		Cilia activity[1]		Phytoplankton change[1]	Carbohydrate[2] (%)	Protein[2] (%)	Digestibility[2]	
		0.30	1.30	Gill	Foot				Carbohydrate	Protein
Ankistrodemus gracilis	3x20	-	-	-	-	-	43.50	13.67	122.22	9.93
Chlamydomounas sp.	8x9	-	-	-	-	-	25.67	20.00	158.74	11.50
Chlorella sp. 1	3	+	+	-	+	+	45.80	22.50	130.74	11.25
Chlorella sp. 2	4	+	+	-	+	+	37.32	25.45	231.77	15.95
Kirchneriella incurvate	4	+	+	+	+	+	21.64	12.27	248.82	16.67
Monoraphidium sp.	12x20	-	+	-	-	+	26.29	17.50	155.08	11.23
Navicular sp.	4x16	+	+	+	+	+	24.2	8.31	-	-
Scenedesmus sp.	16x20	-	+	-	+	-	37.47	16.00	158.73	12.67
Stichococcum sp.	3-4	+	+	-	-	-	82.31	16.33	131.95	11.46
Cocconyxa sp.	3x5	+	+	+	+	+	21.89	25.83	168.47	9.50

Kovitvadhi et al. (2006) identified eleven species of phytoplankton: Ankistrodesmus gracilis, Chlamydomonas sp., Chlorella sp.1, Chlorella sp.2, Kirchneriella incurvata, Monoraphidium sp., Navicula sp., Scenedesmus sp., Stichococcus sp., and Coccomyxa sp. All phytoplankton were then chosen based on 6 criteria outlined by Areekijseree et al. (2006), Kovitvadhi et al. (2006), and Supannapong et al. (2008) for cultivating appropriate juveniles. (1) dimensions Ability to filter phytoplankton in the gastrointestinal tract can be observed under a microscope within 30 minutes and 1.30 hours. Ciliary movement in the gill, mantle, and foot can also be observed under a microscope. The gastrointestinal system should change color from green to yellow or brown and the form should transition from normal to debris. The carbohydrate and protein levels in phytoplankton are being analyzed. Assessing the efficiency of carbohydrate and protein digestion in phytoplankton using crude enzyme extracts from juvenile organisms (in vitro digestibility). The primary phytoplankton species used for rearing juvenile freshwater mussels were Chlorella sp. 2 and K. incurvata (Table 3).

4. JUVENILE CULTURE

After glochidia transformed to juvenile in artificial medium, water of recirculating aquaculture system was added in the ratio of medium to water equaled 3:1, 1:1, 1:3 for 30 min, added *K. incurvata and Chlorella* sp. in ratio 1:1 at density of 1×105 cells per ml in ratio medium to water equaled 1:3. From there, cultured juvenile was transferred to culture units. Culture of juvenile stage could be divided into 3 stages, namely first stage beginning from 0 days old juvenile until two shell mussel completely closed. Kovitvadhi et al. (2008) cultured juvenile 0-120 days old of freshwater mussel, *H. (L.) myersiana* by recirculating aquaculture system (Fig. 2). This system comprised of three filter cabinets made of 6 mm thick acrylic (particulate filter cabinet, macrophytes filter cabinet and biological filter cabinet), one water resting cabinet (Fig. 2E) and nine plastic culture units (Fig. 2F). The size of particulate filter cabinet (Length×Width×Height×Water level = 46×35×51×42 cm) was divided into two equal parts, of which the first part was filled with a 30 cm thick nylon filter (Fig. 2B). Water flowed through this filter and via the second to the macrophytes filter cabinet. The size of macrophytes filter cabinet (Length×Width×Height×Water level = 80×40×51×42 cm) was divided into four equal units. Each unit contained 57 ambulia plants, *Limnophila heterophylla* (Raxb.) Bentham; these, 228 plants in total, were introduced when they were 6 cm in height and had an average weight of 2.69±0.13 g (Fig. 2C). The plants were removed and replaced when their tips reached the water surface. The upper parts of the cabinets were equipped with three fluorescent lamps (each 20 W) 25 cm above the water surface (light intensity at water surface equal to 5320 lux; 24 h) (Fig. 2G). The water then flowed into the biological filter cabinet (60×34×51×42 cm) filled with bioball to full capacity (Fig.

2D) and then to the resting cabinet (46×41×51×42 cm). In the resting cabinet there were two water pumps: The first returned water to the particulate filter cabinet at the rate of 1 l per minute continuously and the second pumped water at 20 ml per minute to nine plastic culture units (each 84×14×15×7 cm). This pump was stopped for 1 h after phytoplankton was introduced into the culture unit. The bottom of the culture unit was filled with sand (<120 μm) at 0.27 g/cm2. The inside of the culture unit was divided into two section, as described previously, but of different sized (section 1-66.1×14×15×7 cm; section 2-17.9×14×15×7 cm). The first section also consisted of five acrylic sheets jutting from the walls on alternate sides. Juveniles were fed *Chlorella* sp. and *K. incurvata*. Each species of algae was collected from the 100 l by being pumped through 0.3 μm ceramic filters and then separated from the water by centrifuging at 8000 ×g. The sediments of the two algal species were mixed at a ratio of 1:1 wet weight and kept in a freezer. When required, the mixture was brought to room temperature then sucked by Pasteur pipette into the all plastic culture unit to an algae density of 1×105 cells per ml. Algae were supplied twice a day (06.00 h and 18.00 h), and the frozen stock was usually used within 7 days of collection.

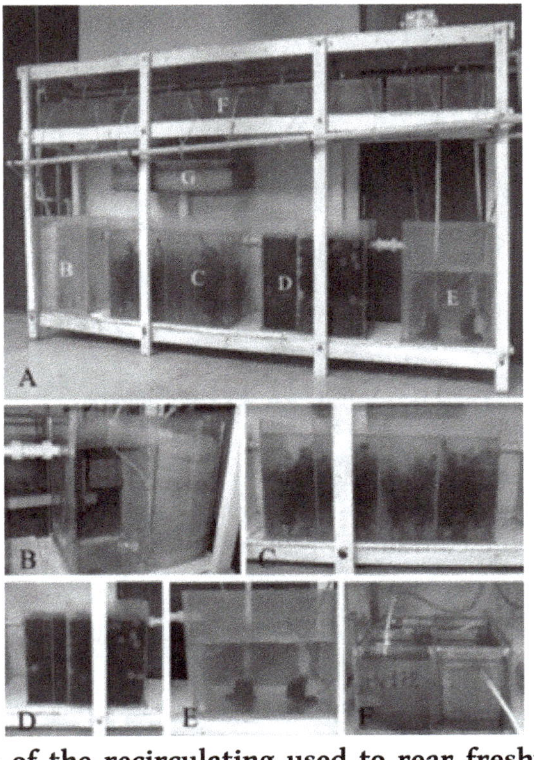

Fig. 2. Photographs of the recirculating used to rear freshwater pearl mussel juveniles. A; Recirculating aquaculture system, B; Particulate filter cabinet, C; Macroplants filter cabinet, D; Biological filter cabinet, E; Water resting cabinet, F; Plastic culture unit, G; Fluorescent box.

Second stage, thirty-five juveniles (120 days old) were transferred to culture units (20×12×72 cm). The culture units had all four vertical sides lined with nylon net (0.42 mm mesh size) and each had a plastic lid with holes to cover the top. The lower part of the culture unit consisted of a section 2 cm in height, which fitted snugly into the culture unit, from which it could be removed (Fig. 3). This lower part contained 400 g of sand (<425 μm in size). The juveniles were placed directly on the sand. The culture unit was then hung in the earthen pond; the base of the culture unit was adjusted to a position approximately 50 cm below the water surface. The juveniles fed by filtering phytoplankton from the water in the earthen pond. All mussels from the culture unit were rinsed every 10 days.

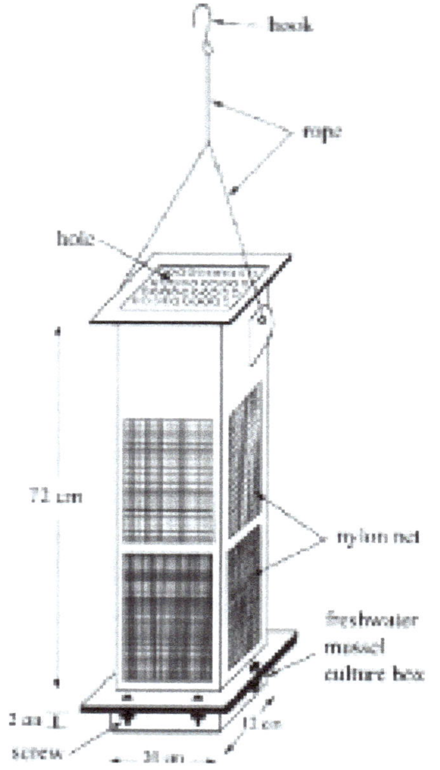

Fig. 3. Rearing container of juveniles (Kovitvadhi et al., 2006).

Third stage, juveniles (180 days old) were transferred to culture into natural habitat or the earthen pond by a cylinder net cage until adult.

5. ADULT CULTURE

At present, adult of some species freshwater pearl mussels had been successfully cultured in an earthen pond and natural habitat. They had high survival and could produce glochidia stage such as *H. (L.) myersiana* (Uthaiwan et al., 2002; Kovitvadhi et al., 2006, 2007, 2008, 2009), *H. (H.) bialatus* (Areekijeree et al., 2006; Supannapong et al., 2008; Srakaew et al., 2010; Chumnanpuen et al., 2011; Kovitvadhi & Kovitvadhi., preparation), *C. hainesiana* (Kovitvadhi et al., submitted). Fifteen female and fifteen male adult mussels, were cultured together in a cylinder net cage (diameter 50 cm × height 50 cm). Then, it was hung under the raft (Fig. 4) at 1.5-2 m deep from water surface which phytoplankton were plenty at this level. The cage was shaken every week for protecting biofouling attachment which could mass mortality.

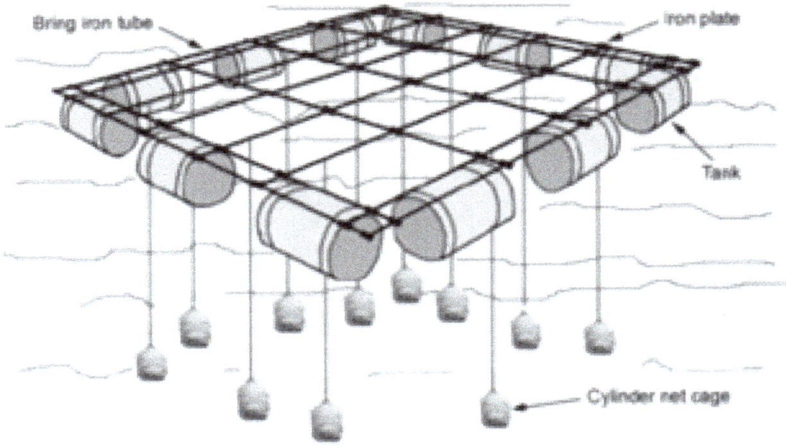

Fig. 4. Raft for adult freshwater pearl mussel culture. (Kovitvadhi, 2008).

6. WATER ANALYSIS

Before culturing mussels, the water quality of the habitat was assessed to ensure it was conducive for development and survival, following the guidelines of Kovitvadhi et al. (2006, 2008). Juveniles were cultured in a laboratory and an earthen pond near their natural habitat, as described by Kovitvadhi et al. (1998) (Table 4). Analyze water quality parameters for freshwater pearl mussel culture, including water temperature, pH, turbidity, conductivity, dissolved oxygen, total alkalinity, free carbon dioxide, total hardness, total ammonia nitrogen, nitrite, nitrate, phosphorus, silica, and calcium. During the juvenile stage, sensitivity to environmental changes was higher compared to other stages, especially in relation to water quality factors such as water temperature, pH, free carbon dioxide, dissolved oxygen, nitrate, phosphorus, total alkalinity, total hardness, total ammonia nitrogen, silica, and calcium. Nitrite was the only factor with increasing values Fig. 5 When averaged water quality value was calculated to relationship with averaged survival value and shell length with equation: $Y=b_0+b_1X+b_2X^2+b_3X^3$ where Y is the survival or shell length, X is age (days), and b_0, b_1, b_2 and b_3 are parameters. It was found that survival of 0-120-day-old juveniles would have direct relationship with pH, total alkalinity, total hardness, silica and calcium with highly significant difference ($P<0.01$) and with reverse relationship to free carbon dioxide and nitrite (Table 5) (Kovitvadhi et al., 2008).

Table 4. Water quality during culturing of 0-360 day - old juveniles and the adult mussel habitat of *H.* (*L.*) *myersiana* in the Mae Klong River, Kanchanaburi Province.

Water quality	Culturing of juvenile in laboratry[1]		Culture of mussel in the earthen pond[1] 120 -360 days old (mean±SD)	Culture of mussel[1] 0 -360 days old (mean±SD)	Mussel Habitat[2] (min. – max.)
	0-60 days old (means±SD)	60-120 day old (means±SD)			
Water tem (°C)	25±0.74	28±0.54	26.5±1.1	24.5-28.5	23.8-31.6
pH	7.03±0.02	7.51±0.04	7.65±0.46	6.85-8.08	6.92-8.14
Dissolved oxygen (ppmO_2)	8.1±0.07	7.5±0.04	5.0±0.6	4.2-8.2	2.5-9.0
Free	10±0.04	4.25±0.35	3.95±2.7	0-10.2	0-6.0
carbondioxide (ppmCO_2)					
Total hardness (ppm$CaCO_3$)	154±2.83	123±9.90	196.8±12.6	121-222	90-133
Ammonia nitrogen (ppmNH_3-N)	0.42±0.02	0.28±0.01	0.44±0.19	0.20-0.82	0.22-0.88
Calcium (ppm$CaCO_3$)	139±9.9	89±4.24	101.1±5.1	86-142	65-105
Phosphorus (ppmP)	0.12±0.06	0.19±0.07	0.17±0.1	0.01-0.45	0.08-0.88
Silica (ppmSiO_2)	4.85±0.6	4.05±0.5	5.75±1.6	3-8	0.2-5.5

[1]Kovitvadhi et al. (2006), 2Kovitvadhi et al. (1998).

Table 5. Coefficient of correlation between average survival rate and water quality; average growth rate and water quality of juvenile *H. (L.) myersiana* cultured in recirculating aquaculture system every 10 days. (Kovitvadhi et al., 2008) (* = *P*<0.05, ** = *P*<0.01, *ns* = not significant difference, *P*>0.05).

Parameter	Survival	Shell length	Shell height
Water temperature	-0.093*ns*	0.075*ns*	0.107*ns*
pH	0.716**	-0.597*	-0.590*
Dissolved oxygen	-0.118*ns*	-0.055*ns*	-0.035*ns*
Total alkalinity	0.841**	-0.849**	-0.827**
Free carbon dioxide	-0.634*	0.481*ns*	0.476*ns*
Total hardness	0.769**	-0.764**	-0.751**
Total ammonia nitrogen	-0.152*ns*	-0.051*ns*	-0.061*ns*
Nitrite	-0.716**	0.709**	0.688**
Nitrate	0.203*ns*	-0.200*ns*	-0.218*ns*
Phosphorus	0.003*ns*	-0.085*ns*	-0.091*ns*
Silica	0.914**	-0.913**	-0.091**
Calcium	0.817**	-0.751**	-0.761**

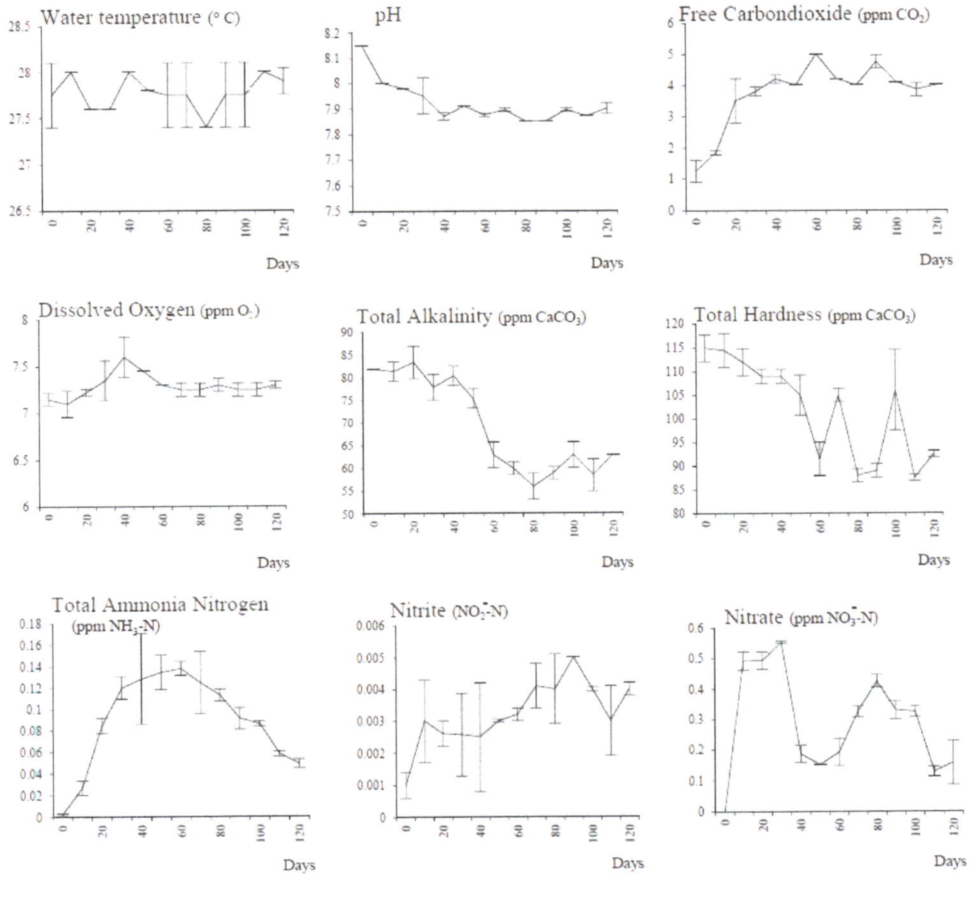

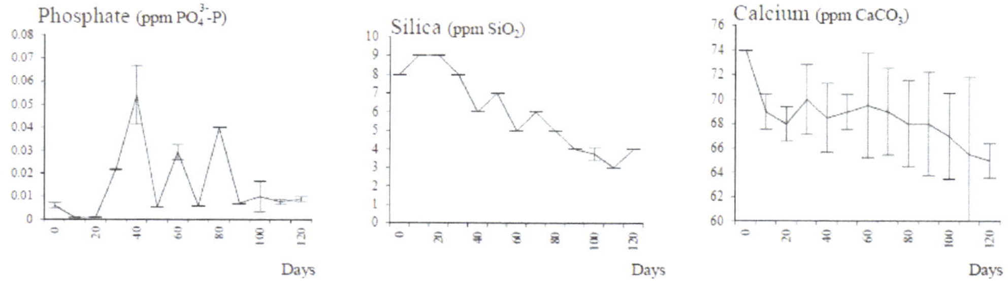

Fig. 5. Water quality during culture for 0-120 days of *Hyriopsis* (*Limnoscapha*) *myersiana* juveniles in recirculating aquaculture system (Kovitvadhi et al., 2008).

7. PHYTOPLANKTON COMMUNITIES

Phytoplankton was discovered in the gastrointestinal tract and served as a nutrient source for various species of freshwater mussels (Hudson and Isom, 1984; Gatenby et al., 1996; Gatenby et al., 1997; ÓBeirn et al., 1998; Kovitvadhi et al., 2000, 2001). This discovery aligns with gut content examinations conducted on other bivalve species by Gale & Lowe (1971), Huca et al. (1983), Paterson (1986), and Parker et al. (1998). The result confirms that freshwater pearl mussels primarily feed on phytoplankton through filtration. To successfully culture freshwater pearl mussels from juvenile to adult stages, it is essential to provide a consistent supply of phytoplankton species as their food source. This is crucial for both laboratory and natural environments as freshwater mussels continuously filter phytoplankton for sustenance. Therefore, an excess density of phytoplankton for feeding juvenile or adult organisms is not required, but a continuous availability of food during the culturing period is essential to enhance survival and growth.

8. DEVELOPMENT OF FRESHWATER MUSSEL MORPHOLOGY

Since the glochidia freshwater pearl mussel of *H.* (*L.*) *myersiana* was cultured in the artificial media that could develop to the adult (Kovitvadhi et al., 2006). Therefore, Kovitvadhi et al. (2007) could study for the morphological development of the juvenile through the adult *H.* (*L.*) *myersiana*. The mussels were collected in sequential developmental stages between 0 and 360 days old. Morphological development was observed by light microscope and SEM. SEM observations were prepared in fixative solution containing 10% neutral buffered formalin for 24 h and stored in 5% neutral buffered formalin for further process. The samples for SEM were thoroughly washed under running water for 30 min and then dehydrated in a graded series of ethanol and dried to critical point. Thereafter, they were mounted on SEM specimen stubs with conductive silver paint and coated with gold and observed with a Jeol Model JSM-5410LV scanning electron microscope operated at 25 KV. All samples before fixation, they were anesthetized in 2% chloral hydrate to observe the internal regions.

The morphological development of *H. (L.) myersiana* juveniles in culture (0-360 days old) is shown in Fig. 6. The early juvenile of *H. (L.) myersiana* at 0 days old after transformation has semi-oval, equivalve shells with an equilateral valve, presenting the same size and shape as the glochidium (Fig. 6A). Anterior shell growth was clearly seen in the first day of juvenile development (Fig. 6B), while posterior shell growth followed afterwards (Fig. 6C). The shells of 0-40-day-old juveniles were thin and transparent as seen under light microscope (Fig. 6D). The inner organs (i.e., stomach, intestine, gills, heart, foot, mantle, and cilia at the gills, mantle and foot) were clearly observed through the shell in this period (Fig. 6E). The shell, however, became thicker during the developmental process and covered all the inner organs (Figs. 6E- L). The first anterior and posterior wings appear in 50-day-old juveniles (Fig. 6E), with the posterior wing becoming dominant relative to the anterior from the 140-day-old stage (Fig. 6H). The mantle lobes of 0-50-day-old juveniles are joined dorsally and are free ventrally (Fig. 6F). The incurrent siphon and excurrent siphon appear after 50 days (Figs. 6G-6L). The complete adult morphology is apparent from 160 days old (Fig. 6I). Males and females reproductive organs are sexually mature about 270-360 days old mussel which depend on environment (Kovitvadhi et al., 2006; Srakaew et al., 2010) (Figs. 6K-6L).

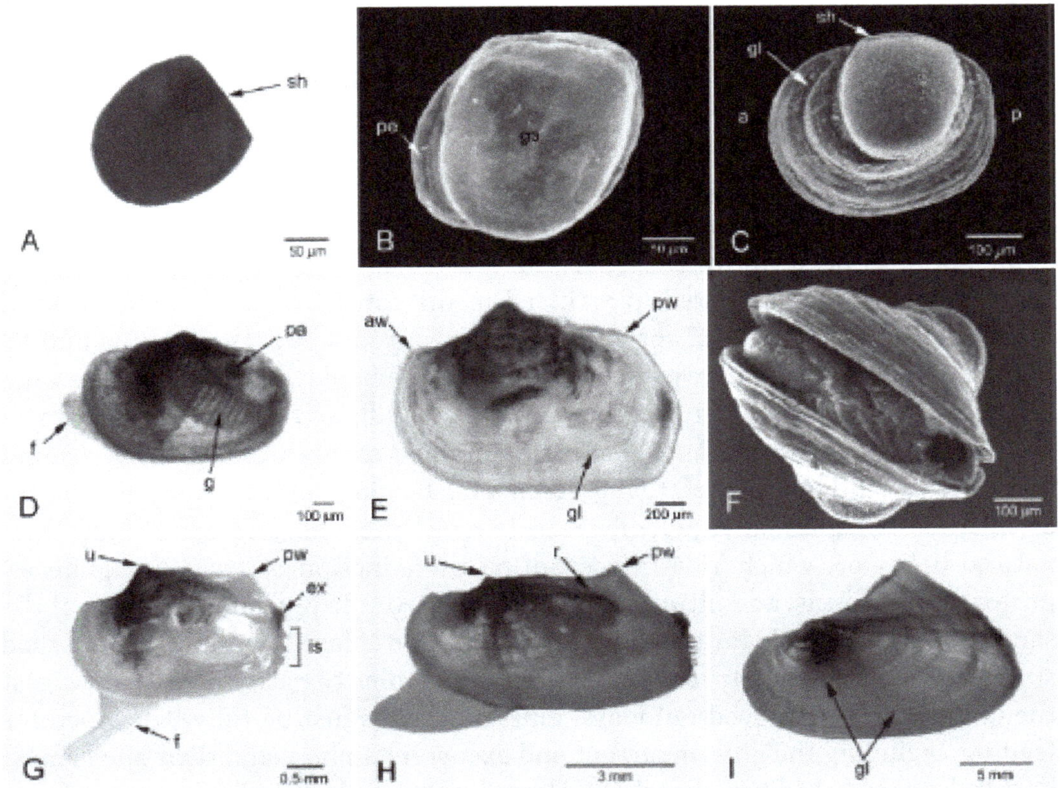

A; Light microscopy of early juvenile (after transformation, 0-day-old) shell, note shell hinge (sh). B; SEM micrograph of 1-day-old juvenile, appearance in anterior region of new soft periostracum (pe), note glochidium shell (gs). C; Juvenile 10 days old, anterior (a) region appears before and grows more than the posterior (p), note growth line (gl). D; Light microscopy of development of shell, 40-day-old juvenile, note foot (f), gill (g), posterior adductor muscle (pa). E; Light microscopy of development of shell, 50-day-old juvenile, note anterior wing (aw); posterior wing (pw). F; SEM of ventral side of 50- day-old juvenile. G; Light microscopy of external morphology of 90-day-old juvenile, note excurrent siphon (es), incurrent siphon (is), umbo (u). H; Light microscopy of external morphology of 140-day-old juvenile, note rectum (r). I-L; External morphology of shell, 160, 180, 270 and 360 days old, respectively.

Fig. 6. Morphological development of 0–360-day juveniles of *Hyriopsis* (*Limnoscapha*) *myersiana*. (Kovitvadhi et al., 2006; 2007).

9. CONCISE SUMMARY

The culture of freshwater pearl mussel involves three sequential steps: (1) cultivating glochidia larvae in artificial medium, (2) raising juveniles, and (3) raising adults. Studies have shown that glochidia from some freshwater mussel species can be successfully cultivated in artificial media consisting of a combination of M199, common carp plasma, antibiotics, and antimycotic agents, resulting in fully mature adults with complete gametogenesis. Key elements in juvenile culture were culturing techniques, water quality, substrate (sand), and food. The laboratory-scale recirculating aquaculture system maintained steady water quality changes and allowed sand to accumulate ingredients for feeding, including organic matter or microbes. Additionally, juveniles can burrow into the sand, mimicking their natural behavior, which aids in preventing the adhesion of feces, pseudofeces, protozoa, flatworms, and ultimately the mortality of the juveniles. The size of the sand should be suitable for each juvenile's size. Phytoplankton is an essential food supply with a size and structure that allows juveniles to easily ingest and digest them. Before being moved outdoors, mussel organs must be fully developed for feeding, including the gills, incurrent and excurrent siphon, and their shells must be totally closed. Additionally, the quality of water and food played a crucial role in the growth and survival of the organism.

REFERENCES

1. Areekijseree, M., Engkagul, A., Kovitvadhi, S., Kovitvadhi, U., Thongpan, A. & Rungruangsak Torrissen, K. (2006). Development of digestive enzymes and *in vitro* digestibility of different species of phytoplankton for culture of early juveniles of the freshwater pearl mussel, *Hyriopsis (Hyriopsis) bialatus* Simpson 1900. *Invertebrate Reproduction and Development* 49, pp. 255-262, ISSN 2157-0272
2. Buddensiek, V. (1995) The culture of juvenile freshwater pearl mussels *Margaritifera margaritifera* L. in cages: a contribution to conservation programmes and the knowledge of habitat requirements. *Biological Conservation* 74, pp. 33-40, ISSN 0006- 3207
3. Chumnanpuen, P., Kovitvadhi, U., Chatchavalvanich, K., Thongpan, A. & Kovitvadhi, S. (2011) Morphological development of glochidia in artificial media through early juvenile of freshwater pearl mussel, *Hyriopsis (Hyriopsis) bialatus*. *Invertebrate Reproduction and Development* 55, pp. 40-52, ISSN 2157-0272
4. Fukuhara, S., Nakai, I. & Nagata, Y. (1990) Development of larvae of *Anodonta woodiana* (Bivalvia) parasitized on host fish. *Venus* 49, pp. 54–61, ISSN 0042-3580
5. Gale, W.F. & Lowe, R.L. (1971) Phytoplankton ingestion by the fingernail clam, *Sphaerium transersum* (Say), in pool 19, Mississippi River. *Ecology* 52, pp. 507-512, ISSN 0012-9658
6. Gatenby, C.M., Neves, R.J. & Parker, B.C. (1996) Influence of sediment and algal food on cultured juvenile freshwater mussels. *Journal of the North American Benthological Society* 15, pp. 597–609, ISSN 0001-4966
7. Gatenby, C.M., Parker, B.C. & Neves, R.J. (1997) Growth and survival of juvenile rainbow mussels *Villosa iris* (Lea, 1829) (Bivalvia: Unionidae), reared on algal diets and sediment. *American Malacological Society* 14, pp. 57–66, ISSN 1000-3207
8. Guillard, R.R.L. & Ryther, J.H. (1962) Studies of marine planktonic diatoms. I *Cyclotella nana* Hustedt and *Detonula confervacea* (Cleve). Gran. *Canadian Journal of Microbiology* 8, pp. 229-239, ISSN 0008-4166
9. Hanlon, S.D. & Neves, R.J. (2006) Seasonal growth and mortality of juveniles of *Lampsilis fasciola* (Bivalvia: Unionidae) released to a fish hatchery raceway. *American Malacological Bulletin* 21, pp. 45–49, ISSN 0740-2783
10. Hoshaw, R.W. & Rosowski, J.R. (1973) *Methods for microscopic algae*. In: Stein, J.R. (ed.), Handbook of Phycological Methods, Culture Methods and Growth Measurements. Cambridge University Press, USA. ISBN 0-521-20049-0
11. Huca, G.A., Brennerand, R.R. & Niveiro, M.H. (1983) A study of the biology of *Diplodon delodontus* (Lamarck, 1819). *Veliger* 25, pp. 51–58, ISSN 0042-3211

12. Hudson, R.G. & Isom, B.G. (1984) Rearing juveniles of the freshwater mussels (Unionidae) in a laboratory setting. *Nautilus* 98, pp. 129–135, ISSN 0028-1344
13. Isom, B.G. & Hudson, R.G. (1982) *In vitro* culture of parasitic freshwater mussel glochidia. *The Nautilus* 96, pp. 147-151, ISSN 0028-1344
14. Isom, B.G. & Hudson, R.G. (1984a) Freshwater mussels and their fish hosts; physiological aspects. *Journal for Parasitology* 70, pp. 318-319, ISSN 0022-3395
15. Isom, B.G. & Hudson R.G. (1984b) Culture of freshwater mussel glochidia in an artificial habitat utilizing complex liquid growth media. U.S. *Patent* 4449480
16. Keller, A.E. & Zam, S.G. (1990) Simplification of *in vitro* culture techniques for Freshwater mussels. *Environment Toxicology and Chemistry* 9, pp. 1291-1296, ISSN 0730-7268
17. Kovitvadhi, S. (2008) *In vitro culture of freshwater mussel juvenile Hyriopsis (Limnoscapha) myersiana (Lea, 1856).* Ph.D. Thesis. University of Porto, Portugal.
18. Kovitvadhi S., Kovitvadhi U. & Meejui O. 2011.Growth and survival of the freshwater pearl mussel *Chamberlainia hainesiana* (Lea, 1856) from juvenile to adult reared in different densities and locations. *Aquaculture* (submitted)
19. Kovitvadhi, S., Kovitvadhi, U., Sawangwong, P. & Machado, J. (2007) Morphological development of the juvenile through to the adult in the freshwater pearl mussel, *Hyriopsis* (*Limnoscapha*) *myersiana*, under artificial culture. *Invertebrate Reproduction and Development* 50, pp. 207–218. ISSN 2157-0272
20. Kovitvadhi, S., Kovitvadhi, U., Sawangwong, P. & Machado, J. (2008) A laboratory-scale recirculating aquaculture system for juveniles of freshwater pearl mussel *Hyriopsis* (*Limnoscapha*) *myersiana* (Lea, 1856). *Aquaculture* 49, pp. 255-262, ISSN 0044-8486
21. Kovitvadhi, S., Kovitvadhi, U., Sawangwong, P., Thongpan, A. & Machado, J. (2006) Optimization of diet and culture environment for larvae and juvenile freshwater pearl mussels, *Hyriopsis* (*Limnoscapha*) *myersiana* (Lea, 1856). *Invertebrate Reproduction and Development* 49, pp. 61–70, ISSN 2157-0272
22. Kovitvadhi, S., Kovitvadhi, U., Sawangwong, P., Trisaranuwatana, P. & Machado, J. (2009) Morphometric relationship of weight and length of cultured freshwater pearl mussel, *Hyriopsis (Limnoscapha) myersiana* (Lea, 1856) under laboratory conditions and the earthen pond phases. *Aquaculture International* 17, pp. 57-67, ISSN 0967- 6120
23. Kovitvadhi, U. (2000) *Culture of glochidia of freshwater pearl mussel Hyriopsis myersiana (Lea, 1856) in artificial media.* Ph.D. Thesis. University of Porto, Portugal.
24. Kovitvadhi, U., Chaopaknam, B., Nagachinta, A., Jongrungwit, K. & Kulayanamit S. (1998) Ecology of freshwater pearl mussel, *Hyriopsis(Limnoscapha) myersiana*

(Lea, 1856) in the Maeklong River, Kanchanaburi Province. *Kasetsart Journal : Natural Science* 32, pp. 1-12, ISSN 0075-5192

25. Kovitvadhi, U., Nagachinta, A. & Aungsirirut, K. (2000) Species composition and abundance of plankton in the gut contents of freshwater pearl mussel, *Hyriopsis (Limnoscapha) myersiana. Journal of Medical and Applied Malacology* 10, 203–209, ISSN 1053-6388
26. Lima, P., Kovitvadhi, U., Kovitvadhi, S. & Machado, J. (2006) *In vitro* culture of glochidia from the freshwater mussel *Anodonta cygnea. Invertebrate Biolology* 125, 34-44, ISSN 1077-8306
27. ÓBeirn, F.X., Neves, R.J. & Steg, M.B. (1998) Survival and growth of juvenile freshwater mussels (Unionidae) in a recirculating aquaculture system. *American Malacological Bulletin* 14, pp. 165–171 ISSN 1000-3207
28. Panha, S. (1992) Infection experiment of the glochidium of a freshwater pearl mussel *Hyriopsis myersiana* (Lea, 1856). *Venus* 51, pp. 303–314, ISSN 0042-3580
29. Parker, B.C., Patterson, M.A. & Neves, R.J. (1998) Feeding interactions between native freshwater mussels (Bivalvia: Unionidae) and zebra mussels (*Dreissena polymorpha*) in the Ohio River. *American Malacological Bulletin* 14, pp. 173–179, ISSN 1000-3207
30. Paterson, C.G. (1986) Particle-size selectivity. I The freshwater bivalve *Elliptio complanata* (Lightfoot). *Veliger*, 29, 235–237, ISSN 0042-3211
31. Srakaew, N., Chatchavalvanich, K., Kovitvadhi, S., Kovitvadhi, U. & Thongpan, A. (2010) Histological structure of developing gonads in the freshwater pearl mussel, *Hyriopsis (Limnoscapha) myersiana. Invertebrate Reproduction and Development* 54, pp. 203-211, ISSN 2157-0272
32. Supannapong, P., Pimsalee, T., A-komol, T., Engkagul, A., Kovitvadhi, U., Kovitvadhi, S., & Rungruangsak Torrissen, K. (2008) Digestive enzymes and *in vitro* digestibility of different species of phytoplankton for culture of the freshwater pearl mussel, *Hyriopsis (Hyriopsis) bialatus. Aquaculture International* 16, pp. 437-453, ISSN 0967- 6120
33. Uthaiwan, K., Noparatnaraporn, N. & Machado, J. (2001) Culture of glochidia of the freshwater pearl mussel *Hyriopsis myersiana* (Lea, 1856) in artificial media. *Aquaculture* 195, pp. 61-69, ISSN 0044-8486
34. Uthaiwan, K., Pakkong, P., Noparatnaraporn, N., Vilarinho, L. & Machado, J. (2002) Study of a suitable fish plasma for *in vitro* culture of glochidia *Hyriopsis myersiana* (Lea, 1856). *Aquaculture* 209, pp. 197-208, ISSN0044-8486
35. Uthaiwan, K., Pakkong, P., Noparatnaraporn, N., Vilarinho, L. & Machado, J. (2003) Studies on the plasma composition of fish hosts of the freshwater

mussel, *Hyriopsis myersiana*, with implications for improvement of the medium for culture of glochidia. *Invertebrate Reproduction and Development* 44, pp. 53-61, ISSN 2157- 0272

13 BARNACLE'S POTENCY IN THE AQUACULTURE SECTOR

1. INTRODUCTION

1.1 Biological characteristics

Barnacles are members of the Cirripedia group, which are mostly sessile crustaceans that permanently attach themselves to various substrates such as inorganic (rock), organic (coral, molluscs, turtles, whales), and even manmade (plastic, wood) surfaces (Southward, 1987). They mostly live in marine habitats, with certain species able to tolerate low salinities in estuaries. They are found at various depths ranging from the upper intertidal zone to depths exceeding 1,000 m. They are social animals that gather in dense groups, generating patches that can cover substrates entirely. The Thoracica superorder is the main group that contains most of the documented Cirripedia species. Unlike other crustaceans, they are distinguished by their underdeveloped abdomen. The group is categorized into three suborders: Lepadomorpha, Balanomorpha, and Verrucomorpha. Typically, they have a delicate, chitinous outer skeleton. Most species also have calcareous plates (mural plates and opercular plates) that encase the individual, offering protection and support. The opercular valves are involved in feeding, respiration, and moulting in the specimen. Additionally, certain species possess a calcareous base according to Southward (1987).

Thoracic barnacles are filter-feeding animals equipped with specialized articulated appendages, including a cirral fan, that allow them to catch suspended particles in the water column. They mostly feed on microscopic algae and the larvae of other invertebrates (Anderson, 1994). These species are simultaneous hermaphrodites that engage in copulation through the extension of a penis for crossover fecundation between adjacent individuals. The ovary is situated at the basal region of the organism. Following fertilization, the egg mass consolidates to create the "ovigerous lamellae," where the embryonic and early larval development of specimens takes place. Barnacles internally develop eggs up to the larval nauplius I stage, which are then released into the sea where they begin a planktonic, free-swimming phase lasting two to four weeks. The larval stage and its movement are affected by nearby oceanographic conditions, as well as larger-scale phenomena as currents, winds, and upwelling. The larvae increase in size during this stage, developing more appendages and progressing through consecutive stages (from nauplius larvae II to nauplius VI) until they transform into the competent cyprid larvae. The cyprid has a bivalve shape, is lecitotrophic, and has a sensory system that helps it find a suitable substrate based on criteria such as texture, color, light amount, and biofilm for attachment. Following cyprid larval settlement, they undergo metamorphosis into adult specimens. Factors like depredation, competition, and physical disturbances are crucial for the early survival of these individuals (Gaines & Roughgarden, 1985; Minchinton and Scheibling, 1991, Thomason et al., 1998). Studies in controlled laboratory settings have shown that parameters including the kind and concentration of food, temperature, and larval density have a crucial role in determining the duration and survival of larval development (Qiu and Qian, 1999; Mishra et al., 2001; Thiyagarajan et al., 2003a; 2003b). Regarding the induction of larval settlement, various chemical signals from conspecific individuals, prey species, and microbial films have been observed to trigger this response (Rodríguez et al., 1992).

Lepadomorph barnacles are identified by their fleshy stalk that attaches them to a surface, together with calcareous plates located in the apical zone known as the capitulum. The species primarily live in the intertidal zone and are classified as neustonic species, attaching themselves to floating structures (Fig.1A). Balanomorph barnacles, also known as acorn barnacles or non-stalked barnacles, do not have a stalk or peduncle. Instead, they attach directly to the substrate using a base that can be membranous or calcareous. They have a more advanced development of the calcareous plates, consisting of 4 to 8 overlapping plates that create a supportive framework. Additionally, two pairs of plates are situated in the top region, creating the operculum consisting of terga and scuta (Fig. 1B). These species are commonly found in the shallow intertidal and subtidal communities along the coast. Verrucomorph barnacles, sometimes known as "wart barnacles," are little asymmetrical animals lacking a peduncle. They are commonly found in deep seas where they attach to other invertebrates as epizoos (Anderson, 1994).

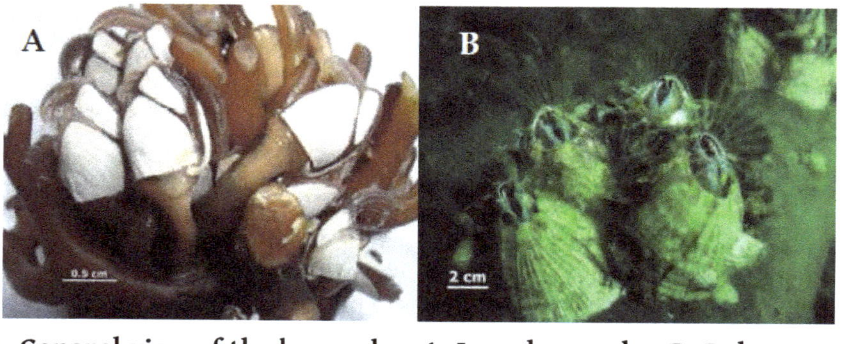

Fig. 1. General view of the barnacles. A. Lepadomorpha. B. Balanomorpha.

1.2 Significant barnacle species with commercial value globally

Table 1. Commercially important barnacle species and species used for human consumption worldwide. Scientific and common names, geographic distribution and production types. Source: Modified from López et al., (2010).

	Scientific name	Common name	Geographic distribution	Production type
"Goose barnacles"	*Pollicipes pollicipes*	"Spanish goose barnacle"	Spain, Portugal, France, Morroco	Fisheries by fishermen´s unions ("perceberos")
	Pollicipes polymerus	"Canadian goose barnacle" "Leaf barnacle"	Canada, USA	Local fisheries
	Pollicipes elegans	"Pacific goose barnacle"	Mexico, Ecuador, Peru	Local fisheries
	Capitulum mitella	"kamenote" or "Japanese goose barnacle"	South of Japan	Artisanal fisheries
"Acorn barnacles"	*Austromegabalanus psittacus*	"picoroco" "giant barnacle"	Perú, Chile, Argentina	Artisanal fisheries/ semi industrial aquaculture
	Balanus nubilus	"giant acorn barnacle"	Alaska, Canada, USA.	Local fisheries
	Balanus rostratus	"mine fujit subo"	Russia, Northern Japan	Local fisheries.
	Megabalanus rosa	"aka fujit subo" "rose barnacle"	Japan	Local fisheries.
	Tetraclita japonica	"kuro fujit subo" "black barnacle"	Japan	Local fisheries.
	Tetraclita kuroshioensis	"hat fujit-subo"	Japan, Indonesia	Coastal fishermen
	Megabalanus azoricus	"craca"	Azores Islands (Portugal)	Local fisheries/ experimental aquaculture
	Megabalanus tintinnabulum	"claca"	Entire tropical zone of the Atlantic and Indian Oceans. Canary Islands (Spain)	Local fisheries

Although there are many barnacle species worldwide, only a select number are commercially significant, primarily because most of these organisms are modest in size. Pollicipes pollicipes, a type of lepadomorph barnacle, commonly known as the "goose neck barnacle," is widely recognized as a gourmet delicacy. The seaweed is harvested by fishermen's organizations, called unions or "perceberos," along the coastline of Galicia, Spain. It can grow up to 10 to 12 cm in height. Between 300 and 500 tons are captured annually, and there is a strong demand for it (Molares & Freire, 2003). Regulation of extraction activities has been implemented due to the overexploitation of natural reserves of this resource. Additional species of the same genus, such P. polymerus (Sowerby, 1833) and P. elegans (Lesson, 1830), have local commercial importance in Canada and Peru. They are also employed to meet the demand in the Iberian market (Bald et al., 2006; Jacinto et al., 2010). The species Capitulum mitella (Linnaeus, 1758), sometimes known as "stalked barnacle," is found along the southern coast of Japan. It is smaller than 5 cm in height and is exclusively sold in the local market.

The most commercially significant species among the balanomorph barnacles included in Table 1 is Austromegabalanus psittacus (Molina, 1782), also known as "picoroco" or "giant barnacle," which is a big species capable of growing up to 30 cm in height. The distribution of this species extends from southern Peru to the austral zone of southern Chile and the southern coast of Argentina, as documented by Pilsbry (1916), Young (2000), and López et al. (2007a). Artisanal fisheries in southern Chile (42°S) exploit this resource on a local scale, with annual catches ranging from 200 to 600 tons (Fig. 2). The average commercial size is 2.2 cm in carino rostral length. Balanus nubilus is a large species found on the Pacific coast from Alaska to North America. It is eaten by indigenous coastal populations. Megabalanus azoricus, sometimes known as "craca," is a subtidal species found exclusively in the Azores Islands archipelago in the Atlantic Ocean. It is harvested locally, and individuals over 1 cm in rostral length are eaten fresh (Lotaçor, 2006; Pham et al., 2008). Enterprise analysis LOTAÇOR S.Data from 1980 to 2010 indicates that recorded landings were small, ranging from 1 to 3.3 tons per year in the 1980s, stabilizing below 1 ton per year in the 1990s. There was a considerable increase between 1999 and 2003, reaching a historic maximum of 7 tons per year. Landings have decreased to approximately 3.7 tons per year in 2010 from 2003. Between 2005 and 2010, the mean initial selling price was around 3.90€/kg, ranging from a minimum of 0.20€/kg to a maximum of 20€/kg. The actual captures and economic worth of the resource might be significantly greater than what has been officially documented. The species is highly valued by the local community and is traditionally caught along the coasts for personal consumption. Dionisio et al. (2009) analyzed LOTAÇOR data to outline certain socio-economic characteristics of this fishery and calculated a per-capita consumption ranging from 88 to 241g/year. The Megabalanus tintinnabulum species, sometimes known as "claca," is harvested in the Madeira archipelago (Portugal) and the Canary Islands (Spain). The species is widespread, found across the tropical zones of the Pacific, Atlantic, and Indian oceans (Young, 1998). On Japan, balanomorph barnacles are known as "fujit-subo" on the market. Balanus rostratus, known as "mine fujit-subo," is commercially harvested in northern Japan, particularly in the Aomori prefecture. This species can grow up to 5 cm in height, with annual landings of approximately 10 tons. Smaller species of balanomorph barnacles, Megabalanus rosa (Pilsbry, 1916) known as "fujit-subo" and Tetraclita japonica (Pilsbry, 1916) called "kuro fujit-subo," are also harvested in the central-southern region of Japan. In Indonesia, the Tetraclita kuroshioensis Chan, Tsang & Chu, 2007, known as "hat fujit-subo," is eaten by coastal residents.

Fig. 2. "Giant barnacle" in a local market in Puerto Montt city, southern Chile.

2. GLOBAL STATUS OF BARNACLE CULTIVATION

Despite the commercial significance of several barnacle species, the cultivation of these species is restricted to only a handful, as noted by López et al. (2010). Previous information on Pollicipes pollicipes in the lepadomorphs group includes details on acquiring young on artificial floating platforms as documented by Goldberg in 1984. Results have been limited due to the fact that larval settling in this species is influenced by the presence of adults of the same species (Molares et al., 1994; Cruz et al., 2010). Similar experiments have been conducted on Megabalanus azoricus, a type of balanomorph barnacle, in the Azores islands. Technologies have been developed for placing collectors in the water column. The findings suggest that larval settling takes place, but it is uncertain if the levels are adequate for large-scale industrial cultures (Pham et al., 2008; Pham & Silva, 2010; López et al., 2010). Progress has been achieved experimentally with these two species in the gathering of spat from the wild. Studies on the "giant barnacle," known as "picoroco" (Austromegabalanus psittacus), have progressed from experimental to semi-industrial cultures solely in this species on the Chilean coast. For this species, research has demonstrated that harvesting spat from the wild at certain levels allows for the establishment of profitable aquaculture operations. The growth rates allow for the rapid attainment of specimens of a commercially viable size (López, 2008; López et al., 2010). Additionally, cost-effective production methods have been created to acquire spat and promote growth. Bedecarratz et al. (2011) also assessed the productivity and financial viability of several products.

Culture activities in Megabalanus azoricus are only beginning. This species of barnacle is exclusively found in intertidal zones with high water flow throughout the Azorean coastline. This clam is highly valued but its native populations are heavily exploited (Santos et al., 1995). The regional authorities have recognized the necessity of conducting trials to develop aquaculture technology for this species (Pham et al., 2008). The preliminary findings for this species show promise according to Pham et al. (2011), however, further extensive investigation is necessary. Understanding the biological properties of barnacle species, including their life cycle, larval dispersion and behavior, and ecological and functional elements during growth, is crucial for their culture. The primary challenges impacting culture based on previous experiences are as follows: Juveniles are either unavailable in the wild or may only be acquired in extremely small amounts. Obtaining young animals through larval cultures in hatcheries may also be an option, but it significantly raises expenses. The limited spat obtained from the wild can be attributed to factors such as insufficient supply of competent cyprid larvae, unsuitable substrates for their behavior, lack of synchronization between larvae quantities and installation of artificial substrates, and variability in collection due to environmental factors. High levels of unpredictability in proficient larval supply operate on a large and intermediate scale in the ocean. It is important to acknowledge that barnacles can create metapopulations and have wide larval dispersion (Lagos et al., 2005); (c) challenges during growth due to predators or competing species for substrate or food (Pham & Silva, 2010); (d) effects influenced by population density linked to mass recruitment (Hills & Thomason, 2003); (e) significant shell weight, particularly in balanomorph species. This also results in reduced harvest yields. The main advantages of the biological characteristics of barnacle species include the ability to obtain spat from the wild, gregarious larval settlement leading to a steady supply of spat with suitable substrates, internal embryonic development in balanomorphs barnacles reducing early mortality, and filter feeding allowing for suspended cultures without external feeding. Future predictions for barnacle farming benefit from the extensive and varied literature on larval development in different species. This literature covers topics such as the length of larval stages, the impact of environmental conditions and feeding, as well as the ecological, physiological, and behavioral factors related to larval settlement and recruitment (Walker, 1995; Jenkins et al., 2000; Dionisio et al., 2007; Tremblay et al., 2007; Pineda et al., 2009). While mostly focused on non-commercially significant species, this data can nevertheless be utilized to enhance barnacle cultivation.

3. SPAT COLLECTION

3.1 Barnacle life cycle

Understanding the barnacle life cycle is crucial for developing cultural techniques. Identifying the unique aspects of embryonic and larval development, including when, where, and to what extent they occur, will allow for the creation of production technologies tailored to these traits. The life cycle of barnacles suitable for semi-intensive cultures, such as the "giant barnacle" and the "craca," have favorable qualities. Internal fertilization and incubation of embryonic stages result in lower mortality during early development compared to invertebrate species like mussels and clams, where fertilization is external and ontogenetic development takes place in the water. However, mortality in planktotrophic larval growth is correlated with the length of this period. Larvae also undergo extensive dispersion, leading to significant variability in the number of competent larvae, both in terms of space and time. This component is crucial when deciding the location of the cultural center, namely in terms of providing resources from the wild. The strong selectivity of the competent larvae influences the choice of artificial substrates, which must meet the biological and physical-chemical parameters for larval settling. The cyprid larva has a sophisticated sensory system that is used to assess the substrate where larval settlement would take place. This stage is essential for the larval's efficiency and influences the future survival of specimens.

3.2 Barnacle spat collection from the wild. Culture technologies: The cases of *Austromegabalanus psittacus* (Chile) and *Megabalanus azoricus* (Portugal)

Methods for collecting spat from the wild have only been created for two species of balanomorphs. With the "giant barnacle" Austromegabalanus psittacus, it is feasible to acquire sufficient spit to support the growth of commercial cultures. The "giant barnacle" is primarily found in the shallow subtidal zone at depths of 5-7 meters, but it can also be infrequently seen in the lower intertidal zone. Moreover, it is a species that grows as a crust and is commonly found on ships, culture systems, and pier support structures, as shown by Brattström (1990) and López et al. (2007a). Cultures have utilized the great geographical heterogeneity and many substrate-types to facilitate larval settlement development. Various types of collectors have been utilized to collect spat from the wild, such as artificial substrates with varying characteristics like texture, size, shape, color, and buoyancy. The materials examined in this species were 900 cm2 rectangular plates made of expanded polystyrene, polythene, and synthetic felts coated with tar. Polythene tubular substrates measuring 70 to 100 cm were tested, providing a usable attachment area of 2,200 to 3,000 cm2 (Fig. 3). The collector system consists of three substrate units connected by polypropylene cables. These cables are positioned vertically in the water column, extending from the surface to a depth of 8 meters. They are suspended from floating systems such as long-lines or culture rafts in shallow, wave-protected, or semi-protected bays. Results indicate that artificial spat collectors with specific properties, such as being benign, rough-textured, dark-colored, and low in buoyancy, guarantee high amounts of spat collecting. The predilection for these properties is attributable to the sensory traits of the competent larvae when they connect to the substrate. They exhibit negative phototaxis, positive geotaxis, and rugophilic behavior. Spat density levels have fluctuated based on collector type, ranging from 0.01 individuals/cm2 to 0.2 individuals/cm2, with nearly full substrate coverage. Research in the Azores involved collecting spit and conducting growth studies on PVC tubes with recruitment densities reaching up to 800 barnacles per square meter per month during late spring and early summer (Pham et al., 2011). Recent tests have demonstrated that selecting materials and their orientation are crucial variables to optimize recruitment values for future spat collector designs (Pham et al., unpublished data). Horizontal placement of material, regardless of its category, results in larger densities compared to vertical placement. The disparity is especially noticeable in the most effective material examined, PVC plates. Consider using PVC plates for designing spat collection systems.

Fig. 3. "Giant barnacle", *Austromegabalanus psittaccus*, spat collection from the wild, southern, Chile.

The private sector has tested a pilot structure for the production of barnacles (Pham & Silva, 2010). The structure tested was designed for both phases of the culture cycle: (a) spat collection and (b) ongrowing.

The device could accommodate 12 PVC tubes in one structural frame and was positioned at a depth of 8 meters for 1.5 years. Over 9,000 barnacles of commercial size were successfully generated with minimal mortality rates, yielding satisfactory results. Efforts are underway to enhance the system's architecture and efficiency, guided by recent findings.

3.3 Spatial and temporal variability

Wild barnacle collections can exhibit spatial and temporal variations due to factors such as latitude, location, depth, substrate type, and time scales ranging from inter-annual to monthly. Barnacle populations function as metapopulations, with larvae being exchanged between nearby populations through passive transport by currents. The number of people at a specific site is influenced not only by the local population of breeders but also by those from farther areas. Local oceanographic forces and variables at meso and macro scales have an influence. Spat collecting for the "giant barnacle" has been assessed by utilizing artificial collectors at various sites around the Chilean coast, at varying depths, and during different times of the year. Larval settlement follows a temporal pattern, with higher spit capture in artificial collectors occurring in the spring-summer period from October to January. A second settlement period is observed in the autumn, specifically in March and April. Gonadal maturity reaches its peak values as water temperature and food supplies increase. An analysis of latitudinal variations was conducted at three locations along the Chilean coast: North Totoralillo (29°29'S; 71°42'W), Tongoy Bay (30°36'S; 71°37'W), and Metri Bay (41°36'S; 72°42'W). The first two are situated in the northern region of Chile, while the third is in the southern part of the country. All three locations are described as shallow bays that are either wave-protected or semi-protected. Distinct climatological and oceanographic variations exist along the Chilean coast, particularly in temperature, with greater temperatures observed in the northern region. After 9 months, the density of specimens recruited in the collectors varied between sites, with significantly larger densities in North Totoralillo compared to Tongoy Bay and Metri Bay. Therefore, the diversity in spit collection does not follow a latitudinal trend, but instead varies depending on the specific site. Differences have been seen in spat collection between tubular plastic collectors and plates, with spat collection being higher in the tubular plastic collectors. This difference may be linked to the varying surface/volume ratio. Differences in water depth also vary between collectors on the surface and those at deeper levels. Cultures are more effective at greater depths, with no differences observed beyond 4 m, perhaps because to the competent larva's positive geotactic and negative phototrophic behavior. Inter-annual fluctuations in spit collection from the wild have been confirmed in certain areas. Oceanographic irregular occurrences, such ENSO (El Niño Southern Oscillation) and local salinity fluctuations, could explain these findings. The fluctuation in spatial and temporal collecting of spat from the wild hinders the advancement of industrial cultures. Therefore, it is recommended to cultivate large populations of the "giant barnacle" in areas suitable for collecting spit, and to designate separate locations for the growth phase. This approach has been employed for numerous years in mussel farming of Mytilus chilensis in southern Chile. Recent records show that the likelihood of finding local "giant barnacles" is linked to factors connected to productivity, such as nitrate concentration and phosphate levels. This could be connected to coastal upwelling processes, which alter the physical and chemical conditions of the coastal water. Chlorophyll content and its change are significant factors influencing the existence of juveniles and adults, perhaps due to their trophic interactions with filtering organisms. Results from a study on M. azoricus, conducted over 2.5 years on settlement panels in two sites near Faial Island, showed recruitment occurring throughout the year, peaking in the summer months from early June to late September, corresponding with rising water temperatures. Newly settled barnacles, less than 15 days old, have a base diameter between 0.79 and 2.80 mm. After two months, their base diameter can increase up to 12 mm, as indicated by the study. The huge size and quick initial growth of M. azoricus recruits enhance their competitive advantage against other barnacle species like Chtamalus montagui and Balanus trigonus settling on the plates.

3.4 Optimization models for spat collection from the wild

Dynamic models were used to study the collection of "giant barnacle" spat from artificial substrates in their natural habitat (Andrade et al., 2011). We can use these models to identify the variables in the spit collection process, the connections between them, and their impact on spat harvest. The dynamic hypothesis suggests that the quantity of "giant barnacle" spat collected from the wild is affected by the abundance of competent larvae (cyprid density) over time, substrate availability, and mortality post larval settling. The multitude of factors influencing their spatial-temporal variability demonstrates the intricacy of the spit collection process and the challenges in predicting outcomes. Simulation experiments were conducted to verify the model's consistency, utilizing empirical data gathered from semi-industrial societies in southern Chile. Within these cultures, the average spat density on artificial collectors varied from 0.1 to 3 spat/cm2, with early mortality rates remaining around 90%. To create 1 gross ton, between 8,000 and 9,000 "giant barnacle" spat are needed when using artificial collectors of 10,000 cm2 strung from 100 m long lines. Ensuring that cyprid abundance levels are synchronized with the placement of artificial substrates in the water is crucial for achieving optimal collector efficiency. Deviation of less than 1 week results in substantial losses (60-70%) in spat production. When the deployment of collectors and the maximum quantity of competent larvae are synchronized, an approximately twofold increase in the number of spit can be achieved. Abundance of capable larvae, together with substrate quantity and quality, are crucial elements for successful spit production from the wild, facilitating the establishment of commercial cultures.

3.5 Spat's production in hatcheries

Currently, there has been no large-scale commercial production of barnacle larval cultures. However, significant knowledge has been acquired regarding the development of various species' larvae. This includes studying the impact of environmental conditions and feeding habits on survival and the length of the development process, as well as the factors affecting larval settlement. Research has been conducted on larval development and its influencing factors in different species of lepadomorphs and balanomorphs (Miller et al., 1989; Molares et al., 1994; Yan & Chan, 2001; Dionisio et al., 2007; Li et al., 2011). Producing spat in hatcheries offers several benefits compared to collecting spat from the wild. Production can be scheduled regardless of the geographical and temporal fluctuations of the larval pool in the environment. Enhancing the energetic characteristics of cyprid larvae can positively impact larval settling percentage, as well as the size and growth rate of juveniles (Thiyagarajan et al., 2002; Thiyagarajan et al., 2003b; Desai & Anil, 2004). Post-settlement mortality can be significantly decreased or entirely eradicated. Research on the cultivation of balanomorph barnacle larvae in laboratory settings has a long history, dating back to studies conducted by Knight-Jones in 1953, Rittschof et al. in 1984, and Brown and Roughgarden in 1985. Cyprid have been collected from several species and used in larval settling experiments (Yan, 2003; Desai and Anil, 2004; Anil et al., 2010; Li et al., 2011). Trials conducted on species like Balanus amphitrite have shown that development times can range from 4 to 18 days depending on the diets and temperatures used, highlighting the impact of these parameters on culture success (Qiu & Quian, 1999). The densities of microalgae from the genera Chaetoceros sp, Skeletonema, and Isochrysis range from 1x105 to 10x105 cells/ml. Survival rates in this species have surpassed 90% according to studies by Qiu & Quian (1999) and Mishra et al. (2001). The larval settlement rate can range from 10% to 80%, influenced by culture conditions, substrate type, and the presence of inductors (Rittschoff et al., 1984; Mishra and Kitamura, 2000; Mishra et al., 2001; Thiyagarajan et al., 2003b). A wide range of chemical signals have been identified as essential for initiating the larval settling response. Natural inductors have been identified through investigations on larval settlement reaction to various natural substrates. These inductors are mostly linked to conspecific individuals, prey species, and microbial films (Rodríguez et al., 1992). Various types of microbial biofilms are recognized as significant factors that trigger larval settling in benthic marine invertebrates. Induction has been documented in diatom and cyanobacteria biofilms (Akashige et al., 1982) and bacterial biofilms (Kirchman et al., 1982; González et al., 1987). Research on bacterial induction has been extensive; the beneficial impact of such biofilms has been seen in barnacles (Maki et al., 1988, Maki et al., 1990). The settlement response linked to these biofilms seems to be caused by the presence of polysaccharides or extracellular glycoproteins on the bacterial wall, as suggested by Kirchman et al. (1982) and Hadfield (1986), or by soluble compounds released by the biofilms, as proposed by Bonar et al. (1990). An escalation in the settlement inductive potential of mature bacterial biofilms was a common factor noted in studies by Maki et al. (1989), Pearce and Scheibling (1990), and Mataxas and Saunders (2009). The opposite impact has also been documented (Maki et al., 1989). Adequate information exists to develop barnacle hatcheries, but the shift from experimental to commercial scale has not yet occurred. There is no information available regarding the associated costs. Implementing commercial cultures will require the advancement of spat production in hatcheries to enhance predictability and quality.

4. GROWTH

4.1 Production technologies

Technological designs in suspended barnacle cultures should prioritize providing optimal access to food and water flow for specimens during the growth period. Adequate space availability is necessary to house animals till they reach the harvest stage. Materials utilized in spat collection systems in the wild must be harmless. The weight must be suitable for suspension from rafts or long-lines, allowing for easy handling and costs that align with commercial activities.

There are two production procedures for the growing phase of the "giant barnacle": "Cultch" systems, which sustain the specimens from spat collection until harvest. The substrates utilized have consisted of plates and polythene tubes ranging from 900 cm2 to 2,200 cm2, which were hung vertically in clusters of three units. One benefit is that collecting spat is unnecessary, making this approach more efficient as specimens are subjected to less mechanical handling, resulting in fewer deaths. Two distinct production technologies are utilized in the "cultchless" system: one for collecting spat from the wild and another for fostering growth. Spat harvest takes place around four months after the larvae have settled. Individuals are then moved to a growing system where they stay until the harvest stage. The surfaces utilized as artificial collectors must allow for the safe removal of young organisms. To detach specimens without destroying the calcareous structures, a surface that avoids harm to the basal zone must be employed since specimens are cemented to the collection substrate. Expanded polystyrene and felts wrapped with tar have been utilized as substrates, ranging in size from 900 cm2 to 1,600 cm2. To promote further development, trays or adhesive might be utilized to connect specimens to a different surface. The most favorable outcomes have been achieved using the "cultch" systems. The "cultch" and "cultchless" systems are situated on floating structures such as long-lines or rafts, suspended from the surface down to a depth of 8 meters. These structures are typically found in shallow bays that are wave-protected or semi-protected. Regular maintenance and cleaning of the systems are essential during the growth phase to prevent the build-up of silt and fouling, particularly from mytilids, ascidians, and barnacles. This helps prevent mortalities caused by suffocation and overgrowth by other species.

4.2 Spatial and temporal variability: Results in the "giant barnacle"

The carino-rostral length or opercular length in balanomorph barnacles is a density-independent measure of growth. It refers to the greatest rectilinear distance between the carinal plate and the rostral plate (López et al., 2007b). Alternative growth estimators like mural plate height or basal diameter are not reliable markers of growth due to their susceptibility to modification by growth density.

In Chile, the average size of the "giant barnacle" is 3.5 cm ± 0.6 carino-rostral length, which is nearly 10 cm in height. Specimens of around 2 cm in carino-rostral length are available in local markets. Specimens from natural banks can grow to over 6 cm in carino-rostral length and more than 30 cm in height. Shell height is influenced by the level of clustering during growth, which leads to the formation of a modified base in the hummocks (López et al., 2007b).

The growth rate changes according on the depth. After twenty months of larval settlement, the average carino-rostral length of "giant barnacle" specimens at depths of 4-6 m was 3.4 ± 0.17 cm, compared to 2.92 ± 0.14 cm for those at depths of 1-2 m (López et al., 2008a).

The cultivation period from larval settlement to harvest can last 7-8 months if the minimum commercial size of 2.1 cm carino-rostral length is considered on the Chilean market. Recent studies in northern Chile indicate a higher growth rate compared to the southern region, likely because to variations in water temperature. In the northern region, specimens larger than 2 cm in carino-rostral length were collected within a timeframe of 4-6 months (Fig. 4).

Fig. 4. "Giant barnacle" culture growth system in northern Chile.

Instantaneous growth rates decrease with age, fluctuating from 0.03 cm/day carino-rostral length at the beginning of growth, to 0.0005 cm/day carino-rostral length after 20 months.

For the Chilean market, "giant barnacle" harvest should occur between 18 – 24 months, depending on the season the collectors are placed in the water. Prior to harvesting, the majority of specimens must have reached over 3 cm carino-rostral length. Nevertheless, the culture period can vary according to the requirements of each market. If, eventually, the "giant barnacle" were to be used as a simil of the "fujit subo" on the Japanese market, the culture period would be substantially less.

Productivity in "giant barnacle" culture systems is extremely high, taking production by extractive fisheries as a point of reference. National average gross production of this resource by small-scale local fisheries is around 200 ton/year; this could be obtained with only 20 to 30 long lines. A double, 100 m floating long-line can produce between 7 and 10 gross ton of total biomass in one season. This can signify approximately 2 ton wet weight of soft parts. Each tubular system of approximately 1 m in length, can produce, on average, 15 ± 8 kg of biomass, equivalent to an average of 100 specimens of commercial size, with a yield corresponding to 20-30 g per individual, depending on the maturity of the female gonad. The results obtained in "giant barnacle" cultures on the Chilean coast indicate that there are spatial differences in growth, both on a latitudinal and local level, as well as bathymetrically and according to the type of culture system. Inter-annual and seasonal temporal differences also exist. Thus, production planning is still subject to a high degree of unpredictability. On the other hand, the heavy weight of the shell, that can fluctuate between 70 to 85% of total weight, together with high variability in the mortality rate during the growth stage, are also limiting factors for cultures. Nevertheless, in view of its rapid growth and productivity, as well as the simple and economic production technologies used during the fattening phase, culture of the "giant barnacle", and other balanomorphs species, constitutes an interesting alternative for diversification of Aquaculture.

4.3 Factors influencing growth

Experiments conducted under controlled conditions have examined the role of environmental variables in the spatial-temporal variability of "giant barnacle" growth (López et al., 2008a). It has been reported that photoperiod and temperature can synergistically affect the metabolism, influencing molt frequency and growth in some barnacle species (Crisp & Patel, 1960; El-Komi & Kajihara, 1991). These variables change seasonally within the geographic distribution range of the "giant barnacle". Temperature and photoperiod affect the instantaneous growth rate, which is greater at 10oC than at 16oC and at photoperiods of 8h light :16 h darkness, than at 16 h light:8 h darkness. Specimens in the natural environment presented higher instantaneous growth rates at depths of between 4 and 6 m than between 1 and 2 m depth, under conditions of greater darkness and lower temperature, in accordance with the results obtained in the laboratory. In other species of barnacles, increased growth has also been reported under darkness conditions (Costlow & Bookhout, 1953; 1956). Futhermore, as occurs in other species of barnacles, growth and molting frequency are not correlated (Costlow & Bookhout, 1956). In the "giant barnacle" molt frequency was greater at 16oC than at 10oC, with no evidence of variation due to photoperiod. This can be associated with a decreasing metabolism as a result of lower temperatures, or to effects at the neuroendocrine system level. However, the inter-molt periods were longer at 10oC than at 16oC and during photoperiods 8: 16 than 16:8 (López et al., 2008a). If this information is applied to suspended "giant barnacle" cultures, the fattening phase should be undertaken at depths of over 4 m and at low temperatures.

5. ECONOMIC ASPECTS

Of the economically significant species of barnacle, only the "goose barnacles" have an established international market; nevertheless, none of them are cultured on a commercial scale. Species of "acorn barnacles" are consumed on very restricted local markets, although progress has been made in the culture of two species, the "picoroco" in Chile and the"craca in Azores Islands. Thus, in spite of the interesting projections associated with barnacle culture, commercialization remains limited.

5.1 Product, yield and market

"Goose barnacles", together with most of the "acorn barnacles", are consumed fresh. Only the "giant barnacle "*Austromegabalanus psittacus* and the "mine fujit subo" *Balanus rostratus*, are available in various types of elaborated product. The "giant barnacle" is consumed fresh, frozen and canned, principally on the local Chilean market (Fig. 5). The "mine fujit subo" is consumed fresh, as tempura and as soup flavouring on the Japanese market.

Fig. 5. Canned products of "giant barnacle".

The yield of cultured "giant barnacle" has been determined in a processing plant as 15-20% of total harvest weight. Loss through processing is approximately 10% (Bedecarratz et al., 2011). At harvest, specimens reach an average total weight of 150 g, most of which is attributable to the calcareous shell, consisting of six mural plates and two opercular plates. Two, double 100 m long-lines can produce an average yield of approximately 420 gross ton/year. This doubles the average annual production of artisanal fisheries activities over the last few years in Chile.

With regard to markets and prices, the "goose barnacles" can reach prices of between €15- 25/Kg on the Iberian market. Of the "acorn barnacle" species, the "giant barnacle" is commercialized on the local Chilean market at prices that range between US$1.5/Kg and 20/Kg. The "mine fujit subo" reaches between US$12-15/Kg on the Japanese market and the "craca", between €2.5-5/Kg on the local market of the Azores Islands, Portugal (López et al., 2010). Market studies of the "giant barnacle" in Japan indicate that some products can reach up to US$25/ Kg. " Giant barnacle" cultures on the Chilean coast have generated an interest in its commercialization in the USA and on the Iberian market.

5.2 Economic feasibility of "giant barnacle" culture. Financial indicators

"Giant barnacle" culture is technically and economically feasible based on spat obtained from the wild (Bedecarratz et al., 2011). Nevertheless, the commercial success of these cultures depends on three factors: (a) appropriate selection of culture locations, according to food availability, as well as environmental, oceanographic and climatic conditions. These variables determine spat supply and growth rate, that in turn define the period from spat collection to harvest; (b) effective culture management in order to reduce mortality; (c) ensuring the best prices on external markets by implementing an adequate commercial marketing strategy and exporting the most profitable products.

The financial indicators obtained for frozen and canned products are adequate. The net present value (NPV) is positive for both products; the internal rate of return (IRR) varies between 31-61% and the discounted payback period (DPBP) varies between 3 and 5 years.

The economic profitability index (EPI) fluctuates between 1.7 and 2.9 and confirms that this aquaculture activity is economically attractive, given that current value of expected cash inflows (at 19.6% discount rate) exceeds initial investments (Bedecarratz et al., 2011).

Greatest cash outlay corresponds to production technology (material and labour costs associated with the construction and installation of production technologies). Sensitivity analysis indicates that positive NPC values are maintained, in spite of significant changes in each critical variable, such as: spat density, gross weight, mortality up to harvest, processing and packing costs, FOB sale prices and exchange rate. For elasticity analysis, the most relevant variables are: gross weight at harvest and sale price FOB, while the variable with lowest impact on NPV is processing and packing costs (Bedecarratz et al., 2011).

6. AT A GLANCE

6.1 Advantages of barnacle culture

The advantages of cultivating "giant barnacles" over other delicate crustacean meats like lobsters or prawns are outlined in López et al., (2010).

- They are omnivorous filter feeders that consume a wide size-range of phytoplankton and zooplankton. Thus, they have a low trophic level and cultures do not require provision of exogenous food, as a result of which costs and environmental impact are moderate.
- Although they are gregarious, they do not create culture problems associated with the territorial behaviour of other species. "Giant barnacles" living in groups do not generate density-dependent effects during growth or reproduction, because of their ability to modify their base, thus limiting demand for areas of adhesion to the substrate (López et al., 2007b).
- Spat can be obtained from the wild in sufficient quantities to permit the development of commercial cultures. This means production costs are low. Spat production in hatcheries ensures reliable supplies, but clearly increases costs (López et al., 2005; 2010; Andrade et al., 2011).
- They have high growth rates, thus specimens can reach a commercial size in periods that, depending on the requirements of the market, fluctuate between 6 and 18 months (López et al., 2010).
- They possess early sexual maturity and high fecundity.
- They have high resistance to environmental variables, particularly hypoxia and to salinity variations, factors that are usually subject to frequent changes, affecting the survival of organisms maintained in suspended cultures (López et al., 2003).
- Culture technologies are simple and economical, and potential for contamination and mass environmental effects, is low. They occupy limited space.
- Although real demand has not been completely defined, commercialization prices on international markets are high.

6.2 Impact of large barnacle farming in expanding aquaculture in Chile

Chile has a strong potential for diversifying aquaculture due to the abundance of economically important local fish, invertebrates, and algae species. Many of them are exported at premium prices, albeit in little quantities. Chile's fisheries and aquaculture exports amount to over US$4 billion and are shipped to over 130 countries. However, constraints resulting from the excessive exploitation of natural populations are present. If the amount of biological knowledge and trial experiences focused at establishing production processes increases, culture alternatives could become a reality. Experimental cultures have been conducted employing approximately twelve species to date (López et al., 2008b). Cultivating native species in industrial settings enhances production predictability and elevates product quality. Aquaculture production currently surpasses 700,000 tons every year. Seventeen species are cultivated, comprising four algae (11.5% of production), five fish species (62.6%), and eight mollusk species (25.9% of production). Approximately 63% of the species are introduced, mostly salmonids, and to a lesser degree, abalones and the Japanese Oyster. Twenty-five percent consists of several types of native mollusks, mainly mytilids, the Chilean oyster, and the northern scallop (Sernapesca, 2009). They are cultured using advanced technology adapted from those used for comparable species in other nations. Approximately 14% are attributed to two kinds of indigenous algae, cultivated with domestic methods.

Traditional fishing activities target 130 species for commercial purposes, including 68 fish, 29 molluscs, 19 crustaceans, 11 algae, and 3 other groupings. Annual landings amount to around 2,000,000 tons per year. According to Sernapesca (2009), fish account for 73% of landings, algae for 19%, molluscs for 5.2%, and crustaceans for 0.7%. The data show diverse and intriguing options for expanding aquaculture by introducing large-scale cultivation of indigenous species. The "giant barnacle" has been extensively studied in relation to various biological aspects relevant to aquaculture, such as growth, density-dependent effects, reproduction, life cycle, environmental factors like salinity and hypoxia, and behavior. Moreover, advancements in production technologies have enabled the cultivation of the "giant barnacle" on a semi-industrial level by acquiring spat from the wild and facilitating growth (López, 2008; López et al., 2010). Cost and worker experience significantly impact the economic viability of aquaculture. Currently, conditions are favorable in southern Chile due to the increased availability of skilled manual labor with aquaculture expertise. This is a result of decreased demand for workers in the salmonid production industry, which has been impacted by the existence of viral illness (Mardones et al., 2009; Vike et al., 2009). This causal element has led to a growth in the availability of maritime concessions, processing plants, transport systems, and other facilities for new aquaculture activities.

REFERENCES

1. Anderson, D.T. (1994). *Barnacles: structure, function, development and evolution.* Chapman and Hall, London. U.K.
2. Andrade, L.I., López, D.A. & López, B. A. (2011). Dynamic models applied to giant barnacle culture. *Aquaculture International* (in press).
3. Akashige, S.T., Kano-noy, H.S. & Nomura, T. (1981). Effects of gamma-aminobutyric acid and certain neurotransmitters on the settlement of the larvae of *Haliotis discus hannai* Ino (Gastropoda). *Bulletin Tohoku Regional Fishery Research Laboratory* 43:37-45.
4. Anil, A.C., Khandeparker, L., Desai, D.V., Baragi, L.V. & Gaonkar, C.A. (2010). Larval development, sensory mechanisms and physiological adaptations in acorn barnacles with special reference to *Balanus amphitrite. Journal of Experimental Marine Biology and Ecology* 392 (1-2): 89-98.
5. Arenas, J. (1971). Distribución de *Elminius kingii* Gray (Cirripedia) en el estuario del río Valdivia. *Beiträge Zur Neotropical Fauna* 6 (3): 199-206.
6. Bald, J., Borja, A. & Muxika, I. (2006). A system dynamics model for the management of the gooseneck barnacle (*Pollicipes pollicipes*) in the marine reserve of Gaztelugatxe (Northern Spain). *Ecological Modelling* 194 (1-3): 306-315.
7. Bedecarratz, P.C., López, D.A., López, B.A. & Mora, O.A. (2011). Economic feasibility of aquaculture of the giant barnacle, *Austromegabalanus psittacus* in southern Chile. *Journal of Shellfish Research* 30 (1): 147-157.
8. Bonar, D., Coon, M., Walch, M., Weiner, R. & Fitt, W. (1990). Control of oyster settlement and metamorphosis by endogenous and exogenous chemical cues. *Bulletin of Marine Sciences* 46: 484-498.
9. Brattström, H. (1990). Intertidal ecology of the northern part of the Chilean Archipelago. *Sarsia* 75: 105-160.
10. Brown, S. & Roughgarden, J. (1985). Growth, morphology, and laboratory culture of *Balanus glandula* larvae (Cirripedia:Thoracica). *Journal of Crustacean Biology* 5: 574-590. Costlow, J.D. & Bookhout, C. G. (1953). Molting and growth in *Balanus improvisus. The Biological Bulletin* 105: 420-433.
11. Costlow, J.D. & Bookhout, C. G. (1956). Molting and shell growth in *Balanus amphitrite niveus. The Biological Bulletin* 110 (2): 107-116.
12. Crisp, D.J. & Patel, B. S. (1960). The moulting cycle in *Balanus balanoides* (L.). *The Biological Bulletin* 118: 31-47.
13. Cruz, T., Castro, J.J. & Hawkins, S.J. (2010). Recruitment, growth and population size structure of *Pollicipes pollicipes* in SW Portugal. *Journal of Experimental Marine Biology and Ecology* 392 (1-2): 200-209.

14. Desai, D.V. & Anil, A.C. (2004). The impact of food type, temperature and starvation on larval development of *Balanus amphitrite* Darwin (Cirripedia: Thoracica). *Journal of Experimental Marine Biology and Ecology* 306; 113-137.

15. Dionisio, M.A., Rodrigues, A. & Costa, A. (2007). Reproductive biology of *Megabalanus azoricus*, (Pilsbry) the Azorean barnacle. *Invertebrate Reproduction & Development* 50(3): 155-162.

16. Dionisio, M.A., Rodrigues, A., Pires, P. & Costa A. (2009). Bases para a gestão e conservação de *Megabalanus azoricus*. *1º Congresso de Desenvolvimento Regional de Cabo Verde / 15º Congresso da APDR*. Cidade da Praia, Cabo Verde, Junho de 2009: 319-337.

17. El-Komi, M.M. & Kajihara, T. (1991). Breeding and molting of barnacles under rearing conditions. *Marine Biology* 108: 83-89.

18. Gaines, S. & Roughgarden, J. (1985). Larval settlement rate: a leading determinant of structure in an ecological community of the marine intertidal zone. *Proceedings of the National Academy of Sciences (USA)* 82: 3703–3711.

19. Gaines, S.D. & Bertness, M.D. (1992). Dispersal of juveniles and variable recruitment in sessile marine species. *Nature* 360: 579-580.

20. Gaines, S.D. & Bertness, M. (1993). The dynamics of juvenile dispersal - why field ecologists must integrate. *Ecology* 74: 2430-2435.

21. Goldberg, H. (1984). Posibilidades de cultivo del percebe, *Pollicipes cornucopia* Leach, en sistemas flotantes. *Informe Técnico. Instituto Español de Oceanografía* 19: 1-13.

22. González, L., Castilla, J. & Guisado, C. (1987). Effect of larval diet and rearing temperature on metamorphosis and juvenile survival of the sea urchin *Loxechinus albus* (Molina, 1782) (Echinodermata: Echinoidea). *Journal of Shellfish Research* 6:109-115.

23. Hadfield, M . (1986). Settlement and recruitment of marine invertebrates: a perspective and some proposals. *Bulletin of Marine Sciences* 39: 418-425.

24. Hills, J. M. & Thomason, J. C. (2003). The 'ghost of settlement past' determines mortality and fecundity in the barnacle, *Semibalanus balanoides*. *Oikos* 101: 529-583.

25. Jacinto, D,. Cruz, T., Silva T. & Castro, J.J. (2010). Stalked barnacle (*Pollicipes pollicipes*) harvesting in the Berlengas Nature Reserve, Portugal: temporal variation and validation of logbook data. *ICES Journal of Marine Science*. 67: 19-25.

26. Jenkins, S.R., Aberg, P., Cerving, G., Coleman, R.A., Delany, J., Della, S.P., Hawkins, S.J., Lacroix, E., Myers, A.A., Lindegarth, M., Power, A.M., Robert, M.F. & Hartnoll, R.G. (2000). Spatial and temporal variation in settlement and

recruitment of the intertidal barnacle *Semibalanus balanoides* (L.) (Crustacea : Cirripedia) over a European scale. *Journal of Experimental Marine Biology and Ecology* 243: 209-225.

27. Kirchman, D., Graham, S., Reish, D. & Mitchell, R. (1982). Bacteria induce settlement and metamorphosis of *Janua (Dexiospira) brasiliensis* Grube (Polychaeta: Spirorbidae). *Journal of Experimental Marine Biology and Ecology* 56: 153-163.
28. Knight-Jones, E.W. (1953). Laboratory experiments on gregariousness during setting in *Balanus balanoides* and other barnacles. *Journal of Experimental Biology* 30: 584– 598.
29. Lagos, N.A., Navarrete, S.A., Veliz, F., Masurero, A. & Castilla, J.C. (2005). Meso-scale spatial variation in settlement and recruitment of intertidal barnacles along the coast of central Chile. *Marine Ecology Progress Series* 290: 165-178.
30. Lagos, N.A., Castilla, J.C. & Broitman, B. (2008). Spatial environmental correlates of intertidal recruitment: a test using barnacles in northern Chile. *Ecological Monographs* 78 (2): 245-261.
31. Li, H.X., Miao, S.Y., Yan, Y., Yu, X.J. & Zhang, L.P. (2011).Larval Development the Barnacle, *Microeuraphia withersi* (Cirripedia,Thoracica,Chthamalidae) reared in the laboratory source. *Crustaceana* 84 (2): 129-152.
32. Lotaçor (2006). *Pescado comercializado nas Lotas de região autónoma dos Açores.* Reviewed April 12. Available from: http://www.lotacor.pt/estatisticas/free/publicacoes_antigas/2006.pdf.
33. López, B.A., & López, D.A.(2005). Moulting frequency and behavioural responses to salinity and diesel oil in *Austromegabalanus psittacus* (Molina) (Cirripedia: Balanidae). *Marine and Freshwater Behaviour and Physiology* 38 (4): 249-258.
34. López, D. (2008) Giant barnacle "picoroco" culture in Chile, *in* C. K. Pham, R. M. Higgins, M. De Girolamo & E. Isidro (Eds.), *Acta of the International Workshop: Developing a Sustainable Aquaculture Industry in the Azores. Arquipelago. Life and Marine Sciences*. Supplement 7: 56-57.
35. López, D. & Toledo, G. (1979). Estudio descriptivo comparado de estados larvales tempranos y de cypris de balanomorfos chilenos. *Acta Zoologica Lilloana* 35: 547-561.
36. López, D.A., Castro, J.M., González, M.L. & Simpfendörfer, R.W. (2003). Physiological responses to hypoxia and anoxia in the giant barnacle, *Austromegabalanus psittacus* (Molina 1782). *Crustaceana* 76: 533-545.
37. López, D.A., López, B.A., González, M.L. & Arriagada, S.E. (2005). Aquaculture diversification in Chile: potential culture of giant barnacles. *Global Aquaculture Advocate* 8 (2): 73-74.

38. López, D.A., López, B.A. & González, M.L. (2007a). Bibliographic index on aquatic biodiversity of Chile: Crustacea, Cirripedia, Thoracica. *Ciencia y Tecnología del Mar* 30 (1):161-165.
39. López, D.A., López, B.A., Burgos, C., Arriagada, S.E. & González, M.L. (2007b). Consequences of base modification in hummocks of the barnacle Austromegabalanus psittacus. *New Zealand Marine and Freshwater Research* 41 (3): 291-298.
40. López, D.A., Espinoza, E.A., López, B.A & Santibañez, A.F. (2008a). Growth and moulting behaviour in the giant barnacle *Austromegabalanus psittacus* (Molina, 1782). *Revista de Biología Marina y Oceanografía* 43 (3): 607-613.
41. López, D. A, López, B.A. & González, M.L. (2008b). Shellfish culture in Chile. *International Journal of Environment and Pollution* 33 (4): 401-431.
42. López, D.A., López, B.A., Pham, C.K., Isidro, E.J. & De Girolamo, M. (2010). Barnacle culture: background, potential and challenges. *Aquaculture Research* 41: e367-e375.
43. Maki, J.S., Rittschof, D., Costlow, J.D. & Mitchell, R. (1988). Inhibition of attachment of larval barnacles, *Balanus amphitrite*, by bacterial surface films. *Marine Biology* 97: 199–206.
44. Maki, J.S., Rittschof, D., Schmidt, A.R., Snyder, A. & Mitchell, R. (1989). Factors controlling attachment of bryozoan larvae: a comparison of bacterial films and unfilmed surfaces. *The Biological Bulletin* 177: 295-302.
45. Maki, J.D., Rittschof, O., Samuelsoon, U., Szewzky, A., Yule, B., Kjelleberg, S., Costlow,J.D.&Mitchell,R.(1990).Effect of marine bacteria and their exopolymers on the attachment of barnacle cypris larvae. *Bulletin of Marine Science* 46 (2):499-511. Mardones, F., Pérez, A. & Carpenter, T. (2009). Epidemiologic investigation of the re- emergence of infection salmon anemia virus in Chile. *Diseases of Aquatic Organisms* 84 (2): 105-114.
46. Metaxas, A. & Saunders, M. (2009). Quantifying the bio-components in biophysical models of larval transport in marine benthic invertebrates: Advances and Pitfalls. *The Biological Bulletin* 216:257-272.
47. Miller, K.M., Blower, S.M., Hedgecock, D. & Roughgarden, J. (1989).Comparison of Larvaland adult stages of *Chthamalus dalli* and *Chthamalus fissus* (Cirripedia, Thoracica). *Journal of Crustacean Biology* 9: 2(242-256).
48. Minchinton, T.E. & Scheibling, R.E. (1991). The influence of larval supply and settlement on the population structure of barnacles. *Ecology* 72: 1867-1879.
49. Mishra, J. K. & Kitamura, H. (2000). The effect of mono-amino acids on larval settlement of the barnacle, *Balanus amphitrite* Darwin. *Biofouling* 14 (4): 299-303.

50. Mishra, J.K., Kitamura, H. & Tomoda, F.I. (2001). Laboratory culture of *Balanus trigonus* larvae by the method of *Balanus amphitrite* for establishing a multi-species settlement assay. *Sessile Organisms* 18 (1):1-6.

51. Molares, J. & Freire, J. (2003). Development and perspectives for community-based management of the goose barnacle *(Pollicipes pollicipes)* fisheries in Galicia (NW Spain*). Fisheries Research* 65 (1-3): 485-492.

52. Molares, J., Tilves, F., & Pascual, C. (1994). Larval development of the pedunculate barnacle *Pollicipes cornucopia* (Cirripedia: Scalpellomorpha) reared in the laboratory. *Marine Biology* 120: 261-264.

53. Morris, R.H., Abbott, D.P. & Haderlie, E.C. (1980). Intertidal invertebrates of California. Stanford University Press, Palo Alto, California, USA.

54. Pearce, C.M. & Scheibiling, R.E. (1990). Induction of settlement and metamorphosis in the sand dollar *Echinarachnius parma*: evidence for an adult-associated factor. *Marine Biology* 107: 363-369.

55. Pham, C.K. & Silva, E. (2010). New species in european shellfish Aquaculture: barnacle farming in the Azores. *Aquaculture Europe* 35 (2): 27-29.

56. Pham, C.K., Higgins, R.M., De Girolamo, M. & Isidro, M. (2008). Acta of the International Workshop: Developing a Sustainable Aquaculture Industry in the Azores. Arquipelago. Life and Marine Sciences Supplement 7: 1- 81.

57. Pham, C.K., De Girolamo, M. & Isidro, E.J. (2011). Recruitment and growth of *Megalabanus azoricus* (Pilsbry, 1916) on artificial substrates: first steps towards commercial culture in the Azores. *Arquipelago. Life and Marine Sciences* 28: 47-56.

58. Pineda, J. (2000). Linking larval settlement to larval transport: assumptions, potentials, and pitfalls. *Oceanography of the Eastern Pacific* 1: 84-105.

59. Pineda, J., Reyns, N.B., Starczak, V.R. (2009). Complexity and simplification in understanding recruitment in benthic populations. *Population Ecology* 51: 17–32.

60. Pilsbry, H. (1916). The sessile barnacles (cirripedia) contained in the collection of the U.S. National Museum including a monograph of the American species. *Bulletin of United States Natural Museum* 93: 1-366.

61. Qiu, J.W. & Qian, P.Y. (1999) .Effects of salinity and temperature on the life-history of *Balanus amphitrite*: effects of past experience. *Marine Ecology Progress Series* 188: 123-132.

62. Regala, J.T. (1999). Contribuição para o estudo da biologia da craca dos Açores, Megabalanus azoricus (Pilsbry 1916). Relatório de Estágio do Curso de Licenciatura em Biologia Marinha. DOP, Universidade do Algarve, Faro. Portugal

63. Rittschof, D., Branscomb ,E.S. & Costlow, J.D. (1984). Settlement and behavior in relation to flow and surface in larval barnacles, *Balanus amphitrite* Darwin. *Journal of Experimental Marine Biology and Ecology* 82: 131– 146.

64. Rodríguez, S.R., Ojeda, F.P. & Inestrosa, N.C. (1992). Inductores químicos del asentamiento de invertebrados marinos bentónicos: importancia y necesidad de su estudio en Chile. *Revista Chilena de Historia Natural* 65: 297-310.

65. Roughgarden, J., Gaines, S. & Possingham, H. (1988). Recruitment dynamics in complex life cycles. *Science* 241: 1460–1466.

66. Santos, R.S., Hawkins, S.J., Monteiro, L.R., Alves, M. & Isidro, E.J. (1995).) Marine research resources and conservation in Azores. *Aquatic Conservation Marine and Freshwater Ecosystems* 5: 311–354.

67. Sernapesca. (2009). *Anuario Estadístico de Pesca. Servicio Nacional de Pesca. Subsecretaria de Pesca, Ministerio de Economía, Fomento y Turismo. Gobierno de Chile.* Reviewed June 2011. Available from http://www.sernapesca.cl/index.php?option=com_remository&Itemid=54&func= select&id=2.

68. Simpfendörfer,R.W., Oelkers, K.B., Nash,D. & López, D.A. (2005). Kinetic properties of the muscular pyruvate kinase from the giant marine barnacle, *Austromegabalanus psittacus* (Molina , 1782) (Cirripedia, Balanomorpha). *Crustaceana* 78 (1): 1203-1218.

69. Simpfendörfer, R.W., Oelkers, K.B & López, D.A. (2006). Phosphofructokinase from muscle of the marine giant barnacle *Austromegabalanus psittacus*: Kinetic characterization and effect of in vitro phosphorylation. *Comparative Biochemistry and Physiology C- Toxicology & Pharmacology* 142 (3-4): 382-389.

70. Southward, A. J. (Ed.). (1987). *Crustacean Issues 5. Barnacle Biology*. A. A. Balkema. Rotterdam, Netherlands.

71. Southward, A. J. (1998). New observations on barnacles (Crustacea: Cirripedia) of the Azores region. *Arquipelago. Life and Marine Sciences* 16a: 11-27.

72. Thiyagarajan, V., Nair, K.V., Subramoniam, T. & Venugopalan, V.P. (2002). Larval settlement behaviour of the barnacle *Balanus reticulatus* in the laboratory. *Journal of the Marine Biological Association of the United Kingdom* 82: 579–582.

73. Thiyagarajan, V., Harder, T. & Qian, P.Y. (2003a). Combined effects of temperature and salinity on larval development and attachment of the subtidal barnacle *Balanus trigonus* Darwin. *Journal of Experimental Marine Biology and Ecology* 287: 223–236.

74. Thiyagarajan, V., Harder, T., Qiu, J.W. & Qian, P.Y. (2003b). Energy content at metamorphosis and growth rate of the early juvenile barnacle *Balanus amphitrite*. *Marine Biology* 143: 543–554.

75. Thomason, J.C., Hills, J.M., Clare, A. & Richardson, A.N. (1998). Hydrodynamic consequences of barnacle colonization. *Hydrobiologia* 375/376: 191-201.

76. Tremblay, R., Oliver, F., Bourget, E. & Rittschof, D. (2007). Physiological condition of *Balanus amphitrite* cyprid larvae determines habitat selection success. *Marine Ecology Progress Series* 340:1-8.

77. Vial, M.V., López, D.A., Simpfendörfer, R.W. & González, M.L. (1999). Responses to environmental hypoxia of balanomorph barnacles. In M. F. Thompson & R. Nagabhushanam (Eds.), *Barnacles. The Biofoulers.* Regency Publications, New Delhi, India, 217–244.

78. Vike, S., Nylund S. & Nylund, A. (2009). ISA virus in Chile: Evidence of vertical transmission. *Archives of Virology* 154 (1): 1-8.

79. Walker, G.(1995). Larval Settlement: Historical and Future Perspective. In F. R. Schram, F. R. & J. T. Hoeg. (Eds), *Crustacean Issues 10, New frontiers in barnacle evolution*, A. A. Balkema, Rotterdam, Netherlands, 69-85.

80. Yan, Y. & Chan, B.K. (2001). Larval development of *Chthamalus malayensis* (Cirripedia: Thoracica) reared in the laboratory. *Journal of the Marine Biological Association of the United Kingdom* 81 (4):623-632.

81. Yan, Y. (2003). Larval development of the barnacle *Chinochthamalus scutelliformis* (Cirripedia: Chthamalidae) reared in the laboratory. *Journal of Crustacean Biology* 23 (3): 513-525.

82. Young, P.S. (1998). Maxillopoda. Thecostraca. *Catalogue of Crustacea of Brazil.* Museo Nacional, Rio de Janeiro, Brazil.

83. Young, P.S. (2000). Cirripedia thoracica (Crustacea) collected during the "Campagne de La Calypso (1961-1962)" from the Atlantic shelf of South America. *Zoosystema* 22 (1): 58-100.

14 ADVANCEMENTS IN BIOTECHNOLOGY AND INTELLECTUAL PROPERTY RIGHTS IN AQUACULTURE – ARE THEY SUSTAINABLE?

1. INTRODUCTION: AN OVERVIEW

This chapter aims to provide a comprehensive study of the latest trends and advancements in biotechnology within aquaculture research and management. Technological advancements and structural changes in the aquaculture industry could impact access and intellectual property rights (IPR) regulations. The discussion will cover a broad range of topics, including immediate and long-term biological repercussions, ethical considerations, and various social factors such as economic, legal, and political difficulties. It will address inherent conflicts that require compromise for sustainable development. The chapter will address current biological difficulties in aquaculture as a developing food production industry, with minimal focus on external factors like environmental impacts. Instances of disease in farmed salmon and cod in Norway, as well as in shrimp and tilapia in Asia, will be emphasized.

2. SUSTAINABLE DEVELOPMENT IN AQUACULTURE

The concept of 'sustainable development' has gained significance in various policy areas with the release of the World Conservation Strategy. A commonly accepted definition of the notion is 'development that fulfills current demands without jeopardizing the capacity of future generations to fulfill their own needs' (WCED, 1987). The Rio declaration of 1992 said that governments bear a worldwide duty to address environmental disputes in a manner that safeguards the interests of both mankind and nature. An example of international commitments that has incorporated this concept is the Convention on Biological Diversity (CBD). Regarding aquaculture, guidelines for promoting sustainability can be located in the Holmenkollen guidelines for sustainable aquaculture (1999), the Norwegian Ministry of Fisheries and Coastal Affairs strategy for an environmentally sustainable seafood industry (2009), and the EU communication titled "A Strategy for the Sustainable Development of European Aquaculture" (2002).

Critics of sustainable development contend that the idea is unclear and there are differing viewpoints among scientists and regulators regarding its definition and ramifications. The primary debated aspects of sustainable development include determining the significant values within sustainable development, establishing priorities among them, and balancing the preservation of nature and biodiversity with the promotion of a fair society and economic growth (Kamara et al., 2006). The swift expansion of aquaculture has caused worry regarding alterations in land use in coastal regions, effects on wild fish due to escapees, environmental contamination, and the substantial utilization of marine resources for fish feed production. A conceptual framework for sustainable aquaculture has been outlined based on three viewpoints: environmental, economic, and sociological (Caffey et al., 1998). This suggests that the implementation of modern biotechnology should be examined in terms of the effectiveness of current methods and the problem-solving capabilities of the new technology. Furthermore, it is necessary to recognize that the use of current biotechnology in aquaculture impacts socio-economic factors such as employment, income, local economic activity, and ethics, all of which are crucial components of sustainability according to most users. Sustainable development necessitates a renewed emphasis on stakeholders and their requirements. It requires a greater comprehension of stakeholders' viewpoints and public concerns, as well as addressing challenges related to institutional structure and representation in decision-making processes.

The following part discusses biological and ecological issues in aquaculture, while section four outlines current technical advancements that may impact these challenges. Section five discusses international and domestic rules related to modern biotechnology and access difficulties, which are relevant to the sustainability topic. We will provide a concise overview of the current structural advancements and managerial patterns in aquaculture. We will now analyze how actors perceive the impact of biotechnology and IPR on sustainability in aquaculture within this comprehensive framework. This part is based on surveys and interviews conducted with important individuals. In section eight, we discuss the significant issues of how intellectual property rights (IPR) and biotechnology could impact sustainability in aquaculture. We will now analyze the consequences and provide suggestions on how advancements in biotechnology and intellectual property rights in aquaculture may support sustainability.

3. PRESENT ISSUES IN THE MANAGEMENT AND BIOLOGY OF AQUACULTURE

The aquaculture business is currently encountering numerous obstacles. These factors encompass animal health and wellbeing, environmental impact, and social implications such as economics, global resource usage, equitable resource distribution, and rural sustainability. Within the structure of a book chapter, only a restricted number of difficulties can be effectively addressed. This part will focus specifically on biological difficulties related to animal health and welfare, as well as the management of fish reproduction, including biotechnological methods.

3.1 Search for improved animal health

The expansion of aquaculture has led to a rise in diseases caused by various pathogens, such as bacteria, viruses, parasites, and fungi. Currently, infections in aquaculture are leading to substantial economic issues and considerably impacting animal welfare. The concentration of fish and efficient pathogen movement in water provides a conducive environment for these pathogens to thrive. Diseases have a tendency to increase in farm conditions, posing possible ecological risks to both the farmed fish and the surrounding environment, which includes wild fish. Antibiotics and pesticides were commonly used for disease prevention in salmon farming for an extended period. The potential pollution from chemicals and overuse of antibiotics, along with the development of antibiotic resistance, have raised concerns and led to a search for alternate solutions, such as selecting for improved disease resistance.

Enhancing disease resistance in fish has primarily relied on conducting challenge tests in controlled environments. Salmon that have been tested in challenges cannot be employed as parents for the following generation of elite salmon, preventing direct application of selection on breeding candidates. Geneticists are seeking genes that regulate the level of resistance to certain diseases to overcome this issue. Genes that serve as markers could be excellent selection criteria as they can be used directly without the need for challenge testing (see to section 4.2 for more details). Therefore, enhancing the precision of selection can minimize the necessity to sacrifice fish in challenge testing.

Norwegian salmon farming has prioritized the selection for genetic disease resistance since 1995. In 2007, Moen et al. (2009) discovered genetic markers for a gene that accounts for 80% of the genetic diversity in resistance to infectious pancreatic necrosis (IPN) in both fry and post-smolts. Aqua Gen has created and implemented a tool that use these markers to choose fish that are resistant to IPN. This method can accurately identify if individual fish has zero, one, or two copies of the gene variation associated with high resistance. This method could potentially be beneficial for pancreas disease (PD), a severe economic disease in farmed Atlantic salmon that results in substantial losses due to death and decreased production (SalmoBreed, 2011).

Salmon lice (Lepeophtheirus salmonis) are a significant health and welfare issue in the salmon industry. Moreover, it poses an ecological issue since the lice proliferate in aquaculture facilities and subsequently transfer to the wild salmon population. Chemical treatment is often used to fight lice, however the usage of biological methods like cleaner fish has been rising because lice have become resistant to the chemicals. Moderate genetic variation has been demonstrated for resistance to the salmon louse, suggesting the potential for mitigating lice-related issues through selective breeding programs (Kolstad et al., 2005). Developing disease and parasite resistance in Norwegian salmon and trout is crucial for the fish, producers, and consumers, as it would enhance the industry's sustainability. This knowledge might also be used to other aquaculture species.

3.2 Genetic diversity and fish breeding strategies

The expansion of aquaculture has led to a rise in diseases caused by various pathogens, such as bacteria, viruses, parasites, and fungi. Currently, infections in aquaculture are leading to substantial economic issues and considerably impacting animal welfare. The concentration of fish and efficient pathogen movement in water provides a conducive environment for these pathogens to thrive. Diseases have a tendency to increase in farm conditions, posing possible ecological risks to both the farmed fish and the surrounding environment, which includes wild fish. Antibiotics and pesticides were commonly used for disease prevention in salmon farming for an extended period. The potential pollution from chemicals and overuse of antibiotics, along with the development of antibiotic resistance, have raised concerns and led to a search for alternate solutions, such as selecting for improved disease resistance.

Enhancing disease resistance in fish has primarily relied on conducting challenge tests in controlled environments. Salmon that have been tested in challenges cannot be employed as parents for the following generation of elite salmon, preventing direct application of selection on breeding candidates. Geneticists are seeking genes that regulate the level of resistance to certain diseases to overcome this issue. Genes that serve as markers could be excellent selection criteria as they can be used directly without the need for challenge testing (see to section 4.2 for more details). Therefore, enhancing the precision of selection can minimize the necessity to sacrifice fish in challenge testing.

Norwegian salmon farming has prioritized the selection for genetic disease resistance since 1995. In 2007, Moen et al. (2009) discovered genetic markers for a gene that accounts for 80% of the genetic diversity in resistance to infectious pancreatic necrosis (IPN) in both fry and post-smolts. Aqua Gen has created and implemented a tool that use these markers to choose fish that are resistant to IPN. This method can accurately identify if individual fish has zero, one, or two copies of the gene variation associated with high resistance. This method could potentially be beneficial for pancreas disease (PD), a severe economic disease in farmed Atlantic salmon that results in substantial losses due to death and decreased production (SalmoBreed, 2011).

Salmon lice (Lepeophtheirus salmonis) are a significant health and welfare issue in the salmon industry. Moreover, it poses an ecological issue since the lice proliferate in aquaculture facilities and subsequently transfer to the wild salmon population. Chemical treatment is often used to fight lice, however the usage of biological methods like cleaner fish has been rising because lice have become resistant to the chemicals. Moderate genetic variation has been demonstrated for resistance to the salmon louse, suggesting the potential for mitigating lice-related issues through selective breeding programs (Kolstad et al., 2005). Developing disease and parasite resistance in Norwegian salmon and trout is crucial for the fish, producers, and consumers, as it would enhance the industry's sustainability. This knowledge might also be used to other aquaculture species.

Genetic diversity is crucial for responding to selection, hence a large and genetically varied breeding population is necessary for starting and maintaining an animal breeding program. An optimal breeding design with suitable family structure is essential for maintaining a sizable effective population size and achieving a prolonged selection response while minimizing inbreeding in fish breeding. Bentsen & Olesen (2002) determined through a simulation research that a minimum of 50 families (pairs of parents) are necessary for preventing inbreeding and achieving a sustained improvement in a mass selection program for aquaculture. Gjerde et al. (1996) introduced optimal designs for fish breeding programs that consider restricted inbreeding and mass selection. Bentsen & Gjerde (1994) show and assess different breeding schemes for between-family, within-family, and combined selection (between- and within-family). Utilizing combined family-based selection designs can significantly enhance selection accuracy, especially for traits with low heritability. To prevent a high rate of inbreeding and maintain genetic diversity, limit the amount of broodstock picked per family due to the increased likelihood of choosing many siblings from a small number of families.

In 2005, less than 10% of the fish used for aquaculture came from family-based selection programs (Gjedrem et al., 2011). Recently, the situation has not shown significant improvement, particularly for aquaculture species like carps that have high production volumes, as only a limited number of effective selection programs are now in operation. Additionally, the effective population sizes are frequently constrained and sometimes insufficient due to the high reproductive capacity enabling the utilization of a small number of broodstock. Improved populations may nevertheless have a significant short-term advantage over non-improved populations and dominate a large portion of the market. Consequently, this deters additional genetic introductions into the breeding nucleus. Long-term inbreeding and reduced genetic variety due to genetic drift might impact performance and hinder long-term genetic advancement. Populations like these require techniques to introduce genetic variety from beyond the breeding nucleus continuously without negative performance effects. Moreover, the commencement of new breeding initiatives tailored to various conditions for key farmed species is anticipated to enhance the current situation.

3.3 Improving fish welfare through genetic enhancement opportunities

Initially, breeding initiatives for farmed fish usually focus on improving productive traits such as growth. During advanced stages, priority is given to disease resistance, survival, and product quality characteristics when selecting to develop a more robust fish. Domesticated fish thrive in captivity and farm environments, leading to less stress, enhanced resilience, and better growth and survival rates. Thus, it is essential to prioritize fish welfare by minimizing stress and fostering their well-being to improve the sustainability and economic viability of fish farming.

Huntingford et al. (2006) pinpoint several challenges to fish welfare in aquaculture, including aggressive interactions, handling and removal from water, infections, enduring unfavorable physical conditions, and potentially increased aggression due to selective breeding for rapid growth. Research on salmon and cod has shown that via multiple generations of selective breeding for development and domestication in hatcheries and farms, a more docile and less hostile carnivorous fish can be produced. Ruzzante (1994) found that domestication can greatly influence behavioral characteristics in fish, altering the intensity of behavior rather than the behavioral pattern itself. Olesen et al. (2011) emphasized the possible linked effects on stress regulation in fast-evolving fish.

Selection of terrestrial animals based on their efficiency in production can have adverse effects on traits such as health and reproduction (Rauw et al., 1998). In Nordic countries, breeding objectives have been selected to include a broader spectrum of traits, including functional and welfare attributes. Olesen et al. (2000) investigated the development of breeding goals for sustainable livestock production by introducing a system that considers non-market values to evaluate characteristics that enhance public benefits, such welfare attributes. Since 1995, farmed salmon in Norway have been selectively bred for disease resistance. This decision will undoubtedly reduce the stress and suffering linked to sickness. Insufficient data exists on genetic variants and correlations between welfare traits, such as behavior (e.g., aggression) and stress management, in farmed fish. We lack awareness of potential adverse relationships in specific fish welfare indicators, like diminished stress management capabilities, because of the current emphasis on productive traits in breeding selection. Further research is needed to examine the wellbeing traits of fish and their genetic components. Differences in genetics between wild and farmed salmon have been documented regarding survival, maturation, behavior, dominance, aggression, and activity level (Fleming & Gross,

1992; McGinnity et al., 1997; Metcalfe et al., 2003; Petersson & Järvi, 2006). Hatchery-raised salmonids showed decreased antipredator behavior (Johnsson et al., 1996) and lower physiological stress levels due to higher stocking densities (Mazur & Iwama, 1993) compared to wild salmonids. Domesticated fish that have acclimated to the farm environment will encounter less adverse consequences in that setting. Key fish welfare indicators for selective breeding encompass growth rate, survival rate, social behavior (e.g., cannibalism in carnivorous species), and frequency of injuries (e.g., fin injuries) (Turnbull et al., 1998).

3.4 Animal welfare and ethics

European Union policy for sustainable aquaculture development emphasizes promoting excellent husbandry techniques and safeguarding the welfare of farmed fish. Conflicts and trade-offs frequently arise between the industry's short-term profit and the desire for inexpensive animal products, on one side, and animal welfare, on the other, which does not benefit the animals. Animal welfare has primarily been focused on research animals, livestock for food production, and pets. Some of these difficulties will be emphasized before we discuss the effects of fish farming on animal welfare.

An essential inquiry in animal husbandry is whether it is ethically acceptable to view animals solely as a resource or tool to fulfill our need, or if there are moral principles that impose limitations on this perspective. Numerous challenging inquiries have emerged concerning the inherent worth of animals. If animals possess intrinsic value, any infringements on their lives by humans raise moral concerns that require thoughtful responses and actions. The Norwegian Animal Welfare Act of 2010 asserts that animals possess an inherent value. This concept emphasizes the importance of prioritizing animal welfare regardless of the animal's perceived value to humans, which helps define the animal's status.

The term 'welfare' originates from the combination of 'well' and 'fare,' referring to the quality of an animal's journey through life. How effectively can an animal adjust its biological functions in response to its surroundings? Animal welfare is defined as the state of an animal in relation to its efforts to adapt to its surroundings, according to Broom (1986). Other definitions emphasize an animal's subjective experience or awareness of its situation and if it can live a natural life. The word 'animal welfare' encompasses the mental/emotional and physical well-being of an individual animal or its ability to adapt to its surroundings. The phrase encompasses behavior, physiological, and immunological variables. In this context, health is defined in a more comprehensive manner than only the lack of sickness. Ensuring animal health relies significantly on the well-being of the animals. The concept also encompasses positive welfare, suggesting that withholding pleasurable experiences and stimuli from animals is an ethical issue concerning animal protection. Animal protection refers to safeguarding the emotional and physical well-being of individual animals.

Most contemporary animal ethicists base direct moral issues on animals' capacity for consciousness. Lund et al. (2007) argued that fish welfare should be seriously considered based on their possessed of ethically important similarities related to sentience. The authors argue that fish are likely to possess sentience and, as a result, should be given significant consideration. They determined through a basic risk analysis that there is a significant likelihood that fish can experience pain, and if they do, the impact might be substantial considering the potentially large number of animals suffering. Therefore, farmed fish should be presumed innocent. From a self-centered perspective, we can make a case for the just treatment of animals. Causing harm to animals goes against human dignity and may lead to the creation of a more brutal society, as suggested by Mahatma Gandhi's quote: "The greatness of a nation and its moral progress can be judged by the way its animals are treated."

4. BIOTECHNOLOGY'S IMPACT ON AQUACULTURE AND ITS CONTRIBUTION TO INNOVATION

Modern biotechnology utilizes innovative tools to address the issues now faced by aquaculture. This chapter will focus on breeding, vaccine development, and the potential advancements through chromosome modification, DNA marker selection, and genetic engineering.

4.1 Technology that manipulates chromosomes for reproduction

External fertilization is common in many aquaculture species and allows for various genetic engineering techniques, such as manipulating chromosome numbers to create haploids, triploids, and tetraploids. Additionally, animals can be generated with chromosomes exclusively from the mother (gynogenesis) or from the father (androgenesis). For a detailed explanation of the methods and uses, see to Refstie & Gjedrem (2005).

Sterile fish production can address the issue of escaped fish breeding with wild fish and the necessity to safeguard enhanced genetic material from unauthorized reproduction. Premature sexual maturation leads to issues in commercial farming such as lower filet quality, increased mortality, and decreased growth. Triploid fish that are sterile will not develop reproductive organs and can keep developing and be harvested at any point. In many animals, one sex reaches sexual maturity early, resulting in decreased development and body weights. Chromosome modifications can be utilized to selectively generate either all male (e.g., tilapia and salmon) or all female fish (e.g., halibut, hake, and angler fish) based on the desired sex. The most prevalent uses of chromosomal modifications in aquaculture are triploid trout and all-male tilapia.

Cost/benefit analysis questions to consider before utilizing reproductive technology include: Is one gender more valuable in the market? Is one gender more productive for the producer? Is gender determination in this species genetically determined or primarily influenced by the environment? Will the 'clean green' market perception and sales pitch be at risk if this technology is utilized (Robinson, 2002)?

4.2 Application of molecular genetics in aquaculture breeding

The two areas of modern biotechnology that has been expected to have most significant impact on genetic improvement of aquaculture species are DNA markers and transgenics (Hayes & Andersen, 2005). A DNA marker is an identifiable physical location on a chromosome whose inheritance can be monitored (Hyperdictionary, 2003). A comprehensive overview of DNA markers and linkage mapping together with a discussion of potential applications of DNA markers in aquaculture breeding programmes is given by Hayes & Andersen (2005). Furthermore, whole genome sequencing and application of genomics in aquaculture breeding programs is discussed by Quinn et al. (2011). As mentioned, DNA markers have already been applied in aquaculture breeding for direct and highly accurate selection of IPN-resistant fish (Aqua Gen, 2010). So called marker assisted selection (MAS) can double genetic gain for traits that can not be measured on selection candidates (e.g. disease resistance), because it utilizes the between family variance, and may also contribute to reduce inbreeding. For MAS, quantitative trait loci (QTL) must be mapped and their effect determined. This is not the case for genomic selection (GS), where the effects of a large number of loci are first estimated using a test group. Selection can then be carried out on genome wide breeding values of the breeding candidates predicted as the sum of the marker effects estimated, assuming an additive genetic model (Meuwissen et al., 2001). However, the high genotyping costs for GS has so far limited its application in aquaculture breeding. Therefore, a scheme with pre-election of parents for growth combined with selective genotyping of large and pooled family groups has been suggested to obtain high accuracies while reducing number of genotypes and costs many fold (Sonesson et al., 2010).

Furthermore, application of new tools of molecular genetics for gaining understanding about genetic regulation of complex traits such as disease resistance may be important (see section 3.1).

4.3 Genetically modified organisms

Modern biotechnology allows for the identification of genes responsible for certain traits and the mapping of genomes, leading to the creation of genetically modified organisms (GMOs). Research and development of transgenic fish, genetically modified (GM) vaccinations (including DNA vaccines), and current and upcoming GM plants for feed are particularly important for aquaculture. Genetic engineering can facilitate meeting patent requirements like the inventive step and reproducibility, thereby enhancing the utilization of intellectual property rights.

4.3.1 Introduction of transgenic fish

Most focus on trangenic fish is on the possibilities for enhancement of the quality of cultured stocks by improving growth rate and increasing resistance to disease and stress (Melamed et al., 2002). Improved growth rate has been possible by the introduction of growth hormone (GH) genes, in species such as Atlantic salmon, coho salmon, Nile tilapia and hybrid tilapia. The most known example of transgenic salmon is the *AquAdvantage*, developed by Aqua Bounty, which contains a gene construct composed of the regulatory elements of an ocean pout antifreeze protein gene controlling a chinook salmon GH gene. The antifreeze promotor enables stimulation of the growth hormone gene also during cold periods with the result that the transgenic salmon grows much faster than its non-GM counterpart. The company is seeking approval for commercial use of this transgenic fish, and the application has been under evaluation in more than ten years by the US Food and Drug Administration (Niiler, 2000). Other highly relevant approaches are development of disease and parasite resistant fish. At present there is a lack of understanding of genes responsible for disease resistance in fish and different strategies are discussed. These strategies include antisense technology for production of complementary RNA for foreign RNA, expression of antimicrobial substances and peptides (as lysozyme) and efforts to increase production of the fish cytokines and other genes involved in immune defence. Moreover transgenic approaches that combine interesting characteristics, as enhanced growth and disease resistance, together with approaches for development of sterile fish or fish where reproductive activity can be down-regulated is also highly relevant since this will minimise the risk of transgenic fish breeding with wild populations after accidental release or escape.

The potential of transgenic fish to escape and enter the natural environment is an important concern for regulators (Le Curie-Belfond et al., 2009). Unless transgenic fish is used in contained facilities, transgenic fish will certainly escape into the environment. The environmental impact is difficult to predict and will depend on the number of escaped fish, their phenotypic characteristics (related to ability for reproduction and survival over time), and the aquatic biodiversity present in the receiving ecosystem (Kapuscinski & Brister, 2001).

Another potential problem related to transgenic fish that is disease or parasite resistant may be similar to what has been experienced with insect resistant crops (Le Curie-Belford et al., 2009). It has for example been reported that insects have developed resistance to insect resistant crops. Hence, the benefit achieved may over time develop into a long-term problem. If the same unexpected events develops with disease or parasite resistant fish the consequence will be a need for more or other antibiotics and chemicals to cope with resistant pathogens and parasites or new emerging pathogens.

Development of transgenic fish does also raise ethical concern. One implication by the genetic modification process itself is that it may affect fish welfare, behaviour and reproduction (see also Le Curie-Belford et al., 2009). It has been reported pleiotropic effects as changes in coloration, cranial, opercula and lower jaw deformation in transgenic coho salmon (Devlin et al., 1995). Concerns have also been raised that genetic modification strategies may affect the animal's integrity (Verhoog, 2001). This is controversial, and for example Sandøe & Holtug (1998) argue that only welfare of animals and humans are relevant ethical considerations, and that these considerations imply to:

1. Clarify realistic alternatives for improving animals welfare through breeding or biotechnology
2. Consider positive and negative effects on animal welfare
3. Find what benefits are at stake for humans
4. If animals welfare is reduced, consider if the costs of animals will weigh up for humans benefits
5. Weigh conflicting concerns, including risk of possible long term side effects.

When weighing conflicting ethical concerns we often have to compromise between efficiency and animal welfare, where different ethical theories of animal ethics may affect acceptance of biotechnology on animals (Sandøe & Christensen, 2008):

- Contractarianism: implies no problem with instrumental use of animals as a biotechnological resource or as a tool for biotechnology.
- Utilitarianism: can accept the use of biotechnology to improve human and animal welfare if the benefits are bigger than the costs in terms of harmful effects.
- The animal rights view: makes the use of biotechnology in animals more problematic. One can for example not sacrifice a live animal and make it suffer from e.g. health problems in order to improve animal health of future generations of an animal population. If there is a risk that e.g. transgenic fish have unintended health problems, this will be particularly troublesome. Use of both animal breeding and biotechnological methods on animals can be considered as instrumental use of animals and does therefore interfere with this view although it may not imply important harm or suffering for the animal.
- Respect for nature often implies moral problems with both animal and plant breeding and biotechnology. Those who sympathise with this view will often be critical to tamper with the nature although it may not cause animal or human harm or suffering. Hence, they will only support a very cautious and restrictive policy to animal breeding and biotechnology. For farmed fish

with close wild relatives in nature this will likely be even more troublesome than for other farm animals. The risk of escaped farmed transgenic fish mating with wild fish is for example particularly problematic in the view of natural populations almost considered as sacred. Traditionally bred farmed salmon with the same alleles as wild salmon, but with other gene frequencies and other combinations can also be troublesome for the same people. When it comes to salmon, the fact that it is often considered as an iconic species may also be a relevant issue.

In the Norwegian Animal Welfare Act (2010) §25, the following is stated about animal breeding: 'Reproduction, including through methods of gene technology, shall not be carried out in such a way that it:

- changes genes in such a way that they influence the animals' physical or mental functions in a negative way, or passes on such genes,
- reduces the animals' ability to practise natural behaviour, or
- stimulates general ethical reactions.

This may indicate a combination of utilitarian and animal rights perspectives. The initial point (animals' usual functions) could also involve a component of reverence for nature. One may argue that this hybrid could form the foundation for the public ethical perspective on utilizing transgenic animals in Norway.

4.3.2 Intellectual Property Rights (IPR) and the implementation of Genetically Modified (GM) and DNA vaccinations

Modern biotechnology offers instruments for quick illness diagnosis and identification, as well as the potential for developing new and enhanced vaccines. There are typically two ways for developing GM vaccines. The first method involves using gene-deleted bacteria or viruses for homologous vaccination to induce protective immunity against the genetically modified vaccine. Another strategy entails creating recombinant vaccines through genetic engineering. This involves inserting a gene into bacteria or yeast to produce proteins for a vaccine, incorporating the gene into a virus or bacteria for use as a vaccine, or using a DNA vaccine.

Currently, the most efficient vaccines in aquaculture are multivalent, targeting salmon. Research is also being conducted to provide vaccines for other species such as seabass, tilapia, and grouper. Current methods are restricted to certain bacterial and viral illnesses, with no available vaccinations for fish parasites. Intracellular pathogens, such viruses and some bacteria, are particularly challenging to eliminate using conventional immunizations. Therefore, DNA vaccines could provide a technical remedy for these issues. A DNA vaccine example is a plasmid containing the infectious haematopoietic necrosis virus (IHNV) glycoprotein gene regulated by a cytomegalovirus promoter (pCMV). This vaccine was administered to Atlantic salmon to confer resistance to IHNV (Traxler et al., 1999). After first trials on DNA vaccination in mammals, various tests have been carried out on fish, showing positive outcomes such full immunity to viral infections (Romøren, 2003 and references therein). A DNA vaccine containing multiple plasmids encoding various antigens of a pathogen has the potential to stimulate a wide range of antibody responses, making it effective for vaccinating against viruses that exhibit antigenic variation, such as infectious pancreas necrosis virus (IPNV) and infectious salmon anaemia virus (ISAV) (Kibenge et al., 2001).

Developing GM and DNA vaccines for infectious fish diseases offers advantages such as cost-effectiveness, simple production, enhanced quality control, heat resistance, uniform production methods for various vaccines, and the ability to create multivalent vaccines (Hew & Fletcher 2002; Kwang 2000). Currently, there is a restricted scientific comprehension of what happens to these vaccines once they are injected into the animal. Research is required to investigate the stability of vectors and DNA constructs, unanticipated immunological implications, and the biological effects of the vaccine post-injection, such as persistence, distribution, expression, and integration (Gillund et al., 2008a, 2008b). It is crucial to study the possibility of recombination with related organisms and dissemination by vectors in the environment for genetically modified vaccines (Myhr & Traavik, 2011).

An instance of the socioeconomic challenges concerning Intellectual Property Rights (IPR) is the situation involving Intervet's patent on the Pancreas Disease (PD) virus and its potential impact on the progress of a PD vaccine. Intervet has obtained a patent for the virus, granting them exclusive rights to produce vaccines for Pancreas Disease (PD). Pharmaq sought a license to manufacture a PD vaccination and requested a compulsory license from the competition authorities, but their request was unsuccessful. They claimed that the existing Intervet vaccine is ineffective and produced in inadequate quantities. They asserted that their proposed vaccine is superior in terms of efficiency, cost-effectiveness, and animal well-being, as it can be administered as part of a multi-vaccination injection. The Ministry of Fisheries and Coastal Affairs (FKD) was not directly involved, but the competition authorities sought their advice. Pharmaq stated in a trial that they use a different variant of the PD virus from the one patented by Intervet. The case remains unresolved at the time of this book's publication.

5. INTERNATIONAL COMMITMENTS AND STATE REGULATIONS: THEIR IMPACT ON PROMOTING SUSTAINABLE DEVELOPMENT IN AQUACULTURE

Sections two to four have shown that biotechnological advancements offer advantages but also pose new concerns in terms of their impact on economic, social, and environmental conditions. This section discusses the regulation of these activities at both the international and domestic levels to promote sustainability in the aquaculture industry. This section offers an overview of two distinct sets of legal requirements. The first group is extensive and focuses on access and benefit sharing legislation, indirectly addressing sustainable development. The second is less advanced on the international and national stages, but these legislative acts and instruments more directly address sustainable development in aquaculture. Both factors are pertinent to the Convention on Biological Diversity (CBD). During its second meeting in November 1995 in Jakarta, Indonesia, the Conference of the Parties (COP) to the CBD established a program of action known as the "Jakarta Mandate on Marine and Coastal Biological Diversity," resulting in the development of a work program in this field.

The Convention on Biological Diversity aims to achieve three interconnected goals: conserving and sustainably using biodiversity, as well as ensuring fair access to and equitable sharing of benefits derived from the utilization of genetic resources. The CBD focuses on conserving and sustainably using wild species and enhancing breeding stocks, while also ensuring fair distribution of benefits from the utilization of the world's genetic resources. Controversial negotiations have surrounded this topic since the formation of the CBD. A Protocol on Access and Benefit Sharing was established during the 10th Conference of the Parties to the CBD in Nagoya in October 2010 through negotiations.

The World Trade Organization (WTO) sets global norms for aligning intellectual property rights, whereas the World Intellectual Property Organization aims to promote cooperation and alignment of IPR among all member countries. Uniform intellectual property rights legislation apply to all technological sectors, such as biotechnology, especially when the invention involves biological material. The conflicting goals of different international treaties create tension and controversy between northern and southern countries.

Two alterations in patent law have caused controversy in the patent system: firstly, the granting of patents on different types of biological material; and secondly, the granting of patents on fundamental research, which enhances the commercialization of research. These inclinations prompt inquiries on the extent to which patents foster innovation. The connection between exclusive rights and access rules is problematic since only a small portion of the profits derived from the use of genetic resources have been shared with the suppliers. The CBD aims to create a compromise between technological availability and genetic resources in biotechnology by managing the growth of patent systems. The clash of various global goals has led to confrontations between nations over the availability of seeds and therapeutic plants compared to proprietary technologies in the agricultural and medical industries. In aquaculture and animal husbandry, breeding material typically does not flow from south to north, unlike in other industries.

Genetic resource access, conservation, fair benefit sharing, and intellectual property rights regimes are globally recognized goals, although they may not always align harmoniously (Rosendal, 2006). Conservation is a fundamental requirement for achieving the other purposes, as stated in the three objectives of the CBD. The main purpose of the CBD is to connect the equilibrium between ABS and IPR with that of conservation. Lack of access to genetic resources hinders innovation. The CBD aims to create a framework that promotes innovation using biodiversity to help conserve variety equitably. Intellectual property rights legislation aims to enhance the motivation for innovation. There are signs that extensive patent claims could hinder innovation by restricting access to technologies and raising transaction costs for other parties. Innovation is essential for maximizing advantages. Lack of revenue sharing from genetic resource utilization may reduce the motivation and capacity to preserve biodiversity in underdeveloped nations. This aspect is less significant in aquaculture compared to the agricultural and pharmaceutical industries (FAO, 2009).

Norway uses domestic norms and regulations to create goals for aquaculture, focusing on safeguarding coastal settlements, increasing value, promoting sustainable management, and fostering innovation (White Paper, 2005:9, 136). Norway takes responsibility for approximately one third of the global wild salmon populations and environmental obligations related to Norwegian-owned salmon farms, including those operating in foreign countries (White Paper, 2009:142). This prompts intriguing inquiries concerning the correlation between Norway's use of this resource and its obligation to manage the natural material in accordance with the CBD.

Norway has recently enacted two significant legal acts at the national level: The Nature Diversity Act of 2009 and the Act on Management of Wild Marine Resources of 6th June 2008. The Wild Marine Resources Act allows the government to set up a process for obtaining government approval before exploring wild marine genetic resources, making it less directly applicable to exporting from breeding programs. The Nature Diversity Act designates genetic material as a common resource in Norway and grants the Ministry the authority to mandate permits for accessing genetic resources. The Marine Resources Act declares that marine resources are a shared resource. Both Acts mandate the ministries to create specific administrative regulations for accessing genetic resources, however there are still obstacles in implementing these regulations. The relationship between access to resources and the right to use them for a patentable innovation is one aspect to consider. Exporting fingerlings and breeding material will include genetic material, which may necessitate particular protection to safeguard the exporter's interests in maintaining these resources as a shared asset.

Now, let's focus on equipment specifically designed to address environmental issues in aquaculture. The primary organization in this field is the Aquaculture Stewardship Council (ASC), anticipated to be fully operational by 2011, serving as the foremost certification and labeling program for ethically farmed seafood worldwide. The Aquaculture Stewardship Council (ASC) is established as a result of the Aquaculture Dialogues. It is tasked with collaborating with independent third-party organizations to approve farms that meet the standards for responsible aquaculture set by the Dialogues. The criteria aim to reduce the primary adverse environmental and social effects associated with 12 aquaculture species. These guidelines, like those of the Forest Stewardship and Marine Stewardship Councils, specify specific performance levels that farmers need to achieve in order to obtain certification. Over 2,000 aquaculture producers, environmentalists, scientists, and other participants are engaged in the process, overseen by the World Wildlife Fund (WWF). The WWF, in collaboration with the Dutch Sustainable Trade Initiative (IDH), contributed to funding the establishment of the ASC, a non-profit organization.

6. ADVANCEMENTS IN AQUACULTURE INFRASTRUCTURE AND CHANGES IN MANAGEMENT PRACTICES

The new promises and threats of the technological developments provide a backdrop also for this section, which is aimed at combining technology and structural trends. The general structural trends in aquaculture are similar to those in agriculture and animal husbandry worldwide, that is to say moving away from public funding and small scale enterprises towards merging, privatization and internationalization. In this section we briefly account for these trends and then in section eight, we point to some of the implications this may have for management and sustainable development in aquaculture.

In Norway, salmon and trout breeding programs were started with public financing in 1971 by a non-profit research institute (Gjøen & Bentsen, 1997). The base populations of these programs were collected from Norwegian rivers (Atlantic salmon) and from Scandinavian farmed populations (rainbow trout) and these breeding populations were transferred in 1985 to a cooperative ownership by salmon farmers' organizations. However, as a result of an economic crisis in the late 1980s, this activity was transferred to a shareholder company in 1992 (at present Aqua Gen AS). At that time it was decided that the value of the breeding material should be secured in a way that took care of the public interest; hence the structure of ownership was divided between private and public shareholders, but that structure was only bound for a five years period. With the next government the largest public shareholder was turned into a private venture company, who in 2007 decided to sell its shares in Aqua Gen AS. The German EW Group, which is also the holding company of the world's leading poultry genetic companies (Aviagen), concluded an agreement to take majority ownership of the shares in Aqua Gen AS. Thus, the Ministry of Fisheries and Coastal Affairs (FKD) gradually lost control over the material from the originally public supported breeding programme for salmon.

Currently, the Norwegian legal system is unclear on regulating genetic material originating in the wild or coming from public breeding programmes and hence a comprehensive management system for aquatic genetic resources is not in place. The sale of Aqua Gen AS to German EW Group is illustrative of the dilemma. This breeding material can now in theory be patented and removed from the public domain. The development has moved from a situation of public control and ownership, via a cooperative situation, to the current situation of increasingly dominating market actors. This raises question about eventual effects of recent Norwegian access regulations in the Nature Diversity Act and the Marine Wild Species Act. The situation has raised questions about the need to regulate access to wild or breeding material in other species, as cod and halibut, because a similar level of international exchange has not taken place for these other species of fish.

Since the early 1990s, public support for Norwegian farm animal breeding has decreased, reflecting a political will to privatize breeding. A counter trend is the initiative from the Ministry of Fisheries and Coastal Affairs to fund the establishment of a breeding program for cod at the research institute, Nofima in Tromsø. In line with the policy goal of safeguarding rural settlements, this may reflect an intact willingness to finance development of breeding programmes in Norway, at least in the districts of Northern Norway. The current official goal is to retain the cod breeding material and associated competence and knowledge bases that are being built up as a Norwegian public good asset. The end goal may be that cod, like salmon, is intended to become profitable and commercialised at some point down the line, but the legal process of how to deal with this has only just started (interview NN1).

7. ACTOR INTERESTS RESPONDING TO STRUCTURAL AND TECHNOLOGICAL CHANGES

In this section we turn to the description of how affected actors perceive the legal, structural and technological developments in the aquaculture sector. The aim is to study their views on whether and how these parallel trends are likely to affect sustainable development in aquaculture.

7.1 Public opinion and attitudes to new biotechnologies

In Europe, surveys have been carried out from 1996 to 2007, to identify Europeans opinion and attitudes to modern biotechnology. Although difficult to draw any conclusion it seems that modern biotechnology used in production for medicines and pharmaceutical products receives highest support while modern biotechnology used in agriculture receives less support (although concern about GM foods have had a small decline since 2001). For example in one recent Eurobarometer poll (GMO compass, 2009), European consumers were asked to identify the environmental themes about which they were most concerned, and on an average, 20% of respondents cited the topic of 'GMOs in farming'. Although these surveys have been on GM crops and medicines, the aspects related to benefits and environmental issues may have similar support and doubt versus the different possibilities for introduction of modern biotechnology in aquaculture (see also 7.1.1 for results from a survey among Norwegian students).

There are several challenges involved in measuring actor perceptions of GMO. A study of the GM debate in the UK, Australia and New Zealand found that access to decision- making and the inability to weigh explicit social value judgements with the broad science consensus were the major obstacles to successful deliberative public debate (Walls et al., 2005). For instance, in the New Zealand experience, non-scientific arguments were implicitly marginalised because the templates (questionnaire) employed for the interest groups made it difficult to use holistic arguments. A 'holistic argument' in this case might imply a consideration of the growing dominance of multinational corporations in the life sciences. These enterprises increasingly decide on options for the development of new medicines and food, they are part and parcel of the GM revolution – but somehow their role seemed to be 'beside the point' in the questionnaire developed to study the public debate (Walls et al., 2005). Most studies on GM plants and products are based on information provided by research laboratories and/or released by industry (Gaskell et al., 2003). It is also important to note that this documentation, along with the GMO applications, is provided by multinational corporations that enjoy little public trust (Gaskell & Bauer, 2001).

7.1.1 Students attitudes to the use of GMO in aquaculture

Knowledge and attitudes regarding the use of GMOs in aquaculture was studied among students at the Norwegian University of Life Sciences (UMB) that can be considered as future stakeholders in food production and management of biological resources (Unpublished data). The survey result showed that most students were well informed about genetic engineering and/or GMOs. Many students were very concerned about environmental impacts from the use of GMOs. For example it was found that most of the respondents disagreed on the statement on whether the use of GMOs results in no negative environmental effects, this was also confirmed by the question on whether the students would buy transgenic salmon if it was more environmentally friendly (showed by an 40 % increase in willingness to buy transgenic salmon if it was more environmentally friendly). However, many students were not willing to buy transgenic salmon if it was more nutritious or more disease resistant. Only 47% of the respondents would buy transgenic salmon if a relevant authority (e.g. Norwegian Food Safety Authority) has approved it as safe.

Many students were totally agreeing that there is a need to reduce the risk by initiating more research on GMOs, that new knowledge about risks must be taken into account and that we need to seek expert advice to get more understanding about the potential risks to health and the environment. The students were also strongly agreeing on the need to reduce the risk of genetic engineering applications by increasing transparency to the public about research and information about the technology. Moreover, the need for improving communication between scientists and the public was recognized.

Spirituality or religion had little or no influence when it came to ethical issues by GMOs in this survey. Students supported use of GMOs for saving human lives (e.g. by producing medicines and vaccines) followed by production of animal feed (e.g. from plants, algae and microorganisms). Surprisingly, it was not found that production of cheaper food could encourage the students to support the use of GMOs. In general, students requested labelling of GM food as well as of salmon fed with GM plant feed.

7.2 Public attitudes to animal welfare and willingness to support and pay for e.g. improved health and welfare of farmed fish

In a recent study by Olesen et al. (2010) it was found that a relatively large percentage of Norwegian consumers agreed (at least partially) that Norwegian animal welfare standards were sufficiently strict (78,1%), and that fish welfare was sufficiently protected in Norwegian fish farming (67.9%). In order to estimate a lower bound for the consumers' willingness to pay for improved welfare for farmed salmon, a real choice experiment with eco-labelled salmon was carried out in Norway (Olesen et al., 2010). It was found that the average respondent preferred eco-labelled salmon to conventional salmon when the colour was the same, and was willing to pay additional 2 euro per kg fillet for eco-labelled salmon. The price premium depends on the conventional and organic salmon being the same colour, and an inferior appearance due to lack of pigmentation significantly reduces consumer interest in organic salmon. This is also consistent with the results in other studies investigating consumer preferences toward organic products with inferior appearance (Thompson & Kidwell, 1998; Yue et al., 2009).

7.3 Actor perceptions of needs and interests in access to aquatic genetic material

Studies of the fish breeding sector for several species and in several countries indicate that they are all prone to rapid structural changes in response to calls for profitability and commercialization (Olesen et al., 2007; Rosendal et al forthcoming). This correspond with a recent survey among fish farmers in Norway where it was found that the most important source of risk for the industry was future salmon prices, institutional risks and diseases (Bergfjord, 2009). In the same study the respondents (farmers) was also asked to identify the most important risk management strategies to the risk issues identified, which was to keep cost low and ensure profitability.

The public breeding programme on cod is currently seen as a public good for Norwegian breeders (interview NN2). The authorities are concerned that the public and private cod breeding programmes can compete on a level playing field, so that the one with public funding is not given unfair competitiveness. At this early stage, it is acknowledged that it is very hard to fund a breeding program, as the economic returns from increased growth may still be a long way off. For salmon, the real growth and economic returns from the breeding program was not apparent until about the fourth generation – and cod is still only in the second or third. There are two major reasons why public funding may be the preferred solution, at least in the early phases: First, during the early phases of breeding, basic mass selection using individual phenotypic information can provide a similar and much cheaper response in growth. This is why more advanced breeding programmes are often less profitable, particularly on a short term, as they are equally costly to start and run the first generations. However, phenotype or mass selection is usually much more limited with respect to selection towards a broader breeding goal with several traits, such as disease resistance. Also, it may be more vulnerable to less control of and rapid increase of inbreeding with resulting genetic erosion. Hence, in the long run, the more advanced family based breeding programmes will become more economically and biologically viable – or sustainable (interview, NN3).

A public or cooperative breeding program typically has a wider variety of breeding aims, such as animal welfare and environmental considerations, in contrast to a private breeding program focused on short-term profit. Therefore, it is anticipated that visible improvements in growth rate and economic benefits will be achieved later compared to a private initiative, but will result in more sustainable fish material in the long term. The private cod program has a short-term competitive advantage, whereas the public program has a long-term competitive advantage. Together, they align with the goals outlined in the 2005 White Paper for rural and coastal settlements, emphasizing improved value, sustainable management, and innovation. Shrimp in India and tilapia in Asia show similar tendencies (Ramanna Pathak, forthcoming, Ponzoni et al., 2010).

Norwegian aquaculture investments and operations in Chile are facing challenges due to widespread outbreaks of infectious salmon anaemia. As a result, the enterprises anticipate stricter regulations concerning biological and environmental risks (Marine Harvest, 2011). New restrictions will need investment in new technology.

7.4 Researchers' perspectives and attitudes towards emerging biotechnologies

There have been few studies investigating scholars' perspectives and opinions on new biotechnologies. Kvakkestad et al. (2007) found that experts from diverse scientific fields and work settings perceived evidence differently in uncertain situations, resulting in varied judgments on the hazards of GM crops. In 2010, Gillund and Myhr conducted a study that identified different perspectives on alternative feed options for salmon among stakeholders in Norwegian aquaculture. The study discussed the sustainability of plant production in industrial agriculture, with a specific focus on the cultivation of genetically modified (GM) plants by the participants. The participants discussed several appraisal criteria concerning health, welfare, economic, environmental, knowledge, and social issues, highlighting the difficulty of finding sustainable alternative feed resources.

8. DISCUSSION

Although the cost ratios and value creation from more efficient fish production and lower fish prices are significant, only approximately 10% of the current global aquaculture production utilizes genetically improved material from modern breeding programs. Aquaculture is significantly falling behind the agriculture sector due to the absence of established practices and expertise in systematic selective breeding, limited collaboration with agriculture or aquaculture research organizations possessing the necessary knowledge and technology, inexpensive prices of roe and fry leading to reduced profit margins for breeders and hatcheries. Genetic improvement has become an unreliable investment because to the simplicity of illegally generating highly fertile fish and selling counterfeit material. Over the past twenty years, research and development dollars have been mostly allocated to molecular genetics and genomics, with less emphasis on advancing and implementing selective breeding, which has demonstrated long-term genetic improvements. Efficient application of genetic information requires a selective breeding program.

There are notable parallels between the ABS argument in the CBD and the perceived lack of immediate utility in breeding and breeding programs. Past research indicates that there have been minimal incentives to capitalize on salmon breeding materials because of low roe prices and limited profit from enhanced breeding material. This trend is expected to persist in the foreseeable future (Olesen et al., 2007; Rosendal et al., 2006). Similarly, there appears to be minimal benefit in gaining more knowledge and enhancing genetic resources and their characteristics, as suggested in the ABS discussion (Grajal, 1999). It is difficult to deny the significant financial gains from biotechnology, including traditional breeding and genetic engineering. There is an increasing commercial interest in obtaining important genetic resources (Laird & Wynberg, 2005). This paradox is evident when considering the beneficial benefit of faster-growing and hence more affordable salmon produced through breeding operations. Here we have a discrepancy where the desire to pay is minimal, while there is a surprisingly strong demand in access. Producing fast-growing, disease-resistant fish is costly, but once achieved, replication is inexpensive. This has resulted in a push for profitability and privatization in the aquaculture industry, which includes public breeding programs. Yet, the expense of upholding a high-quality, disease-free product is considerable. The uncertainty around earnings raises doubts about whether the market will provide this service. Aquaculture breeding operations have proven to be highly beneficial and cost-effective due to the high fertility of aquatic species, resulting in significant value creation for society. Similarly, this applies to family-based strategies that target a comprehensive and enduring breeding objective including multiple features. The paradox highlights the difficulty of achieving governmental objectives related to providing inexpensive access to genetic advancements in breeding and promoting sustainability and innovation in aquaculture. Not supporting public breeding programs could mean losing the goal of offering better breeding material at an inexpensive cost.

Aquaculture is facing pressure to increase production efficiency and prioritize short-term earnings. Therefore, actors encounter challenges related to securing sufficient finance for sustained breeding programs and inexpensive access to enhanced genetic material. Traditionally, aquaculture in India and Norway has primarily relied on public funding to boost production, create and distribute materials widely to a large audience, rather than focusing on generating exclusive products. The original goals of the GIFT tilapia project were also accurate. This demonstrates the characteristics of breeding material as a public good. Increased commercial sector participation necessitates enhanced legal protection of genetic material. Because this restricts access to information, it is believed to have adverse effects on aquaculture. Rosendal et al. (forthcoming) discovered parallels among Norway, India, and the GIFT donors in their normative goal of ensuring inexpensive access to enhanced breeding material. Furthermore, participants interviewed expressed a shared concern about preventing monopolization in a globalized market and ensuring affordable access to aquatic breeding material. The need for profitability is simultaneously weakening these objectives. This could guide cod, shrimp, and carp breeding programs to follow a path similar to that of salmon and GIFT tilapia. Waiving public control in the aquaculture sector may appear to contradict the recommendations and interests of both commercial and public stakeholders.

Market consolidations and privatization are key structural variables acknowledged by the participants as significant in altering the regulations within the salmon industry (Olesen et al., 2007). Privatization and commercialization are likely to shift breeding objectives towards creating products that cater to economically sustainable markets rather than focusing on developing new products based on social, ecological, and biological criteria such as disease resistance and fish welfare traits. Publicly financed breeding programs for salmon and tilapia have transitioned to private ownership, indicating a tendency towards privatization in the farmed cod industry as well. The structural characteristics of the aquaculture industry help explain why it has fewer conflicts between developed and developing countries compared to the plant sector. Although not causing a North-South conflict, the fundamental goals in obtaining breeding material are alike for plant and animal genetic resources. Due to structural changes resulting in fewer and larger enterprises, conflicts over access are more likely to arise between small and large entities inside the sector rather than across nations.

Continuous enhancement and development are crucial in the biology of breeding, and patents can hinder innovation by restricting progress. The expense and duration of acquiring a patent, along with the extended protection period of twenty years in patent legislation, do little to encourage swift innovation in industries where ongoing advancements in a biological dynamic system are the most effective and enduring strategy. It is interesting that AquaGen and Landcatch Natural Selection, the two salmon breeding businesses who successfully mapped the QTL marker for IPN resistance, have decided not to patent it. The expenses associated with pursuing these patents are likely deemed excessive for these enterprises, including AquaGen supported by a large worldwide corporation. However, like agriculture and medicine, the changes in the aquaculture industry's structure appear to have a greater impact on how individuals perceive the importance of access and protection than biological characteristics. This also has implications for the wider international discussion over the Nagoya Protocol on Access and Benefit Sharing (ABS) and the potential need for sector-specific accords tailored to various genetic resource categories. Since monopolization and globalization exhibit comparable structural qualities in other sectors such as agriculture, pharmacy, and aquaculture, it implies that there is less justification for a sector-specific strategy to regulating international transactions involving genetic resources.

9. AT A GLANCE

Aquaculture is a global food source (FAO, 2009). Can its biotechnological advancement be sustainable in both the short and long term? In this chapter, we have outlined many issues that actors in the aquaculture sector are encountering and emphasized the need for decision-making.

The EU statement 'A strategy for the sustainable development of European aquaculture' highlights the importance of maintaining competitiveness, productivity, and durability in the aquaculture sector. Future industry advancements should incorporate farming technologies, socio-economics, natural resource utilization, and governance in an integrated manner to attain sustainability. This vision for the future is ambitious and involves a comprehensive evaluation of environmental, social, economic, and legal factors. We suggest the following:

- Selective breeding procedures are started with a low inbreeding rate to produce domesticated and genetically modified seeds suitable for farming. Additionally, breeding programs are required for farming various species in diverse settings.
- Include breeding practices focused on achieving wide and long-term goals to maintain genetic variety and produce resilient farm animals with high welfare standards. Efforts to enhance breeding programs for disease and parasite resistance are being initiated and enhanced to secure the industry's survival. Molecular genetics applications such as Marker-Assisted Selection (MAS) and Genomic Selection (GS) could enhance the effectiveness of these programs. Efficient utilization of MAS and GS necessitates well-structured selective breeding programs.
- Public support or cooperative ownership and organization are essential to advocate for the solutions mentioned. Support and training are required to establish missing selective breeding programs in order to utilize advanced molecular techniques. Additionally, funding and training are necessary to enhance local capabilities in genetic engineering, biotechnology, and selective breeding initiatives.
- Patent rules and practice should steer clear of too broad patents, such as those related to viruses and bacteria, that impede further innovation.
- The introduction of GMOs and products like transgenic fish, DNA vaccines, and GM vaccinations should incorporate precautionary measures, including studying the potential detrimental consequences of inserted genes on animal health, welfare, and the environment. Additionally, it is essential to implement monitoring techniques to identify any unforeseen consequences. Identifying consumers' and public perceptions and attitudes towards emerging biotechnologies in aquaculture is essential. Research should also focus on determining their readiness to support and pay for advancements such as enhanced health and welfare of farmed fish.
- It is crucial to thoroughly explore the potential impacts of introducing biotechnology to the sector in order to identify how this technology might drive sustainable advances.

REFERENCES

1. Animal Welfare Act. In: *GOVERNMENT.NO, Information from the Government and the Ministries.* 10.08.2011, Available from http://www.regjeringen.no/en/doc/laws/Acts/animal-welfare- act.html?id=571188
2. Aquagen (2010). QTL-rogn - dokumentert IPN-beskyttelse fra første dag. In: *AquaGen Kunnskapsbrev.* No.1. (2010), p. 1, 19.06.2011, Available from: http://aquagen.no/filestore/01-2010QTL-rogn-dokumentertIPN-beskyttelsefrafrstedag.pdf
3. Bentsen, H.B., & Gjerde, B. (1994). Design of fish breeding programs. *Proceedings of the 5th World Congress of Genetics Applied to Livestock Production*, Vol.19, pp. 353-359, Guelph, Canada, August 1994.
4. Bentsen, H.B. & Olesen, I. (2002). Designing aquaculture mass selection programs to avoid high inbreeding rates. *Aquaculture*, Vol.204, pp. 349-359 .
5. Bergfjord, O.J. (2009). Risk perception and risk management in Norwegian aquaculture, *Journal of Risk Research,* Vol.12, No.1, pp.91-104.
6. Broom, D.M. (1986). Indicators of poor welfare. *British Veterinary Journal,* Vol.142, pp. 524– 526.
7. Caffey, R.H., Kazmierczak, R.F., Romaire, R.P. & Avault, J.W. (1998). Indicators of aquaculture sustainability: a Delphi survey, *Book of Abstracts of The international triennial conference and exposition of the World Aquaculture Society, the National Shellfisheries Association and the Fish Culture Section of the American Fisheries Society*, pp.91, Las Vegas, USA, 1998.
8. CBD (2000). *Cartagena Protocol on Biosafety to the Convention on Biological Diversity.* (Montreal: Secretariat of the Convention on Biological Diversity). 16.12.2010, Available from http://bch.cbd.int/protocol/publications/cartagena-protocol-en.pdf
9. Devlin, R.H., Yesaki, T.Y., Donaldson, E.M., Du, S.J., Shao, J.D. & Hew, C.L. (1995). Production of germline transgenic Pacific salmonids with dramatically increased growth performance. *Canadian Journal of Fisheries and Aquatic Sciences,* Vol. 52, pp. 1376-1384.
10. European Commission (2002). *A strategy for the sustainable development of European aquaculture.* 29.4.2010, Available from http://ec.europa.eu/fisheries/cfp/aquaculture_processing/aquaculture/strategy_ en.htm.
11. FAO, (2009). Bartley, D.M., Benzie, J.A.H. et al.: *The Use and Exchange of Aquatic Genetic Resources for Food and Agriculture.* CGRFA Background Study Paper No. 45, FAO, Rome

12. Fleming, I.A. & Gross, M.R. (1992). Reproductive-behavior of hatchery and wild coho salmon (oncorhynchus-kisutch) - does it differ. *Aquaculture*, Vol.103, pp. 101-121.

13. Gaskell, G. & Bauer, M.W. (Eds.) (2001). *Biotechnology, 1996–2000: The years of controversies*, Science Museum Press, London

14. Gaskell, G., Allum, N. & Stares, S. (2003). *Europeans and Biotechnology in 2002: Eurobarometer 58.0.* A report to the EC Directorate General for Research from the project Life Sciences in European Society. March 21, 2003 (2nd edition).

15. Gillund, F., Kjølberg, K., Krayer von Krauss, M. & Myhr, A.I. (2008a). Do uncertainty analyses reveal uncertainties? Using the introduction of DNA vaccines to aquaculture as a case. *Science of the Total Environment* , Vol.407, pp. 185-196.

16. Gillund, F., Tonheim T., Seternes T., Dalmo R.A. & Myhr A.I. (2008b). DNA vaccination in aquaculture –Expert judgements of impact on environment and fish health, *Aquaculture*, Vol. 284, pp. 25-34.

17. Gillund, F. & Myhr, A.I. (2010). Perspectives on salmon feed: A deliberative assessment of several alternative feed resources. *Journal of Agricultural and Environmental Ethics*, Vol. 23, pp. 527-550.

18. Gjedrem, T., Robinson, N., Rye, M. (2011). The importance of selective breeding in aquaculture to meet future demands for animal protein: A review. *Aquaculture.* Submitted

19. Gjerde, B., Gjøen, H.M. & Villanueva, B. (1996). Optimum designs for fish breeding programmes with constrained inbreeding. Mass selection for a normally distributed trait. *Livestock Production Science*, Vol.47, pp. 57-72.

20. Gjøen, H.M. & Bentsen, H.B. (1997). Past, present and future of genetic improvement in salmon aquaculture, ICES Journal of Marine Science, Vol.54, pp. 1009-1014

21. GMO Compass (2009). Opposition decreasing or acceptance increasing?, 28.5.2011, Available from http://www.gmo- compass.org/eng/news/stories/415.an_overview_european_consumer_polls_attit udes_gmos.html

22. Grajal, A. (1999). Biodiversity and the Nation State: Regulating Access to Genetic Resources Limits Biodiversity Research in Developing Countries. *Conservation Biology*, Vol.13, No.4, pp. 6-10.

23. Hayes, B. & Andersen, Ø. (2005). Modern biotechnology in aquaculture. In: *Selection and Breeding Programmes*, T. Gjedrem (Ed.), pp. 301-317, Springer, ISBN -10 1-4020-3341- 9, Dordrecht.

24. Haavind Vislie advokatfirma. (2008). Letter of 19 June to Norwegian competition authorities.

25. Hew, C.L. & Fletcher, G.L. (2002). The role of aquatic biotechnology in aquaculture. *Aquaculture*, Vol.197, pp. 191-204.

26. Holmenkollen Guidelines (1999). Holmekollen Guidelines for sustainable aquaculture 1997. In: *Sustainable Aquaculture*, N. Svennevig, H. Reinertsen, M. New (Eds.), pp. 343- 347, Proceedings of the Second International Symposium on Sustainable Aquaculture: Food for the future? Oslo, Norway, November, 1997, AA. Balkema, Rotterdam/Brookfield

27. Hyperdictionary, (2003). *hyperdictionary.* Available from: http://www.hyperdictionary.com.

28. Huntingford, F.A., Adams, C., Braithwaite, V.A., Kadri, S., Pottinger, T.G., Sandøe, P. & Turnbull, J.F. (2006). Current issues in fish welfare. *Journal of Fish Biology*, Vol.68, pp. 332-372.

29. Johnsson, J.I., Petersson, E., Joensson, E., Bjoernsson, B.T. & Jaervi, T. (1996). Domestication and growth hormone alter antipredator behavior and growth patterns in juvenile brown trout, Salmo trutta. *Canadian Journal of Fisheries and Aquatic Sciences*, Vol.53, pp. 1546-1554

30. Kamara, M., Coff, C. & Wynne, B. (2006). *GMOs and Sustainability: Contested visions, routes and drivers.* Report prepared for the Danish Council of Ethics, Copenhagen.

31. Kapuscinski, A.R. & Brister. D. J. (2001). Genetic impacts of aquaculture. In: *Environmental Impacts of aquaculture*, K.D. Black (Ed.), pp. 128-153, Sheffield, UK, Sheffield Academic press.

32. Kibenge, F.S.B., Garate, O.N., Johnson, G., Arriagada, R., Kibenge, M.J.T. & Wadowska, D. (2001). Isolation and identification of infectious salmon anaemia virus (ISAV) from Coho salmon in Chile. *Diseases of Aquatic Organisms* , Vol.45, pp. 9–18.

33. Koester, V. (1997). The Biodiversity Convention Negotiation Process and Some Comments on the Outcome. *Environmental Policy and Law,* Vol. 27, No.3, pp. 175-192.

34. Kolstad, K., Heuch, P.A., Gjerde, B., Gjedrem, T. & Salte R. (2005). Genetic variation in resistance of Atlantic salmon (Salmo salar) to the salmon louse *Lepeophtheirus salmonis. Aquaculture,* Vol. 247, pp. 145-151.

35. Kvakkestad, V., Gillund, F., Kjølberg, K. & Vatn, A. (2007). Scientists' perspectives on the deliberate release of GM crops. *Environmental Values*, Vol.16, pp. 79-104.

36. Kwang, J. (2000). Fishing for vaccines. *Nature Biotechnology*, Vol.18, pp. 1145-1146.

37. Laird, S. & Wynberg, R. (2005). *The Commercial Use of Biodiversity: An Update on Current Trends in Demand for Access to Genetic Resources and Benefit-Sharing, and Industry Perspectives on ABS Policy and Implementation*. UNEP/CBD/WGABS/4/INF/5.

38. Le Curieux – Belfond, O., Vandelac, L., Caron, J. & Seralini, G.E. (2009). Factors to consider before production and commercialization of aquatic genetically modified organisms: the case of transgenic salmon. *Environmental Science and Policy*, Vol.12, pp. 170-189.

39. Lund, V., Mejdell, C., Röcklinsberg, H., Anthony, R. & Håstein, T. (2007). Expanding the moral circle: farmed fish as objects of moral concern. *Diseases of Aquatic Organisms*, Vol.75, pp. 109-118.

40. Marine Harvest. (2011). *Salmon, a small protein source, but a great alternative in the global market place*. Presentation held by A. Aarskog at the North Atlantic Seafood Conference in Oslo 3.3. 2011, Available from http://hugin.info/209/R/1494191/430048.pdf

41. Mazur, C.F. & Iwama, G.K. (1993). Effect of handling and stocking density on hematocrit, plasma-cortisol, and survival in wild and hatchery-reared chinook salmon (oncorhynchus-tshawytscha). *Aquaculture,* Vol.112, pp. 291-299

42. McGinnity, P., Stone, C., Taggart, J.B. Cooke, D., Cotter, D., Hynes, R. et al. (1997). Genetic impact of escaped farmed Atlantic salmon (Salmo salar L.) on native populations: use of DNA profiling to assess freshwater performance of wild, farmed, and hybrid progeny in a natural river environment. *ICES Journal of Marine Science,* Vol.54, pp. 998-1008

43. Melamed, P., Gong, Z., Fletcher, G. & Hew, C.L. (2002). The potential impact of modern biotechnology on fish aquaculture. *Aquaculture*, Vol. 204, pp. 255-269.

44. Metcalfe, N.B., Valdimarsson, S.K. & Morgan, I.J. (2003). The relative roles of domestication, rearing environment, prior residence and body size in deciding territorial contests between hatchery and wild juvenile salmon. *Journal of Applied Ecology*, Vol.40, pp. 535-544.

45. Meuwissen, T.H.E., Hayes, B.J. & Goddard M.E. (2001). Prediction of total genetic value using genome-wide dense marker maps. *Genetics,* Vol.157, pp. 1819-1829.

46. Moen T., Baranski M., Sonesson A.K. & Kjøglum, S. (2009). Confirmation and fine-mapping of a major QTL for resistance to infectious pancreatic necrosis in Atlantic salmon (Salmo salar): population-level associations between markers and trait. *BMC GENOMICS,* Vol.10, pp. 368.

47. Myhr, A.I. & Traavik, T. (2011). *Genetically Engineered Virus Vaccine Vectors: Environmental Risk Management Challenges*, In: Genetic Engineering, InTech, ISBN 978-953-307-671- 3.

48. Niiler, E. (2000). FOA researchers consider first transgenic fish. *Nature Biotechnology*, Vol. 18, pp. 143.

49. Norwegian Ministry of Fisheries and Costal Affairs (2009). *Strategi for en miljømessig bærekraftig havbruksnæring.* 10.07.2009, Available from http://www.regjeringen.no/...strategier.../strategi-for-en-miljomessig-barekraftig- .html.

50. Olesen, I., Groen, A.F. & Gjerde, B. (2000). Definition of animal breeding goals for sustainable production systems. *Journal of Animal Science*, Vol.78, pp. 570-582.

51. Olesen, I., Rosendal, G.K., Bentsen, H.B., Tvedt, M.W. & Bryde, M. (2007). Access to and protection of aquaculture genetic resources - Strategies and regulations. *Aquaculture.* Vol.272, No.1, pp. S47-S61

52. Olesen, I., Alfnes, F., Røra, M., Navrud, S. & Kolstad, K. (2010). Eliciting consumers' willingness to pay for organic and welfare labelled salmon by a non-hypothetical choice experiment. *Livestock Science*, Vol.127, pp. 218-226.

53. Olesen, I., Myhr, A.I. & Rosendal, G.K. (2011). Sustainable aquaculture: are we getting there? Ethical perspectives on salmon farming. *Journal of Agricultural and Environmental Ethics.* Vol.24, No. 4, pp. 381-408.

54. Petersson, E. & Järvi, T. (2006). Anti-predator response in wild and sea-ranched brown trout and their crosses. *Aquaculture*, Vol.253, pp. 218-228.

55. Ponzoni, R.W., Khaw. H.L. & Yee, H.Y. (2010). GIFT: the story since leaving ICLARM (now WorldFish Centre) – socioeconomic, access and benefit sharing and dissemination aspects. *FNI Report*, 14/2010, Oslo, Norway

56. Ramanna Pathak, A. (fortcoming) Balancing biodiversity, access to genetic resources and profits in India's shrimp sector.

57. Rauw, W.M., Kanis, E., Noordhuizen-Stassen, E.N. & Grommers, F.J. (1998). Undesirable side effects of selection for high production efficiency in farm animals. A review. *Livestock Production Science*, Vol.56, pp. 15-33.

58. Refstie, T. & Gjedrem, T. (2005). Chromosome Engineering. In *selection and Breeding, Programmes, T.* Gjedrem (Ed.), pp 287-299, Springer, ISBN -10 1-4020-3341-9, Dordrecht

59. Robinson, N. (2002). Techniques for advanced genetic improvement of fin fish and shellfish in aquaculture. *Report on an NRE scientific exchange visit to aquaculture and marine science institute in Scotland and Norway*, September 2002. Australia, Victorian Institute of Animal Science

60. Romøren, K. (2003). Liposome- and Chitosan-based formulations used for gene delivery to fish: -*in vivo* and *in vitro* studies of possibilities and limitations. Ph.D. dissertation, ISSN 1501-7710. Faculty of Mathematics and Natural Sciences, Department of Pharmaceutics, School of Pharmacy, University of Oslo, Norway

61. Rosendal, G.K. (2000). *The Convention on Biological Diversity and Developing Countries.* Dordrecht: Kluwer Academic Publishers

62. Rosendal, G. K., Olesen, I., Bentsen, H.B., Walløe Tvedt, M. & Bryde, M. (2006). Access to and legal protection of aquaculture genetic resources – Norwegian perspectives. *Journal of World Intellectual Property,* Vol. 9, No.4, pp. 392-412.

63. Rosendal, G.K. (2006). Regulating the Use of Genetic Resources – Between International Authorities. *European Environment*, Vol.16, No. 5, pp. 265-277.

64. Ruzzante, D.E. (1994). Domestication effects on aggressive and schooling behavior in fish. *Aquaculture*, Vol.120, pp. 1-24.

65. SalmoBreed. (2011). SalmoBreed AS har identifisert genetisk markør for sykdommen pancrease disease (PD). In: *Nyhetsbrev fra SalmoBreed*. Vol.7, No.3, (June 2011), p.1, 19.06.2011, Available from: http://www.salmobreed.no/files/Nyhetsbrev/nyhetsbrev_2011/2011_3_nyhetsbr ev_SalmoBreed_-_Kopi.pdf

66. Sandøe; P. & Holtug, N. (1998). Ethical aspects of gene and biotechnology. *Acta Agriculturae Scandinavia, Section A: Animal Science,* Suppl. 29, pp. 51-58.

67. Sandøe, P. & Christensen, S.B. (2008). *Ethics of animal use.* Blackwell Publishing, ISBN 978-1- 4051-5120-7, Oxford, UK

68. Sonesson, A.K., Meuwissen, T.H.E. & Goddard, M.E. (2010). The use of communal rearing of families and DNA pooling in aquaculture genomic selection schemes. *Genetics Selection Evolution,* Vol.42, No.41, pp. 9

69. Thompson, G.D. & Kidwell, J. (1998). Explaining the choice of organic produce: Cosmetic defects, prices, and consumer preferences. *American Journal of Agriculture Economics,* Vol.80, pp. 277–287.

70. Traxler, G.S., Anderson, E., LaPatra, S.E., Richard, J., Shewmaker, B. & Kurath, G. (1999). Naked DNA vaccination of Atlantic salmon, *Salmon salar* against IHNV. *Diseases of Aquatic Organisms*, Vol. 38, pp. 183–190.

71. Turnbull, J.F., Adams, C.E., Richards, R.H. & Robertson, D.A. (1998). Attack site and resultant damage during aggressive encounters in Atlantic salmon (Salmo salar L.) parr. *Aquaculture*, Vol.159, pp. 345-353. Verhoog, H. (2001). The intrinsic value of animals: its implementation in governmental regulations in the Netherlands and its implications for plants. *Proceedings of Ifgene workshop Intrinsic value and integrity of plants in the context of GE.* pp. 15-18, Dornach, Switzerland, May 9-11, 2001

72. Walls, J., Tee Rogers-Hayden, A.M. & O'Riordan, T. (2005). 'Seeking citizens' views on GM crops: Experiences from the United Kingdom, Australia and New Zealand', *Environment,* Vol.47, No.7, pp. 23–36

73. WCED (1987). *Our Common Future*. Oxford: Oxford University Press.

74. White Paper. (2005). (Stortingsmelding) *Den blå åker (The blue field)* No. 19. 2004-2005.

75. White Paper. (2009). (Stortingsmelding) *Interesser, ansvar og muligheter. Hovedlinjer i norsk utenrikspolitikk (Interests, responsibilities and options. Main avenues in Norwegian foreign politics)* No.15. 2008-2009.

76. Yue, C., Alfnes, F. & Jensen, H.H. (2009). Discounting spotted apples: Investigating consumers' willingness to accept cosmetic damage in an organic product. *Journal of Agriculture & Applied Economics* Vol.41, pp. 1-18

15 OMICS METHODOLOGIES: NEW TOOLS IN AQUACULTURE STUDIES

1. INTRODUCTION

According to the FAO, a growing percentage of world aquatic production is derived from aquaculture, whose importance is increasing dramatically due to commercial overfishing and a growing demand for seafood (FAO, 2010). In 1980, aquaculture production represented 9% of fishery resources; by 2010, it had increased to 43%. It is thought that such a production will need to double in the next 25 years. The FAO is promoting aquaculture because it is an important source of income and employment and also because of its great contribution to food security and the development of many countries. Currently, there are three main challenges for developing productive, feasible and sustainable aquaculture: 1) diversification of the proteins used for the feeds, 2) resolution of problems derived from stressful conditions, diseases and/or deterioration of environmental conditions, and 3) introduction of new species to make this industry less vulnerable to market demand (COM, 2002). The Senegalese sole (*Solea senegalensis*) is a flatfish species with a high potential for use in marine aquaculture diversification. The cultivation of sole has been successful under several husbandry conditions, but the frequent occurrence of opportunistic diseases and its high sensitivity to different stressors, such as manipulation, pollutants, etc., make sole unable to be produced industrially (Cañavate, 2005; Dinis et al., 1999). Consequently, the identification of biomarkers responsive to pathological situations and pollutants will help to prevent health problems and to improve their farming.

Biomarkers provide evidence of alterations by physiological or environmental conditions (López-Barea, 1995a). The so-called "classic" biomarkers are suggested *a priori* by virtue of their biological roles but are rather biased because they concentrate on a small number of proteins, excluding others that are also altered in the same conditions but whose relationship with the physiological or environmental changes is unknown (López-Barea & Gómez-Ariza, 2006). In 1989, a group at the University of Cordoba (UCO) began to develop a battery of biomarkers sensitive to physiological or environmental changes in several bioindicator species, including bivalves, crustaceans, fish, mammals and mammalian cell lines. A variety of biochemical parameters were included, such as phase I (ethoxyresorufin- *O*-deethylase, EROD) or phase II biotransforming enzymes (GSH transferase, GST), antioxidative defences (superoxide dismutases, SOD; catalase, CAT; glutathione peroxidases, GSHPx, glucose-6P and 6P-gluconate dehydrogenases, glutathione reductase, GSSGrase), neurotransmission-linked esterase activities, such as acetylcholine (AcChE) and carboxyl esterases (CbE), oxidative damages to biomolecules, including DNA (8-oxo-dG), proteins (protein-SSG mixed disulphides), lipids (malondialdehyde, MDA), and the glutathione content and redox status (total glutathione, GSSG/GSH). The UCO group also developed new biochemical indicators that are altered by physiological or environmental changes, such as the levels of individual GST and SOD isoenzymes, the activation of promutagens to genotoxins by exposure to extracts of reference or exposed animals –a global measure of biotransforming capacity– and the metallothionein (MT) levels using a new and extremely sensitive HPLC-based fluorescent assay.

The utility of these "classic" biochemical biomarkers was later validated by the UCO group in studies carried out preferentially in natural sites in Spain, Slovakia and Tunisia, and contrasted with experimental exposures to model contaminants carried out under controlled conditions. These studies were reported in the following publications, limited in this review to those made in fish, and listed here by their date of publication: Rodriguez-Ariza et al. (1992, 1993, 1994a, 1994b), Martínez-Lara et al. (1992, 1996, 1997), Pedrajas et al. (1993, 1995, 1998), López-Barea (1995b), Lenartova et al. (1997), López-Barea & Pueyo (1998), Cousinou et al. (1999, 2000), Alhama et al. (2006, 2010), Romero-Ruiz et al. (2003, 2008), and Jebali et al. (2008). These "classic" biochemical biomarkers also responded to physiological changes, including oxidative alterations promoted by different feeding schemes, as described in Pascual et al. (1995a, 1995b, 1997, 2003) and Cánovas-Conesa et al. (2007).

While genes typically exert their functions at the protein level, genetic responses to stress are often regulated at the transcriptional level. Therefore, the determination of transcriptional profiles has become an essential approach in understanding the coordinated gene response to various physiological and pathological variables. The construction of cDNA libraries by *suppression subtractive hybridization* (SSH) (Prieto-Álamo et al., 2009; Williams et al., 2003) is a fundamental methodology used in differential expression studies with non-model species because it enables the identification of genes with no previous knowledge of their sequences. SSH is a PCR-based technique for generating cDNAs enriched in differentially expressed genes, useful for large-scale gene identification in non-model organisms (Diatchenko et al., 1996). Unlike SSH, which only provides qualitative results, *DNA microarrays* give semiquantitative (fold-variation) data, and more importantly, permit, in a single experiment, the analysis of the levels of thousands of transcripts, making them a valuable high-throughput methodology in Functional Genomics. Moreover, heterologous hybridization allows the use of microarrays made from transcripts of one species to probe gene expression in other related species. *Real-time qRT-PCR* has become a reference method to detect and quantify transcripts and to validate the results obtained with other techniques such as subtractive libraries or microarrays.

The UCO team gained wide experience in quantifying changes occurring at the mRNA level by RT-PCR. Of relevance is the devise of new approaches for the quantification of the exact number of transcript molecules and their application to a wide variety of organisms and conditions. This team developed, validated and optimised relative quantifications using complex multiplexed RT-PCR (Gallardo-Madueño et al., 1998; Manchado et al., 2000; Michan et al., 1999; Monje-Casas et al., 2001; Prieto-Álamo et al., 2000; Pueyo et al., 2002) and absolute quantification by real-time RT-PCR (Jiménez et al., 2005; Jurado et al., 2003, 2007; Montes-Nieto et al., 2007; Prieto-Álamo et al., 2003, 2009). They also developed a quantitatively rigorous approach based on a combination of multiplexed and real-time RT-PCR to increase the number of transcripts to be quantified simultaneously without compromising the sensitivity, reliability and repetitiveness of the absolute measurements (Jurado et al., 2003; Michan et al., 2005; Monje-Casas et al., 2004; Ruiz-Laguna et al., 2005, 2006). These studies have demonstrated the potential benefits of absolute transcript quantifications in studies of tissue-specific expression profiles (Jurado et al., 2003, 2007; Prieto-Álamo et al., 2003, 2009; Ruiz-Laguna et al., 2005), of changes associated with growth stages or with the age or sex of an individual and have been particularly useful in studies with free-living animals (Jiménez et al., 2005; Michan et al., 2005; Monje-Casas et al., 2004; Prieto-Álamo et al., 2003; Ruiz-Laguna et al., 2005, 2006). We have demonstrated that the main drawback of relative quantifications is the variability of most popular internal standards. By comparing the differences in the transcript molecules with the conventional fold variations, we have also shown that relative quantifications grossly overestimate changes affecting poorly transcribed genes in comparison with highly abundant mRNAs.

Proteomics addresses the post-genomic challenge of examining the entire complement of proteins (proteome) expressed by a genome in a cell, tissue or organ at a given time under defined conditions (James, 1997). Protein expression is modulated at different levels from transcription to the maturation of the polypeptides produced by the translation of mature mRNAs. Proteins were initially separated by *two-dimensional electrophoresis* (2-DE; Wilkins et al., 1996), and their expression was analysed by 2D software (Melanie, etc.). Proteins were identified by mass spectrometry analysis of their peptide mass fingerprint (MALDI-TOF- PMF) or *de novo* sequencing of some peptides (nESI-MS/MS), comparing the results with public databases (Simpson, 2003). 2-DE, which is labour-intensive and has low reproducibility, requires a large amount of sample, and its narrow dynamic range is problematic with proteins of extreme Mr/pI. Shotgun proteomic methods allow the analysis of complex protein mixtures after full digestion by *multidimensional separation* coupling tandem liquid chromatography (LC/LC) and MS/MS (Washburn et al., 2001). The application of proteomic technology faces the problem of the lack of genomic information on most non-model sentinel organisms. This makes it difficult to identify differentially expressed proteins by high-throughput methods such as MALDI-TOF-PMF (López-Barea & Gómez-Ariza, 2006).

2. CONVENTIONAL AQUACULTURE STUDIES WITH SOLEA SENEGALENSIS

2.1 Early studies

Studies of sole aquaculture began in Faro (S Portugal) and Cádiz (SW Spain) to produce good quality larvae and juveniles (Dinis et al., 1999). Broodstock spawning studies established optimal feeding regimes by combining squid (*Loligo vulgaris*) and polychaetes (*Hediste diversicolor*) at the final maturation stages. Spawning was studied in terms of temperature (stopped <16 ºC), duration (4-6 months), egg fertilisation rate (20-100%) and viable egg rate (72%). Larvae hatch at 2.4 mm and accept *Artemia nauplii* as the first prey two days after hatching (DAH). Metamorphosis spans from 11 to 19 DAH, at which point the fish are fed live *Artemia metanauplii*. They reach 16 mm at 40 DAH and 35 cm/450 g after 1 year, with 8% survival. Pasteurellosis can cause pigmentation abnormalities and malformations associated with eye migration and can progress to death (Dinis et al., 1999).

The potential of sole for aquaculture was reviewed some time later (Cañavate, 2005). Although important progress in reproduction techniques was reached, much basic knowledge remained lacking. Ongrowth was successfully carried out, but progress was limited by opportunistic diseases due to suboptimal rearing conditions resulting in an inability of the sole to achieve an adequate physiological status for resistance. Growth, survival and pigmentation were studied during sole growth in tanks with three bottom types (Rodiles et al., 2005). The final length and weight was similar in the sand, white and dark conditions, but different pigmentation patterns appeared on the sand (clear, dark) and white bottoms (clear, brown, dark). The homogeneous dark pattern, preferred by markets, is only obtained in tanks with a dark bottom. A lower survival rate was found on sand bottoms due to pathologies derived from the difficulties in maintaining the sand bed.

2.2 Organ development and reproductive studies

Digestive tract development was studied in larvae until 30 DAH, which involved the assessment of histology, digestive enzymes, lipids, proteins and carbohydrates in the buccopharyngeal cavity, oesophagus, early stomach, anterior and posterior intestine, pancreas and liver (Ribeiro et al., 1999a). The digestive tract elongates in metamorphosis, increasing absorption. Phosphatases, lipase and aminopeptidase have been detected starting at 2 DAH and the levels increase during development. Proteins abound in the intestinal epithelium and exocrine pancreas, and neutral lipids are found at the yolk sac intestinal epithelium and liver. After 31 DAH larvae ingest, digest and absorb nutrients because they now have a complete digestive tract. A time course of pancreatic and intestinal enzymes was studied in larvae until 31 DAH (Ribeiro et al., 1999b). Digestive enzymes increase until 10 DAH then decrease until 18 DAH, a pattern typical of developing animals. Alkaline phosphatase abounds from 21-27 DAH, during the development of brush border membranes, with a parallel decrease in the cytosolic enzyme, Leu–Ala peptidase.

Thyroid development was studied in sole larvae by histo- and immunohistochemistry to synchronise larval development and improve fish production (Ortiz-Delgado et al., 2006). The first follicle is visible by the first feeding; increases during metamorphosis and has adult characteristics by 30 DAH. Thyroid hormones decrease to undetectable levels at yolk-sac reabsorption. T3 and T4 are detected by 6 DAH and increase during metamorphosis.

Seasonal profiles of sex steroids –17β-estradiol (17β-E), testosterone (T), 11-ketotestosterone (11-KT), and 17,20β-dihydroxy-4-pregnen-3-one (17,20β-P)– were studied in *S. senegalensis* in an attempt to achieve steroid-induced maturation (García-López et al., 2006a). Females have six maturation stages, as follows: early, intermediate and final ovarian development, then partially, mid and spawned out. By summer´s end, a new gonadal cycle starts, as demonstrated by increased reproductive parameters. By mid-autumn some females reach advanced maturation stages, which coincide with a peak of running males. By the start of spring, ovarian development reaches its peak, and plasma steroid levels are maximal at the start of the spawning period, which occurs from March to June. In parallel with oocyte and sperm release, the proportion of spawned out fish and non-running males increases, and steroid levels decline. The high levels of 17,20β-P during spawning make it a candidate for a maturation-inducing steroid.

Testicular development was also studied (García-López et al., 2006b). The spermatogenetic cycle consists of the following five stages: early (I), mid (II), and late (III) spermatogenesis, maturation (IV), and recovery (V). In the summer, stage I and V testes are found with low values of sex steroids and IG (gonadosomatic index). Recrudescence begins in autumn, with an initial increase of IG 11-KT and

T and the appearance of stage II and III testes. In the winter, 11-KT and T peak and soon decrease, and IG slightly declines. In the spring, 11-KT and T decline further, while IG slightly increases and running males peak with stage IV testes. Sperm production and quality was assessed in wild-captured and F1 broodstock fish (Cabrita et al., 2006). Males produce motile sperm from February to November, with specific peaks of high spermiation and fluent males. Sperm volume and cell density is lower in F1 males than in wild-captured broodstock.

Ovarian development was also studied (García-López et al., 2007). In the autumn/winter, oocytes progress to vitellogenic stages in parallel with high levels of K (condition factor), IG, and plasma 17β-E and T. In the late winter/early spring, development is maximal, with females at intermediate and final maturation and K, IG, 17β-E and T peaking. Steroid levels are lower in cultured sole than in naturally spawning females, leading to atresia and lack of oocyte maturation, thus reducing ovary size with declining K, IG, and 17β-El and T levels and many perinucleolar oocytes. The amount of circulating 17,20β-P, the putative maturation-inducing steroid, remains near constant through the period, suggesting that oocytes are unresponsive to its stimulation.

Skeletal development and malformations are a bottleneck in sole aquaculture. Maturation and abnormalities of the vertebral column and caudal skeleton have been studied in sole (Gavaia et al., 2002). Different defects are found in the caudal complex and the vertebral column, and 44% of fish show at least one defect. While the causes are unknown, their high incidence may reflect rearing and/or feeding problems. The tissue distribution and evolution of bone Gla (Bgp) and matrix Gla proteins (Mgp) and Ca2+ deposition were studied in zebrafish during larval development and in adult tissues as well as sole metamorphosis (Gavaia et al., 2006). In zebrafish, Bpg and Mpg accumulate mainly in the matrix of skeletal structures already calcified or under calcification. In sole metamorphosis, Bpg and Mpg increase in parallel to the calcification of the axial skeleton. In both species, Mpg also accumulates in non-mineralised vessel walls.

2.3 Nutrition studies

Studies on the requirements, catabolism and assimilation of amino acids (AAs) were carried out in early larval, metamorphic and post-larval sole. Initial studies on indispensable (IAA) and dispensable (DAA) amino acids (Rønnestad et al., 2001), showed that sole assimilated most (85%) of the dietary IAAs and catabolised most of the DAAs. Such results were confirmed after studying the bioavailability of several AAs in larvae (Conceição et al., 2003). The demand and availability of AAs and proteins in relation to digestive capacity were reviewed, and AAs sources were described, highlighting the regulatory role of cholecystokinin and peristaltic activity (Rønnestad et al., 2003). A balanced AA profile improved amino acid assimilation in post-larval sole (Aragão et al., 2004a). Changes in AA requirements and dietary imbalances were studied in *Sparus aurata* and *S. senegalensis* (Aragão et al., 2004b); the AA profiles of both changed during ontogeny, especially in sole due to its marked metamorphosis. AA imbalances were found during development. In both species, Phe/Tyr addition was studied to assess the effects on metamorphosis after their conversion into thyroid hormones (Pinto et al., 2010). While Phe did not affect sole metamorphosis, dietary Tyr increased the production of thyroid hormones, which was beneficial for sole metamorphosis.

The nutritional physiology of sole development was studied to optimise diets and understand limiting factors in weaning (Conceiçao et al., 2007). Larvae have a high capacity to digest live prey, even at the early stages. Use of inert microdiets in co-feeding with *Artemia* resulted in the development of intestinal activity and enhanced survival, although it was also accompanied by low growth and high size dispersal. Fatty acid absorption increases with their degree of unsaturation, and larvae spare DHE from catabolism. Rotifers and *Artemia* are deficient in one or more AAs, such as His, Lys, Arg, Thr, or those containing sulphur and aromatic rings, depending on the larval stage; balancing the dietary AA profile with dipeptides increases retention and decreases catabolism in *Artemia*-fed larvae.

The effects of non-protein energy levels on growth and oxidative status were studied in sole fed diets with 4 energy levels (Rueda-Jasso et al., 2004). Cellular energy allocation showed differences in liver, but not in muscle. TBARS were higher in fish fed a diet with high lipid content, in parallel to high CAT and SOD activity. Yet, the protein source or energy levels had no major impact on sole growth, nutrient utilisation or fatty acid composition (Valente et al., 2011). Quantitative lipid imbalances and a low protein/neutral lipid ratio increased the accumulation of lipid droplets in the enterocytes and lowered fatty acid absorption in larvae (Morais et al., 2005). The effects of a neutral lipid level and source were studied in marine fish larvae (Morais et al., 2007). A growth-depressing effect of high neutral lipids, as assessed by lower digestive enzyme activity, absorption and/or food intake was reported. In larvae, lipid transport from enterocytes to the body is more critical than lipolytic activities. Phospholipid digestion is more efficient than that of neutral lipids, whose excess leads to the accumulation of large lipid droplets in the enterocytes, reducing fatty acid absorption and growth.

The feed transit, protein and energy digestibility of practical feed were assessed in sole (Dias et al., 2010). Protein digestibility is high for fishmeal and corn gluten, intermediate for soybean meal, and moderate for wheat meal. Energy digestibility varies from 88 to 93% for soybean meal, corn gluten and anchovy fishmeal and is 73% in wheat meal. Thus, flatfish, despite its high dietary protein requirement, digest vegetable ingredients quite well, suggesting that the development of practical feeds with high levels of plant-protein sources would be beneficial.

2.4 Conventional biomarker studies in *S. senegalensis* aquaculture

Nearly 15 years after the studies of the UCO group, the use of biomarkers to follow fish physiology and pollution effects has become popular and is now applied by most groups. Antioxidant enzymes, stress proteins, lipid peroxides and histology were studied in sole larvae (Fernández-Díaz et al., 2006) fed on the following 3 diets: live *Artemia nauplii*, microcapsules, and vitamin A-supplemented microcapsules. Live-fed larvae grow larger and undergo faster metamorphosis than microcapsule-fed larvae, although all groups have near 80% survival. Vitamin A improves the growth and development compared to an inert diet. *Artemia*-fed larvae have organs with normal development, but histological alterations are seen in larvae fed an inert diet. Catalase (CAT), superoxide dismutase (SOD), total GSH- peroxidase (t-GPX), lipid peroxides (MDA) and stress proteins (HSP70, not HSP60) are diet- and age-dependent. Inert diet-fed larvae have similar biomarker responses, but different ($p<0.05$) from *Artemia*-fed larvae. Higher antioxidant defences are attributed to the start of metamorphosis and the use of inert food.

A similar approach was used by Cañavate et al. (2007) to assess the effect of light on the development of sole larvae with or without adding β-carotene-rich *Dunaliella salina*. SOD, CAT, t-GPX and MDA were used as biomarkers. Growth and survival after metamorphosis were unaffected by light or *D. salina*. Light affects CAT and t-GPX throughout development but does not affect MDA, and SOD is only affected in metamorphosis. *D. salina* does not affect SOD, CAT or t-GPX and no interaction with light intensity was found. MDA lowers significantly only when *D. salina* is added, and its effect was found only in metamorphosing larvae, whose MDA levels are much higher than in earlier stages. These results confirm the antiperoxidative effect of β-carotene from live algae in the larval rearing process.

2.5 The Pleurogene project

After the initial studies on the growth and reproduction of the Senegalese sole, "Genoma España" and "Genome Canada" promoted the "Pleurogene" project, in order to develop new technology to assess gene and protein expression during the reproduction and breeding of two flatfish, Senegalese sole and Atlantic halibut. The project [http://www.gen- es.org] aimed to improve basic knowledge of reproduction, larval development and survival, and had de following objectives: 1) Establishment of an EST database and shotgun proteome analysis; 2) Construction of a microarray for high-throughput analysis of gene expression; 3) Construction of genetic linkage map; 4) Development of methods for gene expression profiling through laser capture microdissection RNA; 5) Identification of changes in gene expression during gamete development and maturation; 6) Genomic analysis of sex determination and differentiation; 7) Determination of the pattern of gene expression during larval metamorphosis, the ontogeny of the gastrointestinal tract and the effects of dietary treatments; and 8) Development of E-mold, an integrative bioinformatics platform for genomic, proteomic and morphological information from flatfish.

The Pleurogene project developed genomic and proteomic tools to help achieve these goals (Douglas et al., 2007; Cerdà et al., 2008), and also had the following major research results: 1) Development of genomics tools for the Senegalese sole, including 10 different cDNA libraries from adult and larvae tissues and 1 normalised multi-tissue library, 10,300 new sole EST sequences and nearly 500 peptides, and a sole oligonucleotide microarray with probes to detect 4550 different RNAs; 2) Development of a hormone treatment to increase sperm motility in sole; 3) Generation of a sole genetic linkage map; 4) Development of a progeny test or paternity kit; 5) Development of indirect approaches for sex control; 6) Production of recombinant gonadotropin hormone; 7) Generation of gene and protein expression maps associated to larval development, metamorphosis, and nutrition; 8) Generation of gene and protein expression maps of testes producing high quality sperm; 9) Generation of gene expression maps for sexual differentiation and maturation; and 10) Development of the "Solea-mold" that allows for new data to be included, *in silico* experiments to be performed, and comparative studies to be undertaken.

3. "CLASSIC" BIOMARKERS VERSUS "*OMICS*" METHODOLOGIES

3.1 Assessing pollution effects in fish

After the initial studies of the UCO group, the use of "classic" biomarkers became increasingly popular to assess the effects of pollutants on fish. One example was the "Prestige" oil spill in November 2002 off the Galician coast (NW Spain). Due to the heavy nature of the crude oil and its low solubility in sea water, its dispersion was low and it remained *in situ* in oil patches adhered to rocks and sediments. The biological effects of these oil patches were tested in *S. aurata* (Morales-Caselles et al., 2006), using MT levels, EROD activity and histology as biomarkers, without significant results. Biomarkers including CAT, GSHPx, GSSGrase and DT-diaphorase, AcChE, CbE, GST and MDA were also analysed in *S. senegalensis*, with modest alterations (Solé et al., 2008).

Antioxidative and phase II and III biotransforming enzymes were used to follow the effects of linear alkylbenzene sulphonates in sole via an *in vivo* continuous-flow assay (Alvarez- Muñoz et al., 2007) that was also used in *S. aurata* and *S. senegalensis* to assess the toxicity of sediments from littoral areas in northern and southern Spain (Jiménez-Tenorio et al., 2007), using MTs, EROD and histopathology. Jimenez-Tenorio et al. (2008) assessed the sediment toxicity caused in *S. senegalensis* by acute or chronic spills using histopathology, EROD and GST as biomarkers. Oxidative stress biomarkers and PAH contents were studied in Huelva soles near a petrochemical plant (Oliva et al., 2010). Significant correlations were found among the levels of GST, GPx and CAT, hepatic levels of PAH metabolites and PAH contents in sediments.

The effect of waterborne copper (Cu) was studied in *S. senegalensis* in static conditions (Fonseca et al., 2009) by assessing the following characteristics: biomarkers such as MTs or MDA, mass indices, and biochemical condition indices such as the RNA/DNA ratio and the lipid and protein content. Cu triggers a biomarker response and lowers growth and condition, without changes in morphometric indices. Decreased condition shows that lipid reserves enable fish to respond to toxicity and to maintain growth and protein synthesis, although with lower rates than control fish.

The effects of Mexel®432 and NaClO antifoulings were studied in *S. senegalensis* by assessing osmolality, Na+/K+-ATPase, stress, histology, oxidative damage, antioxidant defences and detoxification (López-Galindo et al., 2010a,b). NaClO increases plasma cortisol, glucose and lactate after an acute stress, with a later recovery. Gill GST and AChE are sensitive to NaClO. Hepatic markers initially respond to NaClO but longer exposures are toxic. Mexel®432 initially increases cortisol, which later returns to basal values, but glucose, lactate and triglycerides decresase. Gills have a lowered Na+,K+-ATPase activity, causing an imbalance in osmoregulation. Moderate changes are found in KAT, GSSGrase and GPX but not in MDA, GST or CbE. A multi-biomarker approach was used in *Dicentrarchus labrax*, *S. senegalensis* and *Pomatoschistus microps*, from the Aveiro and Tejo (Portugal) estuaries (Fonseca et al., 2011), which were affected by anthropic activities, without highly significant results.

Detection of DNA damage by biomarkers for genotoxicity was used in *S. senegalensis* (Costa et al., 2008a) exposed to sediments from three Sado Estuary (W Portugal) sites. The two blood parameters used were erythrocyte nuclear abnormalities (ENA) and DNA strand- breaks (DNA-SB). The levels of metals, PAHs, PCBs and DDTs were determined in the sediments. Scarcely polluted sediments are weaker inducers of genotoxic damage, whereas those under urban, industrial or agricultural influences significantly increase ENA and DNA-SB. A strong correlation exists between PAH and PCB content and genotoxicity, while metals have a weaker correlation. In a parallel study, *S. senegalensis* were exposed to Sado Estuary sediments (Costa et al., 2008b). Livers had more histological lesions than gills, and sediments contaminated by organics caused more damage to both organs than those contaminated by metals. Two "classic" biomarkers, MT and CYP1A, were also assessed. Lethality and biomarker responses do not linearly depend on the cumulative levels of contaminants but rather of their bioavailability and synergistic effects (Costa et al., 2009). In a parallel study, exposure to contaminated sediments induced DNA fragmentation and clastogenesis (Costa et al., 2011). Still, the most contaminated sediment revealed an antagonistic effect between metals and organics, enhanced by higher bioavailability. The laboratory assay caused a more pronounced increase in ENA, whereas a significant increase in DNA-SB exists in field-tested fish exposed to reference sediment.

3.2 The advent of "omics" methodologies

In contrast to the "classic" biomarker strategy, "*omics*" approaches, plus *in vitro* and *in silico* methods, are becoming a powerful multidisciplinary strategy. Their use is still at an early stage because most popular bioindicators are poorly represented in gene/protein sequence databases (Ruiz-Laguna et al., 2006; González-Fernández et al., 2008). *Omics* includes *genomics* to study DNA variations, *transcriptomics* for genome-wide characterisation of gene expression by measuring mRNAs, *proteomics* to assess the cell and tissue-wide expression of proteins, and *metabolomics* for global assessment of metabolite concentrations. These technologies provide detailed molecular information that helps to identify response pathways and to define mechanisms and modes of action without requiring previous knowledge. In 2003, the UCO group began a search for new biomarkers through the use of "omics" methods, which allowed the study of the Aznalcóllar spill and the status of Doñana National Park and its surroundings (SW Spain). These studies were collected in the following publications, listed here by publication date: Rodriguez-Ortega et al. (2002, 2003), Ruiz-Laguna et al. (2005, 2006), Bonilla-Valverde et al. (2004), López-Barea & Gómez-Ariza (2006), Romero-Ruiz et al. (2006), Montes-Nieto et al. (2007, 2010), Vioque-Fernández et al. (2007a, 2007b, 2009a, 2009b), González-Fernández et al. (2008), Abril et al. (2011), and Pueyo et al. (2011).

All organisms are adapted to certain extracellular salinity ranges. Osmoregulatory mechanisms are central to adaptation. Osmotic stress was studied in tilapia *Oreochromis mossambicus*, spiny dogfish shark *Squalus acanthias*, and an intertidal sponge *Tetilla mutabilis*, by genomics and proteomics methods (Kültz et al., 2007). SSH, RACE-PCR and proteomics allowed the identification of genes and proteins involved in adaptation to salinity or other environmental stresses. Algorithms based on sequence homology searches (MSBLASTP2) are powerful tools for protein identification. Gene ontology and pathway analysis can subsequently use identified genes and proteins for modelling molecular mechanisms of environmental adaptation. The dependence on information about biochemical pathways and gene ontology databases for model species is a severe barrier for work with non-model species. To minimise this dependence, focusing on a single biological process is key when applying "omics" methods to non-model organisms.

Environmental metabolomics allows for the characterisation of the metabolism of organisms from the natural environment and of those reared under laboratory conditions. Viant (2007) used this approach to characterise the responses of organisms to natural and anthropogenic stressors, discussing the challenges of measuring metabolites and highlighting the dynamic nature of the metabolome, whose variability is a challenge in environmental studies. The normal metabolic operating range (NMOR) is defined as the region in metabolic space in which 95% of individuals reside, and stress is a deviation from NMOR. The importance of genotypic and phenotypic anchoring (e.g., knowing species, gender, and age) is emphasised to facilitate the interpretation of multivariate metabolomics data.

An NRC-UK sponsored international consortium from government agencies, academia and industry in Canada, Japan, the UK, and the USA was carried out on fish toxicogenomics (Van Aggelen et al., 2010). The following three topics were addressed: progress in ecotoxicogenomics, perspectives on roadblocks for practical implementation of toxicogenomics into risk assessment, and dealing with variability in data sets. Although examples of successful application of "omic" technologies were identified, it is critical to perform studies that relate molecular changes to ecologically adverse outcomes. Although there are hurdles to pass on the road to regulatory acceptance, "omics" are already useful for elucidating modes of action of toxicants and can contribute to the risk assessment process as part of a weight-of-evidence approach.

A qRT-PCR approach was used to assess how *Lactobacillus rhamnosus* IMC 501 added to *Amphiprion ocellaris* larvae, alters development, and also to study the responses after probiotic exposure (Avella et al., 2010). Larvae and juveniles had 2-fold higher weight after probiotics were supplied. Metamorphosis occurs 3 days early, and factors involved in growth and development (I-l GF I/II, myostatin, PPAR α/β, vitamin D receptor α, and retinoic acid receptor γ) have higher gene expression. Probiotics lessen the severity of the general stress response as demonstrated by lower levels of glucocorticoid receptor and 70- kDa HSP expression. Improved development of the skeletal head was also found, with 10– 20% less deformities in probiotic-treated juveniles.

"Omics" have also been used in chemical screening and perturbation studies in zebrafish (Sukardi et al., 2010). Pharmacological efficacy and selectivity have been evaluated by chemical-induced phenotypic effects, although this has limitations in the identification of action mechanisms. "Omics" also facilitates the translatability of zebrafish studies across species by comparing conserved chemically induced responses. Thus, De Wit et al. (2010) characterised the estrogenic and metabolic effects of 17α-ethinylestradiol (EE2) in *D. rerio*, following the concern regarding the effects of endocrine-disrupting compounds. Oligo microarrays, with 3479 zebrafish-specific oligos, were used to generate differential gene expression levels, and proteomic responses were evaluated by DIGE and MALDI-TOF/TOF. Assessment of the differentially expressed transcripts and proteins showed that both individual platforms could profile clear estrogenic interference and multiple metabolism-related effects and stress responses. Cross-comparison of transcriptomics and proteomics datasets have limited concordance, but a revision of the results shows that transcriptional effects project at the protein level as downstream effects of the affected signalling pathways.

Public databases of coexpressed gene sets are valuable resources for many studies, including gene targeting for functional identification and investigations of regulatory mechanisms or protein–protein interactions. While coexpressed gene databases are highly popular in plant biology, those with animal data are limited due to the lower reliability of coexpression data. The COXPRESdb (http://coxpresdb.jp) represents the coexpression relationship in humans and mouse (Obayashi & Kinoshita, 2011). Updates focusing on enhancing the reliability of gene coexpression data in animals have been reported. A new coexpression measure, Mutual Rank, has been implemented, and five other animal species, such as the rat, chicken, zebrafish, fly and nematode, were included to assess the conservation of coexpression. In addition, different layers of "omics" data have been added into the integrated network of genes. Functioning as a gene network representation, COXPRESdb can help researchers to clarify the functional and regulatory networks of genes in a broad array of animal species.

4. TRANSCRIPTOMIC STUDIES IN *SOLEA SENEGALENSIS*

Cultivation of the Senegalese sole is hampered by its sensitivity to different stresses and infectious diseases that can cause high mortality. Consequently, there is a need to identify sole genes responsive to stress, infections and pollutants in order to improve productivity, management and fish welfare. Transcriptomic responses of sole stimulated with lipopolysaccharide (LPS), a mimetic of bacterial infections, and copper sulphate, a zoosanitary compound, were studied by different experimental and methodological approaches, such as SSH libraries, DNA microarrays and real-time qRT-PCR (Prieto-Álamo et al., 2009; Osuna-Jiménez et al., 2009).

4.1 SSH libraries

The construction of subtractive libraries in *S. senegalensis* allowed the identification of differentially expressed genes in response to LPS in the head-kidney, a hematopoietic and lymphoid organ involved in immune response and to CuSO4 in the liver, a central metabolic organ in xenobiotic detoxification and in the defence system (Prieto-Álamo et al., 2009). In both cases, forward (F) and reverse (R) libraries were designed to obtain clones of genes that were up- or down-regulated in response to LPS of CuSO4 relative to the PBS control. To offset inter-individual variations and temporal differences in the responses, the libraries were constructed with total RNA from pooled head-kidney or liver (≥ 10 fish/condition) of soles treated with LPS or CuSO4 for 6 and 24 h. Four hundred sixty clones were sequenced and the products of the ESTs were identified by comparison with the open access databases. A total of 222 unique sequences were detected, and 185 were identified as related to major physiological functions (Table 1).

A high percentage of identified ESTs were related to immune response (Figure 1). Their presence in a sole head-kidney library stimulated with LPS agreed with the immune role of this organ and the immunostimulating effects of LPS (Swain et al., 2008). The number of genes classified as being immune-related was even larger in the liver than in the head- kidney. Most of these immune-related ESTs coded for acute phase proteins (e.g., lysozyme, coagulation factors, proteinase inhibitors, complement components, and Fe transport/homeostasis proteins) according to several genomics studies indicating that, in teleost, the liver is an important source of immune transcripts (Ewart et al., 2005), mediating a powerful acute phase response (Bayne & Gerwick, 2001). In addition, these results supported the capacity of Cu (like other metals) to alter immunological competence, in agreement with a report showing Cu up-regulation of the cytokine TGF-β in striped bass (Geist et al., 2007).

Table 1. Characteristics of the sole SSH libraries and sequences. aMean ± SEM

	LPS (head-kidney)	CuSO4 (liver)
Number of clones sequenced	231	229
Number of clones analysed	222	226
Average sequence length (bp)a	411 ± 168	414 ± 155
Number of unique ESTs	133	89
Up-regulated ESTs	62	48
Identified ESTs	49	44
Non-identified ESTs	13	4
Down-regulated ESTs	71	41
Identified ESTs	58	34
Non-identified ESTs	13	7

Genes coding for products involved in osmoregulation and nitrogen excretion (e.g., liver angiotensinogen, sodium potassium ATPase beta subunit, kininogen 1, angiotensin I converting enzyme 1, and alanine-glyoxylate aminotransferase 2-like) were also identified in libraries from the livers of CuSO4-treated soles, in agreement with the previously reported effects of copper on osmoregulation, acid-base balance and nitrogen excretion (e.g., Blanchard & Grosell, 2006; Evans et al., 2005).

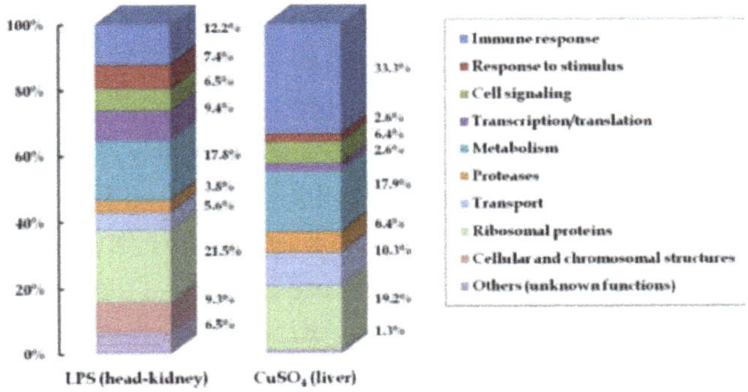

Fig. 1. Functional classification of unique ESTs obtained in SSH libraries from LPS or CuSO4- treated soles. "Others" indicates genes that had significant identities to databases entries of unknown function.

4.2 DNA microarrays

The microarray developed for the European flounder *Platichthys flessus* (GENIPOL platform, Williams et al., 2006) had been used to assess hepatic gene expression of many species of flatfish, confirming that heterologous microarray analyses between closely related species is a suitable approach (Cohen et al., 2007). This platform turned out a very useful tool for analysing the transcriptional expression of *S. senegalensis*.

First, the hepatic response of soles exposed to CuSO4 or LPS was studied using pooled samples. Statistical analyses showed that 405 genes were differentially expressed after Cu treatment at 6 h, 468 with Cu at 24 h, 271 with LPS at 6 h and 664 with LPS at 24 h (Table 2).

Table 2. Number of differentially expressed genes in *S. senegalensis* liver in response to LPS or CuSO4 treatments that were identified by heterologous DNA microarrays.

	LPS		$CuSO_4$	
	6 h	24 h	6 h	24 h
Up-regulated genes	172	251	101	341
Down-regulated genes	233	217	170	323
Total	405	468	271	664

The functional analysis of the results (Blast2GO software) permitted the identification of response-specific genes to CuSO4 (cell junction and cell signalling), to LPS (glutathione transferase and immune response) or to both treatments (immune response, digestive enzymes, unfolded protein binding, intracellular transport and secretion, and proteasome) (Table 3). This way, the functional category cell junction was statistically significantly over- represented amongst the genes induced by copper at 6 h. This term grouped genes related to cellular adhesion, such those coding for claudins (CLDN26) and genes related to cell signalling, such as GIT2 that encodes G-protein-coupled receptor kinase interactor 2, which is in agreement with the ability of copper to alter the tight junction permeability in human intestinal mucosa (Ferruzza et al., 2002). Glutathione-S-transferases were more prevalent amongst the transcripts down-regulated by LPS at 24 h. The capacity of LPS and bacterial infection to down-regulate biotransformation activities such as GSTs has been described in a number of fish species (Reynaud et al., 2008). As shown in Table 3, genes related to the immune response were specifically induced by LPS. These included the antimicrobial peptide hepcidin (HAMP), TNFα-induced protein 9 (TNFAIP9), cytokines (IL8, IL25) and chemotaxins (LECT2).

Other immune-related genes were induced by both LPS and CuSO4 treatments. This is the case for classic piscine acute phase proteins like haptoglobin (HP) or C7 (Bayne & Gerwick, 2001). C7 is a component of the complement system, whose up-regulation by copper is in line with complement proteins being engaged in novel biological functions distinct from their well- established role in innate immunity (Mastellos et al., 2005). Although soles were fasted prior to and during the experiments, digestive enzymes such as trypsin (PRSS2), chymotripsin (CTRB), elastase (ELA4) and carboxypetidase A (CPA1) and B (CPA2) were down-regulated in response to both treatments. This might be due to a general stress caused by the treatment (Auslander et al., 2008). Furthermore, LPS and copper treatments resulted in the up-regulation of genes encoding unfolded protein-binding, which are induced in fish in response to different kinds of stress conditions (e.g., bacterial infection or exposure to heavy metals) (Basu et al., 2002), intracellular transport and secretion, in order to accommodate the rapid onset of cytokine secretion and for membrane traffic associated with the phenotypic changes of immune activation (Pagan et al., 2003), and proteasomal proteins in agreement with previous results in mammalian cells (Qureshi et al., 2003; Fernandes et al., 2006).

Table 3. Selected genes that were significantly differentially expressed in the liver of *S. senegalensis* during LPS or CuSO4 treatments.

	Response to LPS	Response to $CuSO_4$	Selected genes
Cell junctions	---	Up-regulation	CLDN26, GIT2
Glutathione-S- transferases	Down-regulation	---	GST-A, GST1, GST3
Immune response	Up-regulation	---	HAMP, TNFAIP9, IL8, IL25, LECT2
	Up-regulation	Up-regulation	C7, HP
Digestive enzymes	Down-regulation	Down-regulation	PRSS2, CTRB, ELA4, CPA1, CPA2
Unfolded protein binding	Up-regulation	Up-regulation	GP96, HSP70
Intracellular transport/secretion	Up-regulation	Up-regulation	ARF5, TMED7, SEC22
Proteasome	Up-regulation	Up-regulation	PMSD3, MSUG1

Because the GENIPOL cDNA microarray was constructed from ESTs derived from flounder liver and had been used in studies of hepatic expression (Cohen et al., 2007; Williams et al., 2006, 2007, 2008), we investigated whether this platform would be valid for analysing the transcriptional response in the head-kidney. To this end, the GENIPOL microarray was used to compare the basal transcriptional expression in the *Solea senegalensis* head-kidney and liver. We determined that 1004 genes were statistically differentially expressed in both organs: 418 transcripts were more abundant in the liver than in the head-kidney and 586 transcripts were more abundant in the head-kidney than in the liver. Thus, although the microarray was constructed with hepatic ESTs, the number of genes identified as over- expressed is similar in both organs, though slightly higher in the head-kidney than in the liver. The analysis of transcriptional patterns showed that the most represented biological processes amongst the genes up-regulated in the liver in comparison with the head-kidney were those involved in innate immune response, digestion, lipid transport, and monooxygenase activity (Table 4). The category "innate immune response" grouped genes coding for acute phase proteins because, as has been previously discussed, the liver is the main source of these plasmatic proteins. Amongst the genes related to lipid transport that were particularly remarkable were those coding for several apolipoproteins. The term "monooxygenase activity" encompassed genes belonging to the cytochrome P450 family, in agreement with what is known about the detoxifying capacity and the biotransformation activity of the liver in teleosts (Thorgaard et al., 2002).

Functional terms over-represented in the list of transcripts that were more abundant in the head-kidney than in the liver related to cellular division and protein turnover (protein degradation and the proteasome and ribosomal proteins), among others (Table 5). These results agree with the role of the head-kidney as a major haematopoietic organ in teleosts and with its function as a secondary lymphoid organ in the clearance of soluble and particulate antigens from circulation (Whyte, 2007). It is worth noting that when the lists of genes that are up-regulated in both organs were compared, two sequences corresponding to the GAPDH (glyceraldehyde 3-phosphate dehydrogenase) gene, but with different expression patterns, were detected. The first of these sequences was more abundant in the liver, while the second was prevalent in the head-kidney. *Solea senegalensis* possesses two different GADPH paralogous genes that exhibit different tissue expression patterns, with GAPDH1 being more abundant in the liver and the GAPDH2 isoform more prevalent in the head-kidney (Manchado et al., 2007). A detailed analysis of the sequences detected in the microarray study revealed that these sequences match the two described isoforms. Altogether, these results demonstrated that the GENIPOL microarray platform was valid for analysing the transcriptional response of the sole head-kidney, discriminating between genes coding for transcripts with specific patterns of hepatic or renal expression.

Table 4. Selected genes up-regulated in the liver in comparison with the head-kidney in *S. senegalensis*.

Innate immune response
alpha-1-antitrypsin, alpha-2-macroglobulin, anticoagulant protein C precursor, coagulation factor VIIc, chemotaxin, complement component C3, complement component C8, complement component C9, complement regulatory plasma protein, fibrinogen alpha, fibrinogen beta chain precursor, fibrinogen gamma chain precursor, haptoglobin, hepcidin precursor, interleukin 8 precursor, kininogen 1, plasma protease C1 inhibitor precursor, prothrombin precursor, putative complement factor, transferrin
Digestive enzymes
chymotrypsinogen 1, chymotrypsinogen 2, trypsinogen 2 precursor
Lipid transport
apolipoprotein A-I, apolipoprotein A-IV, apolipoprotein C-I precursor, apolipoprotein E, apolipoprotein H, 14kDa apolipoprotein, fatty acid-binding protein
Monooxygenase activity
cytochrome P450 2F2, cytochrome P450 2X, cytochrome P450 3A, cytochrome P450 3A45, cytochrome P450 8B1, cytochrome P450 monooxygenase

Table 5. Selected genes up-regulated in the head-kidney in comparison with the liver in *S. senegalensis*.

Cellular division/cytoskeleton
alpha-tubulin, actin-related protein 3 homolog, actin related protein 2/3 complex subunit 4, beta- actin, cofilin 2, coronin 1A, lamin B1, microtubule-based motor protein, mitotic spindle assembly checkpoint protein, myosin regulatory light chain 2, nuclear movement protein PNUDC, thymosin beta-4
Protein degradation/proteasome
polyubiquitin, proteasome alpha 1 subunit isoform 2, proteasome (prosome, macropain) subunit alpha type 7, proteasome beta-subunit C5, proteasome subunit beta type 3, proteasome (prosome, macropain) subunit beta type 5, proteasome 26S ATPase subunit 5, proteasome subunit N3, ubiquitin carboxyl-terminal hydrolase isozyme L, ubiquitin specific protease 9
Ribosomal proteins
40S ribosomal protein S3a, 40S ribosomal protein S4, 60S ribosomal protein L3, 60S ribosomal protein L4, 60S ribosomal protein L13

Consequently, the GENIPOL platform was used to assess the response to LPS in the head- kidney. After 24 h of LPS treatment, a total of 224 genes was statistically differentially expressed in the head-kidney (117 up-regulated and 107 down-regulated). The functional analysis of the results revealed that the biological processes altered by LPS treatment in the head-kidney were very similar to those detected in the liver. Most notably amongst the up-regulated genes were the functional groups immune response, unfolded protein- binding, intracellular transport/secretion and proteasome, and digestive enzymes in the list of down-regulated genes. In contrast, the glutathione transferases category, whose transcript levels were down-regulated by LPS in the liver, was not affected in the head- kidney.

4.3 Real-time qRT-PCR

Although the determination of absolute transcript abundance (the exact number of molecules of a transcript in all samples of the study) is the only adequate procedure to accurately assess the expression of a gene (Prieto-Álamo et al., 2003), these kinds of studies are infrequent due to the experimental difficulties that impair this type of analysis. A set of more than 20 sole transcripts was selected for the quantification of their levels by this rigorous approach (Table 6) (Prieto-Álamo et al., 2009; Osuna-Jiménez et al., 2009), based on the physiological relevance of their products and on the magnitude and specificity of their responses.

Table 6. Biological functions of proteins encoded by the sole transcripts selected to be validated by real-time qRT-PCR (Prieto-Álamo et al., 2009; Osuna-Jiménez et al., 2009).

Immune Response
complement component C3, complement component C7, transferrin, haptoglobin, ferritin M, natural killer enhancing factor, tumor necrosis factor alpha-induced protein 9, hepcidin, nonspecific cytotoxic cell receptor protein-1, angiotensinogen, sequestosome 1, tumor necrosis factor receptor associated factor
Response to stress
CCAAT/enhancer binding protein, cold inducible RNA binding protein, DNA-damage-inducible transcript 4-like, NHP2 non-histone chromosome protein
Energetic Metabolism
glyceraldehyde-3P-DHase, transketolase, NADH-DHase 1 alpha subcomplex 4
Protein Synthesis, Folding and Degradation
asparaginil-tRNA synthetase, heat shock protein GP96, proteasome 26S non-ATPase subunit 3, cathepsin Z
Transport
α-globin

The absolute quantification of the selected transcripts was not limited to the samples used for global analysis, but rather extended to samples coming from different individuals, organs, treatments and exposure times. Consequently, this gene-to-gene analysis provided valuable additional information on transcriptional expression patterns of the selected genes. Individual quantification is also mandatory to prevent biased interpretations of specimens with abnormal expression levels. The absolute real-time qRT-PCR on individual samples demonstrated that inter-individual variations of most examined transcripts in treated animals was in the range of that in control fish, indicating similar susceptibility to LPS or CuSO4 challenge among individuals. The qRT-PCR quantifications confirm, in general, the results obtained with the subtractive libraries and the DNA microarrays. As expected, substantial differences in abundance were found depending on the transcript and tissue examined. In general, each transcript displayed a characteristic expression profile, distinguishing between constitutive or up-/down-regulated, early or late responsive, stressor-specific or not, etc., as a function of the organ analysed.

5. PROTEOMICS STUDIES OF GBD

Fish reared in earth ponds are eventually affected by Gas Bubble Disease (GBD) outbreaks if ponds are not correctly handled, particularly under high temperature and radiation conditions. GBD is a non-infectious pathology occurring when the partial pressures of atmospheric gases in solution exceed their respective partial pressures in the atmosphere. GBD was initially observed in farmed species, although outbreaks in wild fish, both fresh- water and marine animals, have also been reported. GBD can have serious adverse economic repercussions in fish cultures by reducing productivity and the commercial value of the fish as well as the farm profitability. Thus, the development of biomarkers responsive to hyperoxia stress would be a valuable tool to apply in systems where oxygen supersaturation might be possible, and would also contribute to basic knowledge of oxidative stress. Oxygen supplementation is a common practice in intensive fish farming, in order to allow high density cultivation while reducing the amount of water demanded in aquaculture facilities. It is also required during fish transportation. In intensive aquaculture, the use of oxygen is regulated by sophisticated mechanisms to keep its concentration close to desired values. However, in open ponds, the likelihood of oxygen supersaturation conditions is higher, because primary producers are in a high nutrient environment, occasionally combined with high temperature and radiation. Photosynthetic oxygen overproduction is a factor of concern for pond aquaculture, particularly when species such as sole, which exhibit nocturnal and benthic habits, are considered. These features complicate water management compared to pond operation in pelagic fish farming.

Environmental and physicochemical conditions inducing hyperoxia, such as radiation, temperature and dissolved O2, were monitored in two independent land-based ponds of an aquaculture research centre (IFAPA Centro El Toruño, Puerto de Santa María, Cádiz, Spain) in which *S. senegalensis* were reared, after a GBD outbreak was detected in some of these animals (Salas-Leyton et al., 2009). Fig. 2 (upper) shows the appearance of the earth pond used as control (100 m2, water renewed 4-fold/day) and of that in which GBD developed (900 m2, water unrenewed) in which algal blooms can be observed at its border. As shown in

Fig. 2 (centre), the dissolved oxygen profile detected in the hyperoxic earth pond was typical of that in environments dominated by macroalgal biomass and high photosynthetic activity, where extreme oxygen levels are reached during a great part of the daily cycle (including night hours), always above saturation without a desaturation phase. As shown in Fig. 2 (lower), the following typical GBD symptoms were detected in fish from this pond: exophthalmia caused by retrobulbar bubbles (A), subcutaneous emphysemas, obstruction of gill lamellas (B), big bubbles located at caudal (C) and dorsal fins, haemorrhages, anomalous swimming accompanied by loss of orientation, near- lethargy status and individual isolation were the main effects of O2 supersaturation (Salas- Leyton et al., 2009).

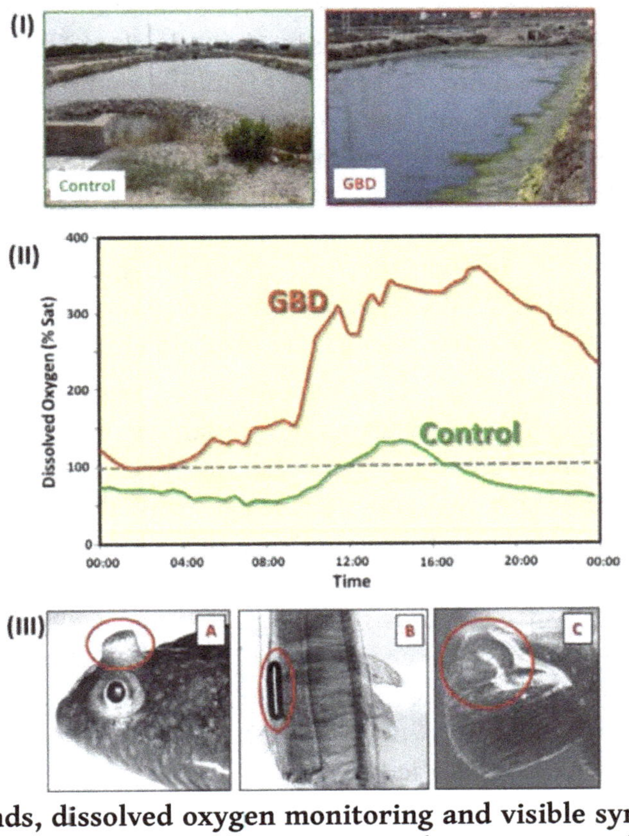

Fig. 2. Earth ponds, dissolved oxygen monitoring and visible symptoms in GBD-affected soles. (I) The experiments were carried out in two independent land-based ponds, one a 900 m2 rectangular pond where oxygen supersaturation developed spontaneously, named the GBD pond (right) and the other a 100 m2 control pond (left) in which water was renewed four times per day. (II) Dissolved oxygen was simultaneously monitored throughout the daily cycle in both the control (green line) and GBD ponds (red line). (III) Visible symptoms observed in GBD-affected soles included the following: exophthalmia caused by retrobulbar bubbles (A), a bubble obstructing a gill lamella (B), and a large bubble located at caudal fin (C).

Under the described aquaculture conditions, a parallel proteomic study was carried out in search of protein alteration patterns that might be used as potential new and unbiased biomarkers of hyperoxic stress (Fig. 3) (Salas-Leyton et al., 2009). The following three health statuses were studied in sole individuals: (1) healthy control fish, (2) GBD-affected but asymptomatic fish, and (3) GBD-affected fish with visible symptoms. Protein expression profiles were studied by 2-DE in cytosolic fractions of gills and livers. A total of 1,525 and 1,632 spots were detected in the four gill and liver gels, respectively, that were run in each of the three situations studied. Fig. 3 (upper) shows the master gels of cytosolic gill (left) and liver (right) proteins. A total of 205 protein spots were differentially expressed in the gills and 498 in the liver in each health status. Fig. 3 (middle) shows the number of spots which are present only, absent, increased or diminished in each of the studied conditions, and the total number of changes found. A significantly higher number of differentially expressed spots were found in GBD-affected soles, mainly in fish with visible symptoms. Of these, 25 spots in the gills and 23 in the liver were selected for identification using tandem mass spectrometry (nESI-IT MS/MS), *de novo* sequencing and a bioinformatics search. Fig. 3 (lower) shows the percentage of the relative intensity of each spot in each health status. Sequence tags were obtained from 9 (gills) and 5 (liver) of the selected spots, resulting in a total of 14 identified spots in the GBD status, which are indicated in Fig. 3 (lower).

Due to the central role of gills in oxygen exchange, the proteins identified in the gills of GBD-affected fish were, unsurprisingly, related to oxidative alteration of the cytoskeleton structure or function (β-tubulin and β-actin), motility (light myosin chain and α- tropomyosin), regulatory pathways (calmodulin, Raf kinase inhibitor protein [RKIP]) and carbohydrate metabolism (glyceraldehyde 3-P dehydrogenase). The hyperoxia-linked effects of these proteins and higher-level responses, related to inflammatory response, apoptosis or cell death, or derived from alterations in the regulatory proteins, have been discussed in depth (Salas-Leyton et al., 2009). In gills, only RKIP was most intense in GBD-affected animals without symptoms, while the other seven identified proteins were most intense in GBD-affected fish with symptoms. Proteins identified in the liver were related to protein oxidative damages (β-globin and fatty acid-binding protein [FABP]), protection from oxidative stress (dicarbonyl/L-xylulose reductase and glycine N- methyltransferase) and inflammatory response (complement component C3), in agreement with the predominant metabolic role of this organ. In the liver, only FABP was most intense in healthy fish, while the other four identified proteins were most intense in GBD-affected fish with visible symptoms. Approximately 50% of the identified proteins corresponded to "unusual" proteins not often found in proteomics screens, while the rest were preferential targets of oxidative stress. Most of these latter proteins were found as truncated forms of oxidised proteins that might trigger cellular defences against hyperoxygenation preceding GBD outbreaks. Some of the identified proteins might be considered to be good hyperoxia stress biomarkers and could be used in the early- warning detection of GBD outbreaks. Massive proteomic approaches have been applied for the first time to the study of this non-infectious pathological condition, gas bubble disease, in fish.

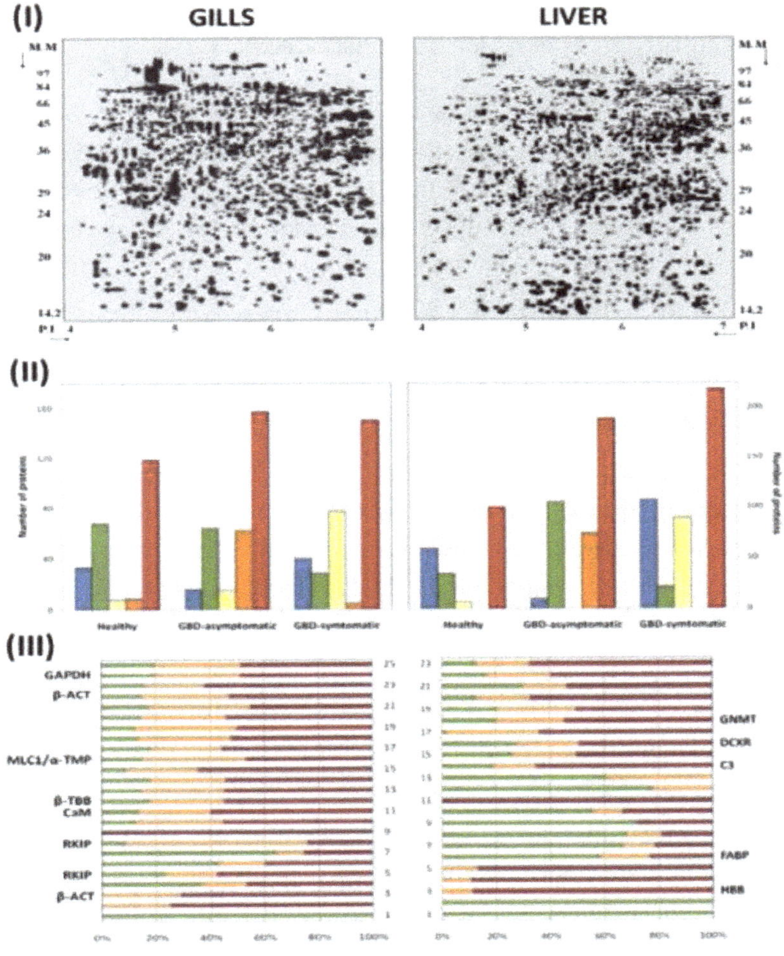

Fig. 3. Proteomic study of GBD-affected juvenile soles after 2-DE of cytosolic gill (left) and liver (right) proteins. (I) Master gels. (II) Number of proteins showing distinct protein expression patterns in healthy, GBD-asymptomatic and GBD-symptomatic fish. Bars represent the number of protein spots present only (■), absent (■), increased (■), diminished (■) and the total number of changes (■) found in each of the conditions. (III) Spots selected for possible identification by mass spectrometry analysis. Bars represent the percentage of the relative intensity of each spot in every status: healthy (■), GBD-asymptomatic (■) and GBD-symptomatic (■) animals. Successfully identified proteins are highlighted.

6. CONCLUSIONS

Given the economic importance of *S. senegalensis*, an aquaculture species that remains largely unexplored at the genomic level, the application of postgenomic methodologies provides results that may be highly relevant at a genetic, immunological and toxicological level, contributing to the improvement of management and welfare of this organism in aquaculture. The utility of high-throughput proteomic methods for unveiling the molecular basis of a cumbersome disease in aquaculture has been also demonstrated.

REFERENCES

1. Abril, N.; Ruiz-Laguna, J.; Osuna-Jiménez, I.; Vioque-Fernández, A.; Fernández-Cisnal, R.; Chicano-Galvez, E.; Alhama, J.; López-Barea, J. & Pueyo, C. (2011). Omic approaches in environmental issues. *Journal of Toxicology and Environmental Health, Part A*, 74: 1-19.
2. Alhama, J.; Romero-Ruiz, A. & López-Barea, J. (2006). Metallothionein quantification in clams (*Chamaelea gallina*) by RP-HPLC and fluorescence detection after monobromobimane derivatization. *Journal of Chromatography* A, 1107, 52-58.
3. Alhama, J.; Romero-Ruiz, A.; Jebali, J. & López-Barea, J. (2010). Total metallothionein quantification by reversed-phase high-performance liquid chromatography coupled to fluorescence detection after monobromobimane derivatization. In: *Cadmium in the Environment*, Parvau, R.G. (Ed.), 389-405, Nova Science Publishers, Inc., ISBN 978-1-60741-934-1, Hauppauge, NY, USA.
4. Álvarez-Muñoz, D.; Lara-Martin, P.A.; Blasco, J.; Gómez-Parra, A. & González-Mazo, E. (2007). Presence, biotransformation and effects of sulfophenylcarboxilic acids in the benthic fish *Solea senegalensis*. *Environment International*, 33, 565-570.
5. Aragão, C.; Conceiçao, L.E.C.; Martins, D.; Rønnestad, I.; Gomes, E. & Dinis, M.T. (2004a). A balanced dietary amino acid profile improves amino acid retention in post-larval Senegalese sole (*Solea senegalensis*). *Aquaculture*, 233, 293-304.
6. Aragão, C.; Conceiçao, L.E.C.; Fyhn, H.J. & Dinis, M.T. (2004b). Estimated amino acid requirements during early ontogeny in fish with different life styles: gilthead seabream (*Sparus aurata*) and Senegalese sole (*Solea senegalensis*). *Aquaculture*, 242, 589-605.
7. Auslander, M.; Yudkovski, Y.; Chalifa-Caspi, V.; Herut, B.; Ophir, R.; Reinhardt, R.; Neumann, P.M. & Tom, M. (2008). Pollution-affected fish hepatic transcriptome and its expression patterns on exposure to cadmium. *Marine Biotechnology, (NY)*, 10(3), 250-261.

8. Avella, M.A.; Olivotto, I.; Silvi, S.; Place, A.R. & Carnevalli, O. (2010). Effect of dietary probiotics on clownfish : a molecular approach to define how lactic bacteria modulate development in a marine fish. *American Journal of Physiology. Regulatory, Integrative and Comparative Physiology*, 298, R359-R371.

9. Basu, N.; Todgham, A.E.; Ackerman, P.A.; Bibeau, M.R.; Nakano, K.; Schulte, P.M. & Iwama G.K. (2002). Heat shock protein genes and their functional significance in fish. *Gene*, 295(2), 173-183.

10. Bayne, C.J. & Gerwick, L. (2001). The acute phase response and innate immunity of fish. *Developmental & Comparative Immunology*, 25(8-9), 725-743.

11. Blanchard, J. & Grosell, M. (2006). Copper toxicity across salinities from freshwater to seawater in the euryhaline fish *Fundulus heteroclitus*: is copper an ionoregulatory toxicant in high salinities? *Aquatic Toxicology*, 80(2), 131-139.

12. Bonilla-Valverde, D.; Ruíz-Laguna, J.; Muñoz, A.; Ballesteros, J.; Lorenzo, F.; Gómez-Ariza, J.L. & López-Barea, J. (2004). Evolution of biological effects of Aznalcollar mining spill in the Algerian mouse (*Mus spretus*) using biochemical biomarkers. *Toxicology*, 197: 123-138.

13. Cabrita, E.; Soares, F. & Dinis, M.T. (2006). Characterization of Senegalese sole, *Solea senegalensis*, male broodstock in terms of sperm production and quality. *Aquaculture*, 261, 967-975.

14. Cánovas-Conesa, B.; Salas, E.; Alhama, J.; López-Barea, J. & Cañavate J.P. (2007). Efecto de la densidad de cultivo sobre los biomarcadores bioquímicos y la carga energética en el lenguado senegalés (*Solea senegalensis*). In: *Libro de Actas del XI Congreso Nacional Acuicultura*, Cerviño, A.; Guerra, A. & Pérez, C. (Eds.), 1463–1466, Universidad de Vigo, ISBN 978-84-611-9085-0, Vigo, España.

15. Cañavate, J.P. (2005). Potential of the Senegal sole *Solea senegalensis* Kaup, 1858 for marine aquaculture diversification. *Boletín del Instituto Español de Oceanografía*, 21, 147-154.

16. Cañavate, J.P.; Prieto, A.; Zerolo, R.; Solé, M.; Sarasquete, C. & Fernández-Díaz, C. (2007). Effects of light intensity and addition of carotene rich *Dunaliella salina* live cells on growth and antioxidant activity of *Solea senegalensis* Kaup (1858) larval and metamorphic stages. *Journal of Fish Biology*, 71, 781-794.

17. Cerdà, J.; Mercadé, J.; Lozano, J.J.; Manchado, M.; Tingaud-Sequeira, A.; Astola, A.; Infante, C.; Halm, S.; Viñas, J.; Castellana, B.; Asensio, E.; Cañavate, P.; Martínez- Rodríguez, G.; Piferrer, F.; Planas, J.V.; Prat, F.; Yúfera, M.; Durany, O.; Subirada, F.; Rosell, E. & Maes, T. (2008). Genomic resources for a commercial flatfish, the Senegalese sole (*Solea senegalensis*): ETS sequencing, oligo microarray design, and development of the Soleamold bioinformatic platform. *BMC Genomics*, 9, 508.

18. Cohen, R.; Chalifa-Caspi, V.; Williams, T.D.; Auslander, M.; George, S.G.; Chipman, J.K. & Tom, M. (2007) Estimating the efficiency of fish cross-species cDNA microarray hybridization. *Marine Biotechnology*, 9, 491-499

19. COM/0511 (2002). Communication from the Commission on *A Strategy for the Sustainable Development of European Aquaculture*. Conceiçao, L.E.C.; Grasdalen, H. & Rønnestad, I. (2003). Amino acid requirements of fish larvae and post-larvae: new tools and recent findings. *Aquaculture*, 227, 221-232. Conceiçao, L.E.C.; Ribeiro, L.; Engrola, S.; Aragão, C.; Morais, S.; Lacuisse, M.; Soares, F. & Dinis, M.T. (2007). Nutritional physiology during development of Senegalese sole (*Solea senegalensis*). *Aquaculture*, 268, 64-81.

20. Costa, P.M.; Lobo, J.; Caeiro, S.; Martins, M.; Ferreira, A.M.; Caetano, M.; Vale, C.; DelValls, T.A. & Costa, M.H. (2008a). Genotoxic damage in *Solea senegalensis* exposed to sediments form the Sado Estuary (Portugal): Effects of metallic and organic contaminants. *Mutation Research*, 654, 29-37.

21. Costa, P.M.; Diniz, M.S.; Caeiro, S.; Lobo, J.; Martins, M.; Ferreira, A.M.; Caetano, M.; Vale, C.; DelValls, T.A. & Costa, M.H. (2008b). Histological markers in liver and gills of juvenile *Solea senegalensis* exposed to contaminated estuarine sediments: A weighted indices approach. *Aquatic Toxicology*, 92, 202-212.

22. Costa, P.M.; Caeiro, S.; Diniz, M.S.; Lobo, J.; Martins, M.; Ferreira, A.M.; Caetano, M.; Vale, C.; DelValls, T.A. & Costa, M.H. (2009). Biochemical endpoints on juvenile *Solea senegalensis* exposed to estuarine sediments: the effects of contaminant mixtures on metallothionein and CYP1A induction. *Ecotoxicology*, 18, 988-1000.

23. Costa, P.M.; Neuparth, T.S.; Caeiro, S.; Lobo, J.; Martins, M.; Ferreira, A.M.; Caetano, M.; Vale, C.; DelValls, T.A. & Costa, M.H. (2011). Assessment of the genotoxic potential of contaminated estuarine sediments in fish peripheral blood: Laboratory versus *in situ* studies. *Environmental Research*, 111, 25-36.

24. Cousinou, M.; Dorado, G. & López-Barea, J. (1999). Amplification and cloning of cDNAs of cytochrome P4501A1 and metallothionein genes from *Sparus aurata* Linnaeus, 1758 and *Liza aurata* (Risso, 1810) by RACE-PCR. *Boletín del Instituto Español de Oceanografía*, 15, 473-484.

25. Cousinou. M.; Nilsen, B.; López-Barea, J. & Dorado, G. (2000). New methods to use of fish cytochrome P4501A to assess marine organic pollutants. *Science of Total Environment*, 247, 213-225.

26. De Wit, M.; Keil, D.; van der Ven, K.; Vandamme, S.; Witters, E. & De Coen, W. (2010). An integrated transcriptomic and proteomic approach characterizing estrogenic and metabolic effects of 17 α-ethinylestradiol in zebrafish (*Danio rerio*). *General & Comparative Endocrinology*, 167, 190-201.

27. Dias, J.; Yúfera, M.; Valente, L.M.P. & Rema, P. (2010). Feed transit and apperent protein, phosphorous and energy digestibility of practical feed ingredients by Senegalese sole (*Solea senegalensis*). *Aquaculture*, 302, 94-99.

28. Diatchenko, L.; Lay, Y.F.; Campbell, A.P.; Chenchik, A.; Moqadam, F.; Huang, B.; Lukyanov, K.; Lukyanov, K.; Gurskaya, N.; Sverdlov, E.D. & Siebert, P.D. (1996). Suppresion subtractive hybridization: A method for generating differentially regulated or tissue-specific cDNA probes and librairies. *Proceedings of the National Academy of Sciences of the USA*, 93, 6025-6030.

29. Dinis, M.T.; Ribero. L.; Soares, F. & Sarasquete, C. (1999). A review on the cultivation potential of *Solea senegalensis* in Spain and Portugal. *Aquaculture*, 176, 27-38 Douglas, S.E.; Knickle, L.C.; Kimball, J. & Reith, M.E. (2007). Comprehensive EST analysis of Atlantic halibut (*Hippoglossus hippoglossus*), a commercially relevant aquaculture species. *BMC Genomics*, 8, 144

30. Evans, D.H.; Piermarini, P.M. & Choe, K.P. (2005). The multifunctional fish gill: dominant site of gas exchange, osmoregulation, acid-base regulation, and excretion of nitrogenous waste. *Physiological Reviews*, 85(1), 97-177.

31. Ewart, K.V.; Belanger, J.C.; Williams, J., Karakach, T.; Penny, S.; Tsoi, S.C.; Richards, R.C. & Douglas, S.E. (2005). Identification of genes differentially expressed in Atlantic salmon (*Salmo salar*) in response to infection by *Aeromonas salmonicida* using cDNA microarray technology. *Developmental & Comparative Immunology*, 29(4), 333-347.

32. FAO: Food and Agriculture Organization of the United Nations (2010). Aquaculture planning. Policy formulation and implementation of sustainable development. FAO Fisheries and Aquaculture Technical Paper 542. Rome FAO.

33. Fernandes, R.; Ramalho, J. & Pereira, P. (2006). Oxidative stress upregulates ubiquitin proteasome pathway in retinal endothelial cells. *Molecular Vision*, 12, 1526-1535.

34. Fernández-Díaz, C.; Kopecka, J.; Cañavate, J.P.; Sarasquete, C. & Solé, M. (2006). Variations on development and stress defences in *Solea senegalensis* larvae fed on live and microencapsulated diets. *Aquaculture*, 251, 573-584.

35. Ferruzza, S.V; Scacchi, M.; Scarino, M.L. & Sambuy, Y. (2002). Iron and copper alter tight junction permeability in human intestinal Caco-2 cells by distinct mechanisms. *Toxicology In Vitro*, 16(4), 399-404.

36. Fonseca, V.; Serafim, A.; Company, R.; Bebianno, M.J.; & Cabral, H.N. (2009) Effect of copper on growth, condition indices and biomarker response in juvenile sole *Solea senegalensis*. *Scientia Marina*, 73, 51-58.

37. Fonseca, V.; Franca, S.; Serafim, A.; Company, R.; Lopes, B.; Bebianno, M.J.; & Cabral, H.N. (2011) Multi-biomarker response to estuarine habitat

contamination in three fish species: *Dicentrarchus labrax, Solea senegalensis* and *Pomatochistus microps*. *Aquatic Toxicology*, 102, 216-277.

38. Gallardo-Madueño, R.; Leal, J.F.; Dorado, G.; Holmgren, A.; López-Barea, J. & Pueyo, C. (1998). *In vivo* transcription of *nrdAB* operon and of *grxA* and *fpg* genes is triggered in *Escherichia coli* lacking both thioredoxin and glutaredoxin 1 or thioredoxin and glutathione, respectively. *The Journal of Biological Chemistry*, 273 (29), 18382-18388.

39. García-López, A.; Fernández-Pasquier, V.; Couto, E.; Canario, A.V.M.; Sarasquete, C. & Martínez-Rodriguez, G. (2006a). Non-invasive assessment of reproductive status and cycle of sex steroid levels in a captive broodstock of Senegalese sole *Solea senegalensis* (Kaup). *Aquaculture*, 254, 583-593.

40. García-López, A.; Anguis, V.; Couto, E.; Canario, A.V.M.; Cañavate, J.P.; Sarasquete, C. & Martínez-Rodriguez, G. (2006b). Testicular development and plasma sex steroid levels in cultured male Senegalese sole *Solea senegalensis* Kaup. *General and Comparative Endocrinology*, 147, 343-351.

41. García-López, A.; Couto, E.; Canario, A.V.M.; Sarasquete, C. & Martínez-Rodriguez, G. (2007). Ovarian development and plasma sex steroid levels in cultured female Senegalese sole *Solea senegalensis*. *Comparative Biochemistry and Physiology - Part A*, 146, 342-354.

42. Gavaia, P.J.; Dinis, M.T. & Cancela, M.L. (2002). Osteological development and abnormalities of the vertebral column and caudal skeleton in larval and juvenile stages of hatchery-reared Senegal sole (*Solea senegalensis*). *Aquaculture*, 211, 305-323.

43. Gavaia, P.J.; Simes, D.C.; Ortiz-Delgado, J.B.; Viegas, C.S.B.; Pinto, J.P.; Kelsh, R.N.; Sarasquete, C. & Cancela, M.L. (2006). Osteocalcin and matrix Gla protein in zebrafish (*Danio rerio*) and Senegal sole (*Solea senegalensis*): Comparative gene and protein expression during larval development through adulthood. *Gene Expression Patterns*, 6, 637-652.

44. Geist, J.; Werner, I.; Eder, K.J. & Leutenegger, C.M. (2007). Comparisons of tissue-specific transcription of stress response genes with whole animal endpoints of adverse effect in striped bass (*Morone saxatilis*) following treatment with copper and esfenvalerate. *Aquatic Toxicology*, 85(1), 28-39.

45. Gonzalez-Fernández, M.; García-Barrera, T.; Jurado, J.; Prieto-Álamo, M.J.; Pueyo, C.; López- Barea, J. & Gómez-Ariza, J.L. (2008). Integrated application of transcriptomics, proteomics, and metallomics in environmental studies. *Pure and Applied Chemistry*, 80, 2609-2626.

46. James, P. (1997). Protein identification in the post-genome era: the rapid rise of proteomics. *Quarterly Review of Biophysics*, 30, 279–331.

47. Jebali, J.; Banni, M.; Gerbej, H.; Boussetta, H.; López-Barea, J. & Alhama, J. (2008). Metallo- thionein induction bu Cu, Cd and Hg in *Dicentrarchus labrax* liver: assessment by RP-HPLC with fluorescent detection and spectrophotometry. *Marine Environmental Research*, 65, 358-363.

48. Jiménez, A.; Prieto-Álamo, M.J.; Fuentes-Almagro, C.A.; Jurado, J.; Gustafsson, J.A.; Pueyo, C. & Miranda-Vizuete, A. (2005) Absolute mRNA levels and transcriptional regulation of the mouse testis-specific thioredoxins. *Biochemical and Biophysical Research Communications*, 330, 56-74.

49. Jiménez-Tenorio, N.; Morales-Caselles, C.; Kalman, J.; Salamanca, M.J.; González de Canales, M.L.; Sarasquete, C. & DelValls, T.A. (2007). Determining sediment quality for regulatory proposes using fish chronic assays. *Environment International*, 33, 474- 480.

50. Jiménez-Tenorio, N.; Salamanca, M.J.; García-Luque, E.; González de Canales, M.L. & DelValls, T.A. (2008). Chronic bioassay in benthic fish for the assessment of the quality of sediments in different areas of the coast of Spain impacted by acute and chronic oilspills. *Environmental Toxicology*, 23, 634-642.

51. Jurado, J.; Prieto-Álamo, M.J.; Madrid-Risques, J. & Pueyo, C. (2003) Absolute gene expression patterns of thioredoxn and glutaredoxin redox systems in mouse. *The Journal of Biological Chemistry*, 278 (46), 45546-45554.

52. Jurado, J.; Fuentes-Almagro, C.A.; Prieto-Álamo, M.J. & Pueyo, C. (2007) Alternative splicing of *c-fos* pre-mRNA : contribution of the rates of synthesis and degradation to the copy number of each transcript isoform and detection of a truncated c-Fos immunoreactive species. *BMC Molecular Biology*, 8, 83.

53. Kültz, D.; Fiol, D.; Valkova, N.; Gomez-Jimenez, S.; Chan, S.Y.; & Lee, J. (2007) Functional genomics and proteomics of the cellular stress response in «non-model» organisms. *The Journal of Experimental Biology*, 210, 1593-1601.

54. Lenartova, V.; Holovska, K.; Pedrajas, J.R.; Martínez-Lara, E.; Peinado, J.; López-Barea, J.; Rossival, I. & Kosuth, P. (1997). Antioxidant and detoxifying fish enzymes as biomarkers of river pollution. *Biomarkers*, 2, 247-252.

55. López-Barea, J. (1995a). Biomarkers in Ecotoxicology: an Overview, In: *Toxicology in Transition*, Degen, G.H.; Seiler, J.P. & Bentely, P. (Eds.), 57-79, Springer, ISBN 978- 3540587811, Berlin, Germany.

56. López-Barea, J. (1995b). Biomarcadores que detectan contaminantes ambientales. In: *Actas del V Congreso Nacional de Acuicultura*, Castelló, F. & Calderer, A. (Eds.), 953-958, Publicaciones de la Universidad de Barcelona, Barcelona, Spain.

57. López-Barea, J. & Pueyo, C. (1998). Mutagen content and metabolic activation of promutagens by molluscs as biomarkers of marine pollution. *Mutation Research*, 399, 3-15.

58. López-Barea, J. & Gómez-Ariza, J.L. (2006). Environmental proteomics and metallomics. *Proteomics*, 6 (Suppl 1), S51-S62.

59. López-Galindo, C.; Vargas-Chacoff, L.; Nebot, E.; Casanueva, J.F.; Rubio, D.; Solé, M. & Mancera, J.M. (2010a). Biomarker responses in *Solea senegalensis* exposed to sodium hypochlorite used as antifouling. *Chemosphere*, 78, 885-893.

60. López-Galindo, C.; Vargas-Chacoff, L.; Nebot, E.; Casanueva, J.F.; Rubio, D.; Solé, M. & Mancera, J.M. (2010b). Sublethal effects or the organic antifoulant Mexel®432 on ormoregulation and xenobiotic detoxification in the flatfish *Solea senegalensis*. *Chemosphere*, 79, 78-85.

61. Manchado, M.; Michán, C. & Pueyo, C. (2000). Hydrogen peroxide activates the *SoxRS* regulon *in vivo*. *The Journal of Bacteriology*, 182(23), 6842-6844.

62. Manchado, M.; Infante, C.; Asensio, E. & Cañavate, J.P. (2007). Differential gene expression and dependence on thyroid hormones of two glyceraldehyde-3-phosphate dehydrogenases in the flatfish Senegalese sole (*Solea senegalensis* Kaup). *Gene*, 400(1- 2), 1-8.

63. Martínez-Lara, E.; Pascual, P.; Toribio, F.; López-Barea, J. & Bárcena, J.A. (1992). Rapid method for the quantification of glutathione transferase isoenzymes in crude extracts. *Journal of Chromatography A*, 609, 141-146.

64. Martínez-Lara, E.; Toribio, F.; López-Barea, J. & Bárcena, J.A. (1996). Glutathione-S- transferase isoenzyme patterns in the gilthead seabream (*Sparus aurata*) exposed to environmental contaminants. *Comparative Biochemistry and Physiology - Part C: Toxicology & Pharmacology*, 113, 215-220.

65. Martínez-Lara, E.; George, S.G.; López-Barea, J. & Bárcena, J. (1997). Purification and characterization of multiple glutathione transferase isoenzymes from grey mullet liver. *Cellular and Molecular Life Sciences*, 53, 759-768.

66. Mastellos, D.; Andronis, C.; Persidis, A. & Lambris, J.D. (2005). Novel biological networks modulated by complement. *Clinical Immunology*, 115(3), 225-235.

67. Michan, C.; Manchado, M.; Dorado, G. & Pueyo, C. (1999). *In vivo* transcription of the *Escherichia coli oxyR* regulon as a function of growth phase and in response to oxidative stress. *The Journal of Bacteriology.*, 181(9), 2759-2764.

68. Michan, C.; Monje-Casas, F. & Pueyo, C. (2005). Transcript copy number of genes for DNA repair and translesion synthesis in yeast: contribution of transcription rate and mRNA stability to the steady-state level of each mRNA along with growth in glucose-fermentative medium. *DNA Repair*, 4(4), 469-478.

69. Monje-Casas, F.; Jurado, J.; Prieto-Álamo, M.J.; Holmgren, A. & Pueyo, C. (2001) Expression analysis of the *nrdHIEF* operon from *Escherichia coli*. Conditions that trigger the transcript level *in vivo*. *Journal of Biological Chemistry*, 276 (21), 18031-18037.

70. Monje-Casas, F.; Michan, C. & Pueyo, C. (2004). Absolute transcript levels of thioredoxin- and glutathione-dependent redox systems in *Saccharomyces cerevisiae*: response to stress and modulation with growth. *Biochemical Journal*, 383(Pt 1), 139-147.

71. Montes-Nieto, R.; Fuentes-Almagro, C.A.; Bonilla-Valverde, D.; Prieto-Álamo, M.J.; Jurado, J.; Carrascal, M.; Gómez-Ariza, J.L.; López-Barea, J. & Pueyo, C. (2007). Proteomics in free-living *Mus spretus* to monitor terrestrial ecosystems. *Proteomics*, 7, 4376-4387.

72. Montes-Nieto, R.; García-Barrera, T.; Gómez-Ariza, J.L. & López-Barea, J. (2010). Environ- mental monitoring of Domingo Rubio stream (Huelva Estuary, SW Spain) by combining conventional biomarkers and Proteomic analysis in *Carcinus maenas*. *Environmental Pollution*, 158, 401-408.

73. Morales-Caselles, C.; Jiménez-Tenorio, N.; González de Canales, M.L.; Sarasquete, C. & DelValls, T.A. (2006). Ecotoxicity of sediments contaminated by the oil spill associated with the tanker "Prestige" using juveniles of the fish *Sparus aurata*. *Archives of Environmental Contamination and Toxicology*, 51, 652-660.

74. Morais, S.; Koven, W.; Rønnestad, I.; Dinis, M.T. & Conceiçao, L.E.C (2005). Dietary protein/lipid ratio affects growth and amino acid and fatty acid absorption and metabolism in Senegalese sole (*Solea senegalensis* Kaup 1858) larvae. *Aquaculture*, 246, 347-357.

75. Morais, S.; Conceiçao, L.E.C.; Rønnestad, I.; Koven, W.; Cahu, C.; Zambonino Infante, J.L. & Dinis, M.T. (2007). Dietary neutral lipid level and source in marine fish larvae: Effects on digestive physiology and food intake. *Aquaculture*, 268, 106-122.

76. Obayashi, T. & Kinoshita, K. (2011). COEXPRESdb: a database to compare gene coexpression in seven model animals. *Nucleic Acids Research*, 39 (Suppl 1), D1016- D1022.

77. Oliva, M.; González de Canales, M.L.; Gravato, C.; Guilhermino, L. & Perales, J.A. (2010). Biochemical effects and polycyclic aromatic hydrocarbons (PAHs) in senegal sole (*Solea senegalensis*) from a Huelva estuary (SW Spain). *Ecotoxicology and Environmental Safety*, 73, 1842-1851.

78. Ortiz-Delgado, J.B.; Ruane, N.M.; Pousao-Ferreira, P.; Dinis, M.T. & Sarasquete, C. (2006). Thyroid gland develoment in Senegalese sole (*Solea senegalensis* Kaup 1858) during early life stages: A histochemical and immunohistochemical approach. *Aquaculture*, 260, 346-356.

79. Osuna-Jiménez, I.; Williams, T.D.; Prieto-Álamo, M.-J.; Abril, N.; Chipman, J.K. & Pueyo, C. (2009). Immune- and stress-related transcriptomic responses of *Solea senegalensis* stimulated with lipopolysaccharide and copper sulphate using heterologous cDNA microarrays. *Fish & Shellfish Immunology*, 26, 699-706.

80. Pagan, J.K.; Wylie, F.G.; Joseph, S.; Widberg, C.; Bryant, N.J.; James, D.E. & Stow, J.L. (2003). The t-SNARE syntaxin 4 is regulated during macrophage activation to function in membrane traffic and cytokine secretion. *Current Biology*, 13(2), 156-160.

81. Pascual, P.; Rodríguez-Ariza, A.; Navas, J.I.; Toribio, F. & López-Barea, J. (1995a). Cambios bioquímicos estacionales durante el cultivo de doradas (*Sparus aurata*). In: *Actas del V Congreso Nacional de Acuicultura*, Castelló, F. & Calderer, A. (Eds.), 582-587, Publicaciones de la Universidad de Barcelona, Barcelona, Spain.

82. Pascual, P.; Navas, J.I.; Toribio, F. & López-Barea, J. (1995b). Indicadores bioquímicos de esters en doradas (*Sparus aurata*). In: *Actas del V Congreso Nacional de Acuicultura*, Castelló, F. & Calderer, A. (Eds.), 754-759, Publicaciones de la Universidad de Barcelona, Barcelona, Spain.

83. Pascual, P.; Navas, J.I.; Toribio, F. & López-Barea, J. (1997). Cambios clínicos estacionales en la carga energética adenílica y guanílica de doradas (*Sparus aurata*). In: *Actas del VI Congreso Nacional de Acuicultura*, de Costa, J.; Abellán, E.; García, B.; Ortega, A. & Zamora, S. (Eds.), 832-837, Publicaciones de la Universidad de Murcia, Murcia, Spain.

84. Pascual, P.; Pedrajas, J.R.; Toribio, F.; López-Barea, J. & Peinado, J. (2003). Effect of food deprivation on oxidative stress biomarkers in fish (*Sparus aurata*). *Chemico-Biological Interactions*, 145, 191-199.

85. Pedrajas, J.R.; Peinado, J. & López-Barea J. (1993). Purification of Cu,Zn-superoxide dismutase isoenzymes from fish liver: appearance of new isoforms as consequence of pollution. *Free Radical Research Communications*, 19, 29-41.

86. Pedrajas, J.R.; Peinado, J. & López-Barea J. (1995). Oxidative stress in fish exposed to model xenobiotics. Oxidatively modified forms of Cu,Zn-superoxide dismutase as potential biomarkers. *Chemico-Biological Interactions*, 98, 267-282.

87. Pedrajas, J.R.; Gavilanes, F.; López-Barea, J. & Peinado, J. (1998). Incubation of superoxide dismutase with malondialdehyde and 4-hydroxy-2-nonenal forms new active isoforms and adducts. An evaluation of xenobiotics in fish. *Chemico-Biological Interactions*, 116, 1-17.

88. Pinto, W.; Rodrigues, V.; Dinis, M.T. & Aragão, C. (2010). Can dietary amino acid supplementation be beneficial during fish metamorphosis?. *Aquaculture*, 310, 200- 205.

89. Prieto-Álamo, M.J.; Jurado, J.; Gallardo-Madueño, R.; Monje-Casas, F.; Holmgren, A. & Pueyo, C. (2000). Transcriptional regulation of glutaredoxin and thioredoxin pathways and related enzymes in response to oxidative stress. The *Journal of Biological Chemistry*, 275(18), 13398-13405.

90. Prieto-Álamo, M.J.; Cabrera-Luque, J.M. & Pueyo C. (2003). Absolute quantitation of normal and ROS-induced patterns of gene expression: an *in vivo* real-time PCR study in mice. *Gene Expression*, 11, 23-34.

91. Prieto-Álamo, M.J.; Abril, N.; Osuna-Jimenez, I. & Pueyo, C. (2009). *Solea senegalensis* genes responding to lipopolysaccharide and copper sulphate challenges: Large-scale identification by suppression subtractive hybridization and absolute quantification of transcriptional profiles by real-time RT-PCR. *Aquatic Toxicology*, 91, 312-319.

92. Pueyo, C.; Jurado, J.; Prieto-Álamo, M.J.; Monje-Casas, F. & López-Barea, J. (2002). Multiplex reverse transcription-polymerase chain reaction for determining transcriptional regulation of thioredoxin and glutaredoxin pathways. *Methods in Enzymology*, 247, 441-451.

93. Pueyo, C.; Gómez-Ariza, J.L.; Bello-López, M.A.; Fernández-Torres, R.; Abril, N.; Alhama, J.; García-Barrera, T. & López-Barea, J. (2011). New methodologies for assessing the presence and ecological effects of pesticides in Doñana National Park (SW Spain). In: *Pesticides in the Modern World / Book 1*, Bakic, S. (Ed.), 165-196, InTech Open Access Publisher, ISBN 978-953-307-437-5, Rijeka, Croatia.

94. Qureshi, N.; Perera, P.Y.; Shen, J.; Zhang, G.; Lenschat, A.; Splitter, G.; Morrison, D.C. & Vogel, S.N. (2003). The proteasome as a lipopolysaccharide-binding protein in macrophages: differential effects of proteasome inhibition on lipopolysaccharide- induced signaling events. *The Journal of Immunology*, 171(3), 1515-1525.

95. Reynaud, S.; Raveton, M. & Ravanel, P. (2008). Interactions between immune and biotransformation systems in fish: a review. *Aquatic Toxicology*, 87(3), 139-145.

96. Ribeiro, L.; Sarasquete, C. & Dinis, M.T. (1999a). Histological and histochemical development of the digestive tract system of *Solea senegalensis* (Kaup, 1858) larvae. *Aquaculture*, 171, 293-308.

97. Ribeiro, L.; Zambonino-Infante, J.L.; Cahu, C. & Dinis, M.T. (1999b). Development of digestive enzymes in larvae of *Solea senegalensis*, Kaup 1858. *Aquaculture*, 179, 293- 308.

98. Rodiles, A.; Herrera, M.; Hachero, I.; Rosano, M.; Ferrer, J.R.; Márquez, J.M. & Navas, J.I. (2005). Influence of different bottom types on the Senegal sole *Solea senegalensis* Kaup, 1858 ongrowing. *Boletín del Instituto Español de Oceanografía*, 21, 195-199.

99. Rodríguez-Ariza, A.; Abril, N.; Navas, J.I.; Dorado, G.; López-Barea, J. & Pueyo, C. (1992). Metal, mutagenicity and biochemical studies on bivalve molluscs from Spanish coasts. *Environmental and Molecular Mutagenesis*, 19, 112-124.

100. Rodríguez-Ariza, A.; Peinado, J.; Pueyo, C. & López-Barea, J. (1993). Biochemical indicators of oxidative stress in fish from polluted littoral areas. *Canadian Journal of Fisheries and Aquatic Sciences*, 50, 2568-2573.

101. Rodríguez-Ariza, A.; Dorado, G.; Navas, J.I.; Pueyo, C. & López-Barea, J. (1994a). Promutagen activation by fish liver as a biomarker of littoral pollution. *Environmental and Molecular Mutagenesis*, 24, 116-123.

102. Rodríguez-Ariza, A.; Toribio, F. & López-Barea, J. (1994b). Rapid determination of glutathione status in fish liver using high-performance liquid chromatography and electrochemical detection. *The Journal of Chromatography B*, 656, 311-318.

103. Rodríguez-Ortega, M.J.; Alhama, J.; Funes, V.; Romero-Ruíz, A.; Rodríguez-Ariza, A. & López-Barea, J. (2002). Biochemical biomarkers of pollution in the clam *Chamaelea gallina* from South-Spanish littoral. *Environmental Toxicology & Chemistry*, 21, 542- 549.

104. Rodríguez-Ortega, M.J.; Grosvik, B.E.; Rodríguez-Ariza, A.; Goksoyr, A. & López-Barea, J. (2003). Changes in protein expression profiles in bivalve molluscs (*Chamaelea gallina*) exposed to four environmental pollutants. *Proteomics*, 3, 1535-1543.

105. Romero-Ruíz, A.; Amezcua, O.; Rodríguez-Ortega, M.J.; Muñoz, J.L.; Alhama, J.; Rodríguez- Ariza, A.; Gómez-Ariza, J.L. & López-Barea, J. (2003). Oxidative stress biomarkers in the bivalve *Scrobicularia plana* transplanted to the Guadalquivir Estuary (S Spain) after Aznalcóllar mining spill. *Environmental Toxicology & Chemistry*, 22, 92-100.

106. Romero-Ruiz, A.; Carrascal, M.; Alhama, J.; Gómez-Ariza, J.L.; Abian, J. & López-Barea, J. (2006). Environmental proteomic studies in clams from the Doñana bank of Guadalquivir Estuary (SW Spain). *Proteomics*, 6 (Suppl 1), S245-S255.

107. Romero-Ruiz, A.; Alhama, J.; Blasco, J.; Gómez-Ariza, J.L. & López-Barea, J. (2008). New metallothionein assay in *Scrobicularia plana*: heating effect and correlation with other biomarkers. *Environmental Pollution*, 156, 1340-1347.

108. Rønnestad, I.; Conceiçao, L.E.C.; Aragão, C. & Dinis, M.T. (2001). Assimilation and catabolism of dispensable and indispensable free amino acids in post-larval Senegal sole (*Solea senegalensis*). *Comparative Biochemistry and Physiology - Part C: Toxicology & Pharmacology*, 130, 461-466.

109. Rønnestad, I.; Tonheim, S.K.; Fyhn, H.J.; Rojas-García, C.R.; Kamisaka, Y.; Koven, W.; Finn, R.N.; Terjesen, B.F.; Barr, Y. & Conceiçao, L.E.C. (2003). The supply

of amino acids during early feeding stages of marine fish larvae: a review of recent findings. *Aquaculture*, 227, 147-164.

110. Rueda-Jasso, R.; Conceiçao, L.E.C.; Dias, J.; De Coen, W.; Gomes, E.; Rees, J.F.; Soares, F.; Dinis, M.T. & Sorgeloos, P. (2004). Effect of dietary non-protein energy levels on condition and oxidative status of Senegalese sole (*Solea senegalensis*) juveniles. *Aquaculture*, 231, 417-433.
111. Ruiz-Laguna, J.; Abril, N.; Prieto-Álamo, M.J.; López-Barea, J. & Pueyo, C. (2005). Tissue, species and enviromental differences in absolute quantities of murine mRNAs coding for α, μ, ω, π and θ glutathione S-transferases. *Gene Expression*, 12, 165-176.
112. Ruiz-Laguna, J.; Abril, N.; García-Barrera, T.; Gómez-Ariza, J.L.; López-Barea, J. & Pueyo, C. (2006). Absolute transcript expression signatures of *Cyp* and *Gst* genes in *Mus spretus* to detect environmental contamination. *Environmental Science & Technology*, 40, 3646-3652.
113. Salas-Leiton, E.; Cánovas-Conesa, B.; Zerolo, R.; López-Barea, J.; Cañavate, J.P. & Alhama, J. (2009). Proteomics of juvenile Senagal sole (*Solea senegalensis*) affected by gas bubble disease in hyperoxygenated ponds. *Marine Biotechnology*, 11, 473-487.
114. Simpson, R.J. (2003). *Proteins and Proteomics. A Laboratory Manual.* Cold Spring Harbor, USA: Cold Spring Harbor Laboratory Press.
115. Solé, M.; Lima, D.; Reis-Henriques, M.A. & Santos, M.M. (2008). Stress biomarkers in juvenile Sole, *Solea senegalensis*, exposed to water-accommodated fraction of the «Prestige» fuel oil. *Bulletin of Environmental Contamination and Toxicology*, 80, 19-23. Sukardi, H.; Ung, C.Y.; Gong, Z. & Lam, S.H. (2010). Incorporating zebrafish Omics into Chemical Biology and Toxicology. *Zebrafish*, 7, 41-52.
116. Swain, P.; Nayak S.K.; Nanda P.K. & Dash S. (2008). Biological effects of bacterial lipopolysaccharide (endotoxin) in fish: A review. *Fish & Shellfish Immunology*, 25(3), 191-201.
117. Thorgaard, G.H.; Bailey, G.S.; Williams, D.; Buhler, D.R.; Kaattari, S.L.; Ristow, S.S.; Hansen, J.D.; Winton, J.R.; Bartholomew, J.L.; Nagler, J.J.; Walsh, P.J.; Vijayan, M.M.; Devlin, R.H.; Hardy, R.W.; Overturf, K.E.; Young, W.P.; Robison, B.D.; Rexroad, C. & Palti, Y. (2002). Status and opportunities for genomics research with rainbow trout. *Comparative Biochemistry and Physiology*, 133(4), 609-646.
118. Valente, L.M.P.; Linares, F.; Villanueva, J.L.R.; Silva, J.M.G.; Espe, M.; Escórcio, C.; Pires, M.A.; Saavedra, M.J.; Borges, P.; Medale, F.; Alvarez-Blázquez, B. & Peleteiro, J.B. (2011). Dietary protein source or energy levels have no major impact on growth performance, nutrient utilisation or flesh fatty acids composition of market-sized Senegalese sole. *Aquaculture*, 318, 128-137.

119. Van Aggelen, G.V.; Ankley, G.T.; Baldwin, W.S.; Bearden, D.W.; Benso, W.H.; Chipman, J.K.; Collette, T.W.; Craft, J.A.; Denslow, N.D.; Embry, M.R.; Falciani, F.; George, S.G.; Helbing, C.C.; Hoekstra, P.F.; Iguchi, T.; Kagami, Y.; Katsiadaki, I.; Kille, P.; Liu, L.; Lord, P.G.; McIntyre, T.; O´Neill, A.; Osachoff, H.; Perkins, E.J.; Santos, E.M.; Skirrow, R.C.; Snape, J.R.; Tyler, C.R.; Versteeg, D.; Viant, M.R.; Volz, D.C.; Williams, T.D. & Yu, L. (2010). Integrating Omic technologies into aquatic Ecological Risk Assessment and Environmental Monitoring: hurdles, achievements, and future outlook. *Environmental Health Perspectives*, 118, 1-5.

120. Viant, M.R. (2007). Metabolomics of aquatic organisms: the new «omics» on the block. *Marine Ecology Progress Series*, 332, 301-306.

121. Vioque-Fernández, A.; Alves-de-Almeida, E.; Ballesteros, J.; García-Barrera, T.; Gómez- Ariza, J.L. & López-Barea, J. (2007a). Doñana National Park survey using *Procambarus clarkii* as bioindicator: esterase inhibition and pollutant levels. *Toxicology Letters*, 168, 260-268.

122. Vioque-Fernández, A.; Alves-de-Almeida, E. & López-Barea, J. (2007b). Esterases as pesticide biomarkers in crayfish (*Procambarus clarkii*, Crustacea): Tissue distribu- tion, sensitivity to model compounds and recovery from inactivation. *Comparative Biochemistry and Physiology - Part C: Toxicology & Pharmacology*, 145, 404-412.

123. Vioque-Fernández, A.; Alves-de-Almeida, E. & López-Barea, J. (2009a). Assessment of Doñana National Park contamination in *Procambarus clarkii*: integration of conven- tional biomarkers and proteomic approaches. *Science of the Total Environment*, 407, 1784-1797.

124. Vioque-Fernández, A.; Alves-de-Almeida, E. & López-Barea, J. (2009b). Biomarker responses and protein expression profile changes in *Procambarus clarkii* after chlorpyrifos or carbaryl exposure. *Biomarkers*, 14, 299-310.

125. Washburn, M.P.; Wolters, D. & Yates, J.R. (2001). Large-scale analysis of the yeast proteome by multidimensional protein identification technology. *Nature Biotechnology*, 19 (3), 242-247.

126. Whyte, S.K. (2007). The innate immune response of finfish--a review of current knowledge. *Fish & Shellfish Immunology*, 23(6), 1127-1151.

127. Wilkins, M.R.; Pasquali, C.; Appel, R.D.; Ou, K.; Golaz, O.; Sanchez, J.-C.; Yan, J.X.; Gooley, A.A.; Hughes, G.; Humphery-Smith, I.; Williams, K.L. & Hochstrasser, D.F. (1996). From proteins to proteomes: large scale protein identification by two-dimensional electrophoresis and amino acid analysis. *Nature Biotechnology*, 14, 61–65.

128. Williams, T.D.; Gensberg, K.; Minchin, S.D. & Chipman, J.K. (2003). A DNA expression array to detect toxic stress response in European flounder (*Platichthys flesus*). *Aquatic Toxicology*, 65, 141-157.

129. Williams, T.D.; Diab, A.M.; George, S.G.; Godfrey, R.E.; Sabine, V.; Conesa, A.; Minchin, S.D.; Watts, P.C. & Chipman, J.K. (2006). Development of the GENIPOL European Flounder (*Platichthys flesus*) microarray and determination of temporal transcriptional responses to cadmium at low dose. *Environmental Science & Technology*, 40, 6479-6488.

130. Williams, T.D.; Diab, A.M.; George, S.G.; Sabine, V. & Chipman, J.K. (2007). Gene expression responses of European flounder (*Platichthys flesus*) to 17-beta estradiol. *Toxicology Letters*, 168(3), 236-248.

131. Williams, T.D.; Diab, A.; Ortega, F.; Sabine, V.S.; Godfrey, R.E.; Falciani., F.; Chipman, J.K. & George, S.G. (2008). Transcriptomic responses of European flounder (*Platichthys flesus*) to model toxicants. *Aquatic Toxicology*, 90(2), 83-91.

Index

www.ingramcontent.com/pod-product-compliance
Lightning Source LLC
LaVergne TN
LVHW062217200926
848484LV00006B/1414
* 9 7 8 1 7 8 7 1 5 2 8 1 6 *